Data Mining
For Business Intelligence

SECOND
EDITION

Data Mining
For Business Intelligence

Concepts, Techniques, and Applications in
Microsoft Office Excel® with XLMiner®

GALIT SHMUELI
University of Maryland
Robert H. Smith School of Business
College Park, MD

NITIN R. PATEL
Massachusetts Institute of Technology
Center for Biomedical Innovation
Cambridge, MA

PETER C. BRUCE
Statistics.com
Arlington, VA

A JOHN WILEY & SONS, INC., PUBLICATION

For general information on our other products and services or for technical support, please contact our Customer
Care Department within the United States at (800) 762-2974, outside the United States at (317) 572-3993 or fax
(317) 572-4002.

Wiley also publishes its books in a variety of electronic formats. Some content that appears in print may not be
available in electronic formats. For more information about Wiley products, visit our web site at www.wiley.com.

Library of Congress Cataloging-in-Publication Data:

Shmueli, Galit, 1971-
 Data mining for business intelligence : concepts, techniques, and
applications in Microsoft Office Excel with XLMiner / Galit Shmueli, Nitin R.
Patel, Peter C. Bruce. – 2nd ed.
      p. cm.
 Includes bibliographical references and index.
  ISBN 978-0-470-52682-8 (cloth)
 1. Business–Data processing. 2. Data mining. 3. Microsoft Excel (Computer
file) I. Patel, Nitin R. (Nitin Ratilal) II. Bruce, Peter C., 1953- III.
Title.
 HF5548.2.S44843 2010
 005.54–dc22
                                                    2010005152

Printed in the United States of America

10

To our families
Boaz and Noa
Tehmi, Arjun, and in
memory of Aneesh
Liz, Lisa, and Allison

Contents

PART II DATA EXPLORATION AND DIMENSION REDUCTION

CHAPTER 3 Data Visualization — 43

Foreword

Data mining—the art of extracting useful information from large amounts of data—is of growing importance in today's world. Your e-mail spam filter relies at least in part on rules that a data mining algorithm has learned from examining millions of e-mail messages that have been classified as spam or not spam. Real-time data mining methods enable Web-based merchants to tell you that "customers who purchased x are also likely to purchase y." Data mining helps banks determine which applicants are likely to default on loans, helps tax authorities identify which tax returns are most likely to be fraudulent, and helps catalog merchants target those customers most likely to purchase.

And data mining is not just about numbers—text mining techniques help search engines like Google and Yahoo find what you are looking for by ordering documents according to their relevance to your query. In the process they have effectively monetized search by ordering sponsored ads that are relevant to your query.

The amount of data flowing from, to, and through enterprises of all sorts is enormous, and growing rapidly—more rapidly than the capabilities of organizations to use it. Successful enterprises are those that make effective use of the abundance of data to which they have access: to make better predictions, better decisions, and better strategies. The margin over a competitor may be small (they, after all, have access to the same methods for making effective use of information), hence the need to take advantage of every possible avenue to advantage.

At no time has the need been greater for quantitatively skilled managerial expertise. Successful managers now need to know about the possibilities and limitations of data mining. But at what level? A high-level overview can provide a general idea of what data mining can do for the enterprise but fails to provide the intuition that could be attained by actually building models with real data. A very technical approach from the computer science, database, or statistical standpoint can get bogged down in detail that has little bearing on decision making.

It is essential that managers be able to translate business or other functional problems into the appropriate statistical problem before it can be "handed off" to a technical team. But it is difficult for managers to do this with confidence unless they have actually had hands-on experience developing models for a variety of real problems using real data. That is the perspective of this book—the use of real data, actual cases, and an Excel-based program to build and compare models with a minimal learning curve.

DARYL PREGIBON

Google Inc, 2006

Preface to the Second Edition

Since the book's appearance in early 2007, it has been used in many classes, ranging from dedicated data mining classes to more general business intelligence courses. Following feedback from instructors teaching both MBA and undergraduate courses, as well as students, we revised some of the existing chapters as well as covered two new topics that are central in data mining: data visualization and time series forecasting.

We have added a set of three chapters on time series forecasting (Chapters 15–17), which present the most commonly used forecasting tools in the business world. They include a set of new datasets and exercises, and a new case (in Chapter 18).

The chapter on data visualization provides comprehensive coverage of basic and advanced visualization techniques that support the exploratory step of data mining. We also provide a discussion of interactive visualization principles and tools, and the chapter exercises include assignments to familiarize readers with interactive visualization in practice.

In the new edition we have created separate chapters for the k-nearest-neighbor and naive Bayes methods. The explanation of the naive Bayes classifier is now clearer, and additional exercises have been added to both chapters.

Another addition are brief chapter summaries at the beginning of each chapter.

We have also reorganized the order of some chapters, following readers' feedback. The chapters are now grouped into seven parts: Preliminaries, Data Exploration and Dimension Reduction, Performance Evaluation, Prediction and Classification Methods, Mining Relationships Among Records, Forecasting Time Series, and Cases. The new organization is aimed at helping instructors of various types of courses to choose subsets of topics to teach.

Two-semester data mining courses could cover in detail data exploration and dimension reduction and supervised learning in one term (choosing the type and amount of prediction and classification methods according to the course flavor and the audience interest). Forecasting time series and unsupervised learning can be covered in the second term.

Single-semester data mining courses would do best to concentrate on the first parts of the book, and only introduce time series forecasting as time allows. This is especially true if a dedicated forecasting course is offered in the program.

General business intelligence courses would best focus on the first three parts, then choose a small number of prediction/classification methods for illustration, and present the mining relationships chapters. All these can be covered via a few cases, where students read the relevant chapters that support the analysis done in the case.

A set of data mining courses that constitute a concentration can be built according to the sequence of parts in the book. The first three parts (Preliminaries, Data Exploration and Dimension Reduction, and Performance Evaluation) should serve as requirements for the next courses. Cases can be used either within appropriate topic courses or as project-type courses.

In all courses, we strongly recommend including a project component, where data are either collected by students according to their interest or provided by the instructor (e.g., from the many data mining competition datasets available). From our experience and other instructors' experience, such projects enhance the learning and provide students with an excellent opportunity to understand the strengths of data mining and the challenges that arise in the process.

Preface to the First Edition

This book arose out of a data mining course at MIT's Sloan School of Management and was refined during its use in data mining courses at the University of Maryland's R. H. Smith School of Business and at statistics.com. Preparation for the course revealed that there are a number of excellent books on the business context of data mining, but their coverage of the statistical and machine-learning algorithms that underlie data mining is not sufficiently detailed to provide a practical guide if the instructor's goal is to equip students with the skills and tools to implement those algorithms. On the other hand, there are also a number of more technical books about data mining algorithms, but these are aimed at the statistical researcher or more advanced graduate student, and do not provide the case-oriented business focus that is successful in teaching business students.

Hence, this book is intended for the business student (and practitioner) of data mining techniques, and its goal is threefold:

1. To provide both a theoretical and a practical understanding of the key methods of classification, prediction, reduction, and exploration that are at the heart of data mining.
2. To provide a business decision-making context for these methods.
3. Using real business cases, to illustrate the application and interpretation of these methods.

The presentation of the cases in the book is structured so that the reader can follow along and implement the algorithms on his or her own with a very low learning hurdle.

Just as a natural science course without a lab component would seem incomplete, a data mining course without practical work with actual data is missing a key ingredient. The MIT data mining course that gave rise to this book followed an introductory quantitative course that relied on Excel—this made its practical work universally accessible. Using Excel for data mining seemed a natural progression. An important feature of this book is the use of Excel, an environment familiar to business analysts. All required data mining algorithms (plus illustrative datasets) are provided in an Excel add-in, XLMiner. Data for both the cases and exercises are available at www.dataminingbook.com.

Although the genesis for this book lay in the need for a case-oriented guide to teaching data mining, analysts and consultants who are considering the application of data mining techniques in contexts where they are not currently in use will also find this a useful, practical guide.

Acknowledgments

The authors thank the many people who assisted us in improving the first edition and improving it further in the second edition. Anthony Babinec, who has been using drafts of this book for years in his data mining courses at statistics.com, provided us with detailed and expert corrections. Similarly, Dan Toy and John Elder IV greeted our project with enthusiasm and provided detailed and useful comments on earlier drafts. Boaz Shmueli and Raquelle Azran gave detailed editorial comments and suggestions on both editions; Bruce McCullough and Adam Hughes did the same for the first edition. Ravi Bapna, who used an early draft in a data mining course at the Indian School of Business, provided invaluable comments and helpful suggestions. Useful comments and feedback have also come from the many instructors, too numerous to mention, who have used the book in their classes.

From the Smith School of Business at the University of Maryland, colleagues Shrivardhan Lele, Wolfgang Jank, and Paul Zantek provided practical advice and comments. We thank Robert Windle, and MBA students Timothy Roach, Pablo Macouzet, and Nathan Birckhead for invaluable datasets. We also thank MBA students Rob Whitener and Daniel Curtis for the heatmap and map charts. And we thank the many MBA students for fruitful discussions and interesting data mining projects that have helped shape and improve the book.

This book would not have seen the light of day without the nurturing support of the faculty at the Sloan School of Management at MIT. Our special thanks to Dimitris Bertsimas, James Orlin, Robert Freund, Roy Welsch, Gordon Kaufmann, and Gabriel Bitran. As teaching assistants for the data mining course at Sloan, Adam Mersereau gave detailed comments on the notes and cases that were the genesis of this book, Romy Shioda helped with the preparation of several cases and exercises used here, and Mahesh Kumar helped with the material on clustering. We are grateful to the MBA students at Sloan for stimulating discussions in the class that led to refinement of the notes as well as XLMiner.

Chris Albright, Gregory Piatetsky-Shapiro, Wayne Winston, and Uday Karmarkar gave us helpful advice on the use of XLMiner. Anand Bodapati provided both data and advice. Suresh Ankolekar and Mayank Shah helped develop several cases and provided valuable pedagogical comments. Vinni Bhandari helped write the Charles Book Club case.

We would like to thank Marvin Zelen, L. J. Wei, and Cyrus Mehta at Harvard, as well as Anil Gore at Pune University, for thought-provoking discussions on the relationship between statistics and data mining. Our thanks to Richard Larson of the Engineering Systems Division, MIT, for sparking many stimulating ideas on the role of data mining in modeling complex systems. They helped us develop a balanced philosophical perspective on the emerging field of data mining.

Our thanks to Ajay Sathe, who energetically shepherded XLMiner's development over the years and continues to do so, and to his colleagues on the XLMiner team: Suresh Ankolekar, Poonam Baviskar, Kuber Deokar, Rupali Desai, Yogesh Gajjar, Ajit Ghanekar, Ayan Khare, Bharat Lande, Dipankar Mukhopadhyay, S. V. Sabnis, Usha Sathe, Anurag Srivastava, V. Subramaniam, Ramesh Raman, and Sanhita Yeolkar.

Steve Quigley at Wiley showed confidence in this book from the beginning and helped us navigate through the publishing process with great speed. Curt Hinrichs' vision, tips, and encouragement helped bring this book to the starting gate. We are also grateful to Ashwini Kumthekar, Achala Sabane, Michael Shapard, and Heidi Sestrich who assisted with typesetting, figures, and indexing, and to Valerie Troiano who has shepherded many instructors through the use of XLMiner and early drafts of this text.

We also thank Catherine Plaisant at the University of Maryland's Human-Computer Interaction Lab, who helped out in a major way by contributing exercises and illustrations to the data visualization chapter, Marietta Tretter at Texas A&M for her helpful comments and thoughts on the time series chapters, and Stephen Few and Ben Shneiderman for feedback and suggestions on the data visualization chapter and overall design tips.

## PART ONE

Preliminaries

Introduction

1.1 WHAT IS DATA MINING?

The field of data mining is still relatively new and in a state of evolution. The first International Conference on Knowledge Discovery and Data Mining (KDD) was held in 1995, and there are a variety of definitions of data mining.

A concise definition that captures the essence of *data mining* is:

Extracting useful information from large data sets.

(Hand et al., 2001)

A slightly longer version is:

Data mining is the process of exploration and analysis, by automatic or semi-automatic means, of large quantities of data in order to discover meaningful patterns and rules.

(Berry and Linoff, 1997, p. 5)

Berry and Linoff later had cause to regret the 1997 reference to "automatic and semi-automatic means," feeling that it shortchanged the role of data exploration and analysis analysis (Berry and Linoff, 2000).

Another definition comes from the Gartner Group, the information technology research firm:

[Data Mining is] the process of discovering meaningful correlations, patterns and trends by sifting through large amounts of data stored in repositories. Data mining employs pattern recognition technologies, as well as statistical and mathematical techniques.

(http://www.gartner.com/6_help/glossary, accessed May 14, 2010)

A summary of the variety of methods encompassed in the term *data mining* is given at the beginning of Chapter 2.

1.2 WHERE IS DATA MINING USED?

Data mining is used in a variety of fields and applications. The military use data mining to learn what roles various factors play in the accuracy of bombs. Intelligence agencies might use it to determine which of a huge quantity of intercepted communications are of interest. Security specialists might use these methods to determine whether a packet of network data constitutes a threat. Medical researchers might use it to predict the likelihood of a cancer relapse.

Although data mining methods and tools have general applicability, most examples in this book are chosen from the business world. Some common business questions that one might address through data mining methods include:

1. From a large list of prospective customers, which are most likely to respond? We can use classification techniques (logistic regression, classification trees, or other methods) to identify those individuals whose demographic and other data most closely matches that of our best existing customers. Similarly, we can use prediction techniques to forecast how much individual prospects will spend.

2. Which customers are most likely to commit, for example, fraud (or might already have committed it)? We can use classification methods to identify (say) medical reimbursement applications that have a higher probability of involving fraud and give them greater attention.

3. Which loan applicants are likely to default? We can use classification techniques to identify them (or logistic regression to assign a "probability of default" value).

4. Which customers are most likely to abandon a subscription service (telephone, magazine, etc.)? Again, we can use classification techniques to identify them (or logistic regression to assign a "probability of leaving" value). In this way, discounts or other enticements can be proffered selectively.

1.3 ORIGINS OF DATA MINING

Data mining stands at the confluence of the fields of statistics and machine learning (also known as artificial intelligence). A variety of techniques for exploring data and building models have been around for a long time in the world of statistics: linear regression, logistic regression, discriminant analysis, and principal components analysis, for example. But the core tenets of classical statistics—computing is difficult and data are scarce—do not apply in data mining applications where both data and computing power are plentiful.

This gives rise to Daryl Pregibon's description of data mining as "statistics at scale and speed" (Pregibon, 1999). A useful extension of this is "statistics at scale, speed, and simplicity." Simplicity in this case refers not to the simplicity of algorithms but, rather, to simplicity in the logic of inference. Due to the scarcity of data in the classical statistical setting, the same sample is used to make an estimate and also to determine how reliable that estimate might be. As a result, the logic of the confidence intervals and hypothesis tests used for inference may seem elusive for many, and their limitations are not well appreciated. By contrast, the data mining paradigm of fitting a model with one sample and assessing its performance with another sample is easily understood.

Computer science has brought us *machine learning techniques*, such as trees and neural networks, that rely on computational intensity and are less structured than classical statistical models. In addition, the growing field of database management is also part of the picture.

The emphasis that classical statistics places on inference (determining whether a pattern or interesting result might have happened by chance) is missing in data mining. In comparison to statistics, data mining deals with large datasets in open-ended fashion, making it impossible to put the strict limits around the question being addressed that inference would require.

As a result, the general approach to data mining is vulnerable to the danger of *overfitting*, where a model is fit so closely to the available sample of data that it describes not merely structural characteristics of the data but random peculiarities as well. In engineering terms, the model is fitting the noise, not just the signal.

1.4 RAPID GROWTH OF DATA MINING

Perhaps the most important factor propelling the growth of data mining is the growth of data. The mass retailer Wal-Mart in 2003 captured 20 million transactions per day in a 10-terabyte database (a terabyte is 1 million megabytes). In 1950, the largest companies had only enough data to occupy, in electronic form, several dozen megabytes. Lyman and Varian (2003) estimate that 5 exabytes of information were produced in 2002, double what was produced in 1999 (1 exabyte is 1 million terabytes); 40% of this was produced in the United States.

The growth of data is driven not simply by an expanding economy and knowledge base but by the decreasing cost and increasing availability of automatic data capture mechanisms. Not only are more events being recorded, but more information per event is captured. Scannable bar codes, point-of-sale

(POS) devices, mouse click trails, and global positioning satellite (GPS) data are examples.

The growth of the Internet has created a vast new arena for information generation. Many of the same actions that people undertake in retail shopping, exploring a library, or catalog shopping have close analogs on the Internet, and all can now be measured in the most minute detail. In marketing, a shift in focus from products and services to a focus on the customer and his or her needs has created a demand for detailed data on customers.

The operational databases used to record individual transactions in support of routine business activity can handle simple queries but are not adequate for more complex and aggregate analysis. Data from these operational databases are therefore extracted, transformed, and exported to a *data warehouse*, a large integrated data storage facility that ties together the decision support systems of an enterprise. Smaller *data marts* devoted to a single subject may also be part of the system. They may include data from external sources (e.g., credit rating data).

Many of the exploratory and analytical techniques used in data mining would not be possible without today's computational power. The constantly declining cost of data storage and retrieval has made it possible to build the facilities required to store and make available vast amounts of data. In short, the rapid and continuing improvement in computing capacity is an essential enabler of the growth of data mining.

1.5 WHY ARE THERE SO MANY DIFFERENT METHODS?

As can be seen in this book or any other resource on data mining, there are many different methods for prediction and classification. You might ask yourself why they coexist and whether some are better than others. The answer is that each method has advantages and disadvantages. The usefulness of a method can depend on factors such as the size of the dataset, the types of patterns that exist in the data, whether the data meet some underlying assumptions of the method, how noisy the data are, and the particular goal of the analysis. A small illustration is shown in Figure 1.1, where the goal is to find a combination of *household income level* and *household lot size* that separate buyers (solid circles) from nonbuyers (hollow circles) of riding mowers. The first method (left panel) looks only for horizontal and vertical lines to separate buyers from nonbuyers, whereas the second method (right panel) looks for a single diagonal line.

Different methods can lead to different results, and their performance can vary. It is therefore customary in data mining to apply several different methods and select the one that is most useful for the goal at hand.

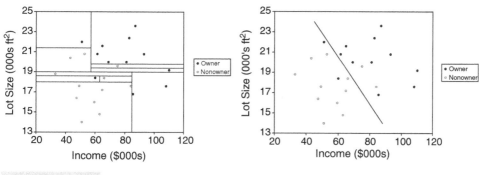

FIGURE 1.1   **TWO METHODS FOR SEPARATING BUYERS FROM NONBUYERS**

1.6 TERMINOLOGY AND NOTATION

Because of the hybrid parentry of data mining, its practitioners often use multiple terms to refer to the same thing. For example, in the machine learning (artificial intelligence) field, the variable being predicted is the output variable or target variable. To a statistician, it is the dependent variable or the response. Here is a summary of terms used:

Algorithm Refers to a specific procedure used to implement a particular data mining technique: classification tree, discriminant analysis, and the like.

Attribute See **Predictor**.

Case See **Observation**.

Confidence Has a specific meaning in association rules of the type "IF A and B are purchased, C is also purchased." Confidence is the conditional probability that C will be purchased IF A and B are purchased.

Confidence Also has a broader meaning in statistics (*confidence interval*), concerning the degree of error in an estimate that results from selecting one sample as opposed to another.

Dependent Variable See **Response**.

Estimation See **Prediction**.

Feature See **Predictor**.

Holdout Sample Is a sample of data not used in fitting a model, used to assess the performance of that model; this book uses the term *validation set* or, if one is used in the problem, *test set* instead of *holdout sample*.

Input Variable See **Predictor**.

Model Refers to an algorithm as applied to a dataset, complete with its settings (many of the algorithms have parameters that the user can adjust).

Observation Is the unit of analysis on which the measurements are taken (a customer, a transaction, etc.); also called *case*, *record*, *pattern*, or *row*. (Each row typically represents a record; each column, a variable.)

Outcome Variable See **Response**.

Output Variable See **Response**.

P (A | B) Is the conditional probability of event A occurring given that event B has occurred. Read as "the probability that A will occur given that B has occurred."

Pattern Is a set of measurements on an observation (e.g., the height, weight, and age of a person).

Prediction The prediction of the value of a continuous output variable; also called *estimation*.

Predictor Usually denoted by X, is also called a *feature*, *input variable*, *independent variable*, or from a database perspective, a *field*.

Record See **Observation**.

Response usually denoted by Y, is the variable being predicted in supervised learning; also called *dependent variable*, *output variable*, *target variable*, or *outcome variable*.

Score Refers to a predicted value or class. *Scoring new data* means to use a model developed with training data to predict output values in new data.

Success Class Is the class of interest in a binary outcome (e.g., *purchasers* in the outcome *purchase/no purchase*).

Supervised Learning Refers to the process of providing an algorithm (logistic regression, regression tree, etc.) with records in which an output variable of interest is known and the algorithm "learns" how to predict this value with new records where the output is unknown.

Test Data (or **test set**) Refers to that portion of the data used only at the end of the model building and selection process to assess how well the final model might perform on additional data.

Training Data (or **training set**) Refers to that portion of data used to fit a model.

Unsupervised Learning Refers to analysis in which one attempts to learn something about the data other than predicting an output value of interest (e.g., whether it falls into clusters).

Validation Data (or **validation set**) Refers to that portion of the data used to assess how well the model fits, to adjust some models, and to select the best model from among those that have been tried.

Variable Is any measurement on the records, including both the input (X) variables and the output (Y) variable.

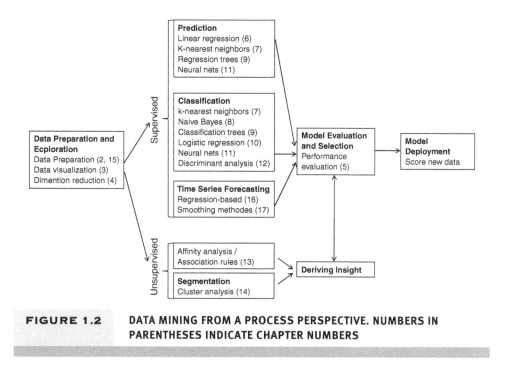

FIGURE 1.2   DATA MINING FROM A PROCESS PERSPECTIVE. NUMBERS IN PARENTHESES INDICATE CHAPTER NUMBERS

1.7 ROAD MAPS TO THIS BOOK

The book covers many of the widely used predictive and classification methods as well as other data mining tools. Figure 1.2 outlines data mining from a process perspective and where the topics in this book fit in. Chapter numbers are indicated beside the topic. Table 1.1 provides a different perspective: It organizes data mining procedures according to the type and structure of the data.

TABLE 1.1   ORGANIZATION OF DATA MINING METHODS IN THIS BOOK, ACCORDING TO THE NATURE OF THE DATA[a]

	Continuous Response	Categorical Response	No Response
Continuous Predictors	Linear regression (6) Neural nets (11) *k* Nearest neighbors (7)	Logistic regression (10) Neural nets (11) Discriminant analysis (12) *k* Nearest neighbors (7)	Principal components (4) Cluster analysis (14)
Categorical Predictors	Linear regression (6) Neural nets (11) Regression trees (9)	Neural nets (11) Classification trees (9) Logistic regression (10) Naive Bayes (8)	Association rules (13)

[a] Numbers in parentheses indicate chapter number.

Order of Topics

The book is divided into five parts: Part I (Chapters 1–2) gives a general overview of data mining and its components. Part II (Chapters 3–4) focuses on the early stage of data exploration and dimension reduction in which typically the most effort is expended.

Part III (Chapter 4) discusses performance evaluation. Although it contains a single chapter, we discuss a variety of topics, from predictive performance metrics to misclassification costs. The principles covered in this part are crucial for the proper evaluation and comparison of supervised learning methods.

Part IV includes eight chapters (Chapters 5–12), covering a variety of popular supervised learning methods (for classification and/or prediction). Within this part, the topics are generally organized according to the level of sophistication of the algorithms, their popularity, and ease of understanding.

Part V focuses on unsupervised learning, presenting association rules (Chapter 13) and cluster analysis (Chapter 14).

Part VI includes three chapters (Chapters 15–17), with the focus on forecasting time series. The first chapter covers general issues related to handling and understanding time series. The next two chapters present two popular forecasting approaches: regression-based forecasting and smoothing methods.

Finally, Part VII includes a set of cases.

Although the topics in the book can be covered in the order of the chapters, each chapter stands alone. It is advised, however, to read Parts I–III before proceeding to the chapters in Parts IV–V, and similarly Chapter 15 should precede other chapters in Part VI.

USING XLMINER SOFTWARE

To facilitate hands-on data mining experience, this book comes with access to XLMiner, a comprehensive data mining add-in for Excel. For those familiar with Excel, the use of an Excel add-in dramatically shortens the software learning curve. XLMiner will help you get started quickly on data mining and offers a variety of methods for analyzing data. The illustrations, exercises, and cases in this book are written in relation to this software. XLMiner has extensive coverage of statistical and data mining techniques for classification, prediction, affinity analysis, and data exploration and reduction. It offers a variety of data mining tools: neural nets, classification and regression trees, k-nearest neighbor classification, naive Bayes, logistic regression, multiple linear regression, and discriminant analysis, all for predictive modeling. It provides for automatic partitioning of data into training, validation, and test samples and for the deployment of the model to new data. It also offers association rules, principal components analysis, k-means clustering, and hierarchical clustering, as well as visualization tools and data-handling utilities. With its short learning curve, affordable price, and reliance on

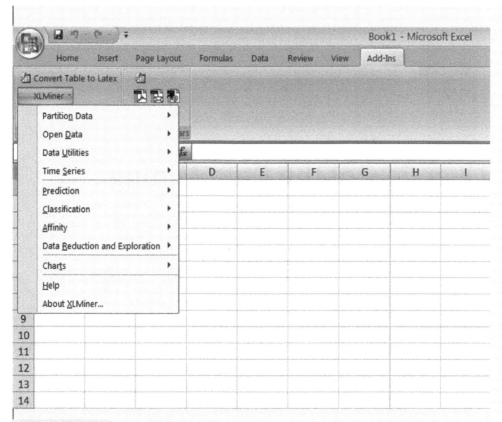

FIGURE 1.3   XLMINER SCREEN

the familiar Excel platform, it is an ideal companion to a book on data mining for the business student.

Download   To download the XLMiner setup program, visit www.xlminer.com/ wiley and follow the instructions there.

Installation   Close any Excel windows, then run the XLMiner setup program. Dialog boxes will guide you through the installation procedure. The final dialog box gives you an option to start Excel and open a "Getting Started" workbook. You'll also find XLMiner options under *Start > All Programs > Frontline Systems > XLMiner*.

Use   XLMiner is loaded when you start Excel, and appears as a pull-down menu on the Add-ins tab of the Excel Ribbon, as shown in Figure 1.3, or in the top toolbar in Excel 2003. By choosing the appropriate menu item, you can run any of XLMiner's procedures on the dataset that is open in the Excel worksheet.

## CHAPTER 2

Overview of the Data Mining Process

In Kåre this chapter we give an overview of the steps involved in data mining, starting from a clear goal definition and ending with model deployment. The general steps are shown schematically in Figure 2.1. We also discuss issues related to data collection, cleaning, and preprocessing. We explain the notion of data partitioning, where methods are trained on a set of training data and then their performance is evaluated on a separate set of validation data, and how this practice helps avoid overfitting. Finally, we illustrate the steps of model building by applying them to data.

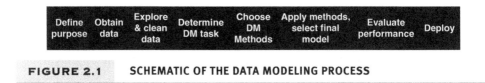

FIGURE 2.1   **SCHEMATIC OF THE DATA MODELING PROCESS**

2.1 INTRODUCTION

In Chapter 1 we saw some very general definitions of data mining. In this chapter we introduce the variety of methods sometimes referred to as *data mining*. The core of this book focuses on what has come to be called *predictive analytics*, the tasks of classification and prediction that are becoming key elements of a

Data Mining for Business Intelligence, By Galit Shmueli, Nitin R. Patel, and Peter C. Bruce
Copyright © 2010 John Wiley & Sons Inc.

"business intelligence" function in most large firms. These terms are described and illustrated below.

Not covered in this book to any great extent are two simpler database methods that are sometimes considered to be data mining techniques: (1) OLAP (online analytical processing) and (2) SQL (structured query language). OLAP and SQL searches on databases are descriptive in nature ("find all credit card customers in a certain zip code with annual charges > $20,000, who own their own home and who pay the entire amount of their monthly bill at least 95% of the time") and do not involve statistical modeling.

2.2 CORE IDEAS IN DATA MINING

Classification

Classification is perhaps the most basic form of data analysis. The recipient of an offer can respond or not respond. An applicant for a loan can repay on time, repay late, or declare bankruptcy. A credit card transaction can be normal or fraudulent. A packet of data traveling on a network can be benign or threatening. A bus in a fleet can be available for service or unavailable. The victim of an illness can be recovered, still be ill, or be deceased.

A common task in data mining is to examine data where the classification is unknown or will occur in the future, with the goal of predicting what that classification is or will be. Similar data where the classification is known are used to develop rules, which are then applied to the data with the unknown classification.

Prediction

Prediction is similar to classification, except that we are trying to predict the value of a numerical variable (e.g., amount of purchase) rather than a class (e.g., purchaser or nonpurchaser). Of course, in classification we are trying to predict a class, but the term *prediction* in this book refers to the prediction of the value of a continuous variable. (Sometimes in the data mining literature, the term *estimation* is used to refer to the prediction of the value of a continuous variable, and *prediction* may be used for both continuous and categorical data.)

Association Rules

Large databases of customer transactions lend themselves naturally to the analysis of associations among items purchased, or "what goes with what." *Association rules*, or *affinity analysis*, can then be used in a variety of ways. For example, grocery stores can use such information after a customer's purchases have all been scanned to print discount coupons, where the items being discounted are determined by mapping the customer's purchases onto the association rules. Online merchants such as Amazon.com and Netflix.com use these methods as the heart of a "recommender" system that suggests new purchases to customers.

Predictive Analytics

Classification, prediction, and to some extent, affinity analysis constitute the analytical methods employed in *predictive analytics*.

Data Reduction

Sensible data analysis often requires distillation of complex data into simpler data. Rather than dealing with thousands of product types, an analyst might wish to group them into a smaller number of groups. This process of consolidating a large number of variables (or cases) into a smaller set is termed *data reduction*.

Data Exploration

Unless our data project is very narrowly focused on answering a specific question determined in advance (in which case it has drifted more into the realm of statistical analysis than of data mining), an essential part of the job is to review and examine the data to see what messages they hold, much as a detective might survey a crime scene. Here, full understanding of the data may require a reduction in its scale or dimension to allow us to see the forest without getting lost in the trees. Similar variables (i.e., variables that supply similar information) might be aggregated into a single variable incorporating all the similar variables. Analogously, records might be aggregated into groups of similar records.

Data Visualization

Another technique for exploring data to see what information they hold is through graphical analysis. This includes looking at each variable separately as well as looking at relationships between variables. For numerical variables, we use histograms and boxplots to learn about the distribution of their values, to detect outliers (extreme observations), and to find other information that is relevant to the analysis task. Similarly, for categorical variables we use bar charts. We can also look at scatterplots of pairs of numerical variables to learn about possible relationships, the type of relationship, and again, to detect outliers. Visualization can be greatly enhanced by adding features such as color, zooming, and interactive navigation.

2.3 SUPERVISED AND UNSUPERVISED LEARNING

A fundamental distinction among data mining techniques is between supervised and unsupervised methods. *Supervised learning algorithms* are those used in classification and prediction. We must have data available in which the value of the outcome of interest (e.g., purchase or no purchase) is known. These *training data* are the data from which the classification or prediction algorithm "learns," or is "trained," about the relationship between predictor variables and the outcome variable. Once the algorithm has learned from the training data, it is then applied to another sample of data (the *validation data*) where the outcome is known, to see how well it does in comparison to other models. If many different models are being tried out, it is prudent to save a third sample of known outcomes (the *test data*) to use with the model finally selected to predict how well it will do. The model can then be used to classify or predict the outcome of interest in new cases where the outcome is unknown. Simple linear regression analysis is an example of supervised learning (although rarely called that in the introductory statistics course where you probably first encountered it). The Y variable is the (known) outcome variable and the X variable is a predictor variable. A regression line is drawn to minimize the sum of squared deviations between the actual Y values and the values predicted by this line. The regression line can now be used to predict Y values for new values of X for which we do not know the Y value.

Unsupervised learning algorithms are those used where there is no outcome variable to predict or classify. Hence, there is no "learning" from cases where such an outcome variable is known. Association rules, dimension reduction methods, and clustering techniques are all unsupervised learning methods.

2.4 STEPS IN DATA MINING

This book focuses on understanding and using data mining algorithms (steps 4–7 below). However, some of the most serious errors in data analysis result from a poor understanding of the problem—an understanding that must be developed before we get into the details of algorithms to be used. Here is a list of steps to be taken in a typical data mining effort:

1. *Develop an understanding of the purpose of the data mining project* (if it is a one-shot effort to answer a question or questions) or application (if it is an ongoing procedure).

2. *Obtain the dataset to be used in the analysis.* This often involves random sampling from a large database to capture records to be used in an analysis. It may also involve pulling together data from different databases. The

databases could be internal (e.g., past purchases made by customers) or external (credit ratings). While data mining deals with very large databases, usually the analysis to be done requires only thousands or tens of thousands of records.

3. *Explore, clean, and preprocess the data.* This involves verifying that the data are in reasonable condition. How should missing data be handled? Are the values in a reasonable range, given what you would expect for each variable? Are there obvious outliers? The data are reviewed graphically: for example, a matrix of scatterplots showing the relationship of each variable with every other variable. We also need to ensure consistency in the definitions of fields, units of measurement, time periods, and so on.

4. *Reduce the data, if necessary, and* (where supervised training is involved) *separate them into training, validation, and test datasets.* This can involve operations such as eliminating unneeded variables, transforming variables (e.g., turning "money spent" into "spent > \$100" vs. "spent $\leq$ \$100"), and creating new variables (e.g., a variable that records whether at least one of several products was purchased). Make sure that you know what each variable means and whether it is sensible to include it in the model.

5. *Determine the data mining task* (classification, prediction, clustering, etc.). This involves translating the general question or problem of step 1 into a more specific statistical question.

6. *Choose the data mining techniques to be used* (regression, neural nets, hierarchical clustering, etc.).

7. *Use algorithms to perform the task.* This is typically an iterative process— trying multiple variants, and often using multiple variants of the same algorithm (choosing different variables or settings within the algorithm). Where appropriate, feedback from the algorithm's performance on validation data is used to refine the settings.

8. *Interpret the results of the algorithms.* This involves making a choice as to the best algorithm to deploy, and where possible, testing the final choice on the test data to get an idea as to how well it will perform. (Recall that each algorithm may also be tested on the validation data for tuning purposes; in this way the validation data become a part of the fitting process and are likely to underestimate the error in the deployment of the model that is finally chosen.)

9. *Deploy the model.* This involves integrating the model into operational systems and running it on real records to produce decisions or actions. For example, the model might be applied to a purchased list of possible customers, and the action might be "include in the mailing if the predicted amount of purchase is > \$10."

The foregoing steps encompass the steps in SEMMA, a methodology developed by SAS:

Sample   Take a sample from the dataset; partition into training, validation, and test datasets.

Explore   Examine the dataset statistically and graphically.

Modify   Transform the variables and impute missing values.

Model   Fit predictive models (e.g., regression tree, collaborative filtering).

Assess   Compare models using a validation dataset.

IBM Modeler (previously SPSS Clementine) has a similar methodology, termed CRISP-DM (cross-industry standard process for data mining).

2.5   PRELIMINARY STEPS

Organization of Datasets

Datasets are nearly always constructed and displayed so that variables are in columns and records are in rows. In the example shown in Section 2.6 (the Boston housing data), the values of 14 variables are recorded for a number of census tracts. The spreadsheet is organized such that each row represents a census tract—the first tract had a per capital crime rate (CRIM) of 0.00632, had 18% of its residential lots zoned for over 25,000 square feet (ZN), and so on. In supervised learning situations, one of these variables will be the outcome variable, typically listed at the end or the beginning (in this case it is median value, MEDV, at the end).

Sampling from a Database

Quite often, we want to perform our data mining analysis on less than the total number of records that are available. Data mining algorithms will have varying limitations on what they can handle in terms of the numbers of records and variables, limitations that may be specific to computing power and capacity as well as software limitations. Even within those limits, many algorithms will execute faster with smaller datasets.

From a statistical perspective, accurate models can often be built with as few as several hundred records (see below). Hence, we will often want to sample a subset of records for model building.

Oversampling Rare Events

If the event we are interested in is rare, however (e.g., customers purchasing a product in response to a mailing), sampling a subset of records may yield so few events (e.g., purchases) that we have little information on them. We would end up with lots of data on nonpurchasers but little on which to base a model that distinguishes purchasers from nonpurchasers. In such cases we would want our sampling procedure to overweight the purchasers relative to the nonpurchasers so that our sample would end up with a healthy complement of purchasers. This issue arises mainly in classification problems because those are the types of problems in which an overwhelming number of 0's is likely to be encountered in the response variable. Although the same principle could be extended to prediction, any prediction problem in which most responses are 0 is likely to raise the question of what distinguishes responses from nonresponses (i.e., a classification question). (For convenience below, we speak of responders and nonresponders as to a promotional offer, but we are really referring to any binary—0/1—outcome situation.)

Assuring an adequate number of responder or "success" cases to train the model is just part of the picture. A more important factor is the costs of misclassification. Whenever the response rate is extremely low, we are likely to attach more importance to identifying a responder than to identifying a nonresponder. In direct-response advertising (whether by traditional mail or via the Internet), we may encounter only one or two responders for every hundred records—the value of finding such a customer far outweighs the costs of reaching him or her. In trying to identify fraudulent transactions, or customers unlikely to repay debt, the costs of failing to find the fraud or the nonpaying customer are likely to exceed the cost of more detailed review of a legitimate transaction or customer.

If the costs of failing to locate responders were comparable to the costs of misidentifying responders as nonresponders, our models would usually be at their best if they identified everyone (or almost everyone, if it is easy to pick off a few responders without catching many nonresponders) as a nonresponder. In such a case, the misclassification rate is very low—equal to the rate of responders—but the model is of no value.

More generally, we want to train our model with the asymmetric costs in mind so that the algorithm will catch the more valuable responders, probably at the cost of "catching" and misclassifying more nonresponders as responders than would be the case if we assume equal costs. This subject is discussed in detail in Chapter 5.

Preprocessing and Cleaning the Data

Types of Variables   There are several ways of classifying variables. Variables can be numerical or text (character). They can be continuous (able to assume any real numerical value, usually in a given range), integer (assuming only integer values), or categorical (assuming one of a limited number of values). Categorical variables can be either numerical (1, 2, 3) or text (payments current, payments not current, bankrupt). Categorical variables can also be unordered (called *nominal variables*) with categories such as North America, Europe, and Asia; or they can be ordered (called *ordinal variables*) with categories such as high value, low value, and nil value.

Continuous variables can be handled by most data mining routines. In XLMiner, all routines take continuous variables, with the exception of the naive Bayes classifier, which deals exclusively with categorical variables. The machine learning roots of data mining grew out of problems with categorical outcomes; the roots of statistics lie in the analysis of continuous variables. Sometimes, it is desirable to convert continuous variables to categorical variables. This is done most typically in the case of outcome variables, where the numerical variable is mapped to a decision (e.g., credit scores above a certain level mean "grant credit," a medical test result above a certain level means "start treatment"). XLMiner has a facility for this type of conversion.

Handling Categorical Variables   Categorical variables can also be handled by most routines but often require special handling. If the categorical variable is ordered (age category, degree of creditworthiness, etc.), we can often use it as is, as if it were a continuous variable. The smaller the number of categories, and the less they represent equal increments of value, the more problematic this procedure becomes, but it often works well enough.

Categorical variables, however, often cannot be used as is. In those cases they must be decomposed into a series of dummy binary variables. For example, a single variable that can have possible values of "student," "unemployed," "employed," or "retired" would be split into four separate variables:

Student—Yes/No
Unemployed—Yes/No
Employed—Yes/No
Retired—Yes/No

Note that only three of the variables need to be used; if the values of three are known, the fourth is also known. For example, given that these four values are the only possible ones, we can know that if a person is neither student, unemployed, nor employed, he or she must be retired. In some routines (e.g., regression and logistic regression), you should not use all four variables—the redundant information will cause the algorithm to fail. XLMiner has a utility to convert categorical variables to binary dummies.

Variable Selection   More is not necessarily better when it comes to selecting variables for a model. Other things being equal, parsimony, or compactness, is a desirable feature in a model. For one thing, the more variables we include, the greater the number of records we will need to assess relationships among the

TABLE 2.1

Advertising	Sales
239	514
364	789
602	550
644	1386
770	1394
789	1440
911	1354

variables. Fifteen records may suffice to give us a rough idea of the relationship between Y and a single predictor variable X. If we now want information about the relationship between Y and 15 predictor variables $X_1 \cdots X_{15}$, 15 records will not be enough (each estimated relationship would have an average of only one record's worth of information, making the estimate very unreliable).

Overfitting The more variables we include, the greater the risk of over-fitting the data. What is overfitting?

In Table 2.1 we show hypothetical data about advertising expenditures in one time period and sales in a subsequent time period (a scatterplot of the data is shown in Figure 2.2). We could connect up these points with a smooth but complicated function, one that explains all these data points perfectly and leaves no error (residuals). This can be seen in Figure 2.3. However, we can see that such a curve is unlikely to be accurate, or even useful, in predicting future sales on the basis of advertising expenditures (e.g., it is hard to believe that increasing expenditures from \$400 to \$500 will actually decrease revenue).

A basic purpose of building a model is to describe relationships among variables in such a way that this description will do a good job of predicting future

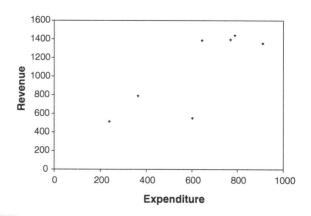

FIGURE 2.2   SCATTERPLOT FOR ADVERTISING AND SALES DATA

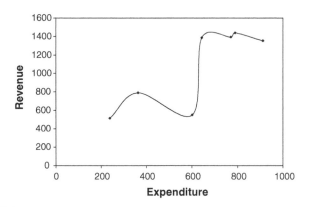

FIGURE 2.3   **SCATTERPLOT SMOOTHED**

outcome (dependent) values on the basis of future predictor (independent) values. Of course, we want the model to do a good job of describing the data we have, but we are more interested in its performance with future data.

In the example above, a simple straight line might do a better job than the complex function does of predicting future sales on the basis of advertising. Instead, we devised a complex function that fit the data perfectly, and in doing so, we overreached. We ended up "explaining" some variation in the data that was nothing more than chance variation. We mislabeled the noise in the data as if it were a signal.

Similarly, we can add predictors to a model to sharpen its performance with the data at hand. Consider a database of 100 individuals, half of whom have contributed to a charitable cause. Information about income, family size, and zip code might do a fair job of predicting whether or not someone is a contributor. If we keep adding additional predictors, we can improve the performance of the model with the data at hand and reduce the misclassification error to a negligible level. However, this low error rate is misleading because it probably includes spurious "explanations."

For example, one of the variables might be height. We have no basis in theory to suppose that tall people might contribute more or less to charity, but if there are several tall people in our sample and they just happened to contribute heavily to charity, our model might include a term for height—the taller you are, the more you will contribute. Of course, when the model is applied to additional data, it is likely that this will not turn out to be a good predictor.

If the dataset is not much larger than the number of predictor variables, it is very likely that a spurious relationship like this will creep into the model. Continuing with our charity example, with a small sample just a few of whom are tall, whatever the contribution level of tall people may be, the algorithm is tempted to attribute it to their being tall. If the dataset is very large relative to the number of predictors, this is less likely. In such a case, each predictor must help predict the outcome for a large number of cases, so the job it does is much less dependent on just a few cases, which might be flukes.

Somewhat surprisingly, even if we know for a fact that a higher degree curve is the appropriate model, if the model-fitting dataset is not large enough, a lower degree function (that is not as likely to fit the noise) is likely to perform better. Overfitting can also result from the application of many different models, from which the best performing is selected (see below).

How Many Variables and How Much Data?   Statisticians give us procedures to learn with some precision how many records we would need to achieve a given degree of reliability with a given dataset and a given model. Data miners' needs are usually not so precise, so we can often get by with rough rules of thumb. A good rule of thumb is to have 10 records for every predictor variable. Another, used by Delmaster and Hancock (2001, p. 68) for classification procedures, is to have at least $6 \times m \times p$ records, where m is the number of outcome classes and p is the number of variables.

Even when we have an ample supply of data, there are good reasons to pay close attention to the variables that are included in a model. Someone with domain knowledge (i.e., knowledge of the business process and the data) should be consulted, as knowledge of what the variables represent can help build a good model and avoid errors.

For example, the amount spent on shipping might be an excellent predictor of the total amount spent, but it is not a helpful one. It will not give us any information about what distinguishes high-paying from low-paying customers that can be put to use with future prospects because we will not have the information on the amount paid for shipping for prospects that have not yet bought anything.

In general, compactness or parsimony is a desirable feature in a model. A matrix of scatterplots can be useful in variable selection. In such a matrix, we can see at a glance scatterplots for all variable combinations. A straight line would be an indication that one variable is exactly correlated with another. Typically, we would want to include only one of them in our model. The idea is to weed out irrelevant and redundant variables from our model.

Outliers   The more data we are dealing with, the greater the chance of encountering erroneous values resulting from measurement error, data entry error, or the like. If the erroneous value is in the same range as the rest of the data, it may be harmless. If it is well outside the range of the rest of the data (e.g., a misplaced decimal), it may have a substantial effect on some of the data mining procedures we plan to use.

Values that lie far away from the bulk of the data are called *outliers*. The term *far away* is deliberately left vague because what is or is not called an outlier is basically an arbitrary decision. Analysts use rules of thumb such as "anything over 3 standard deviations away from the mean is an outlier," but no statistical rule can tell us whether such an outlier is the result of an error. In this statistical sense, an outlier is not necessarily an invalid data point; it is just a distant data point.

The purpose of identifying outliers is usually to call attention to values that need further review. We might come up with an explanation looking at the data—in the case of a misplaced decimal, this is likely. We might have no explanation but know that the value is wrong—a temperature of 178°F for a sick person. Or, we might conclude that the value is within the realm of possibility and leave it alone. All these are judgments best made by someone with *domain knowledge*, knowledge of the particular application being considered: direct mail, mortgage finance, and so on, as opposed to technical knowledge of statistical or data mining procedures. Statistical procedures can do little beyond identifying the record as something that needs review.

If manual review is feasible, some outliers may be identified and corrected. In any case, if the number of records with outliers is very small, they might be treated as missing data. How do we inspect for outliers? One technique in Excel is to sort the records by the first column, then review the data for very large or very small values in that column. Then repeat for each successive column. Another option is to examine the minimum and maximum values of each column using Excel's min and max functions. For a more automated approach that considers each record as a unit, clustering techniques could be used to identify clusters of one or a few records that are distant from others. Those records could then be examined.

Missing Values   Typically, some records will contain missing values. If the number of records with missing values is small, those records might be omitted. However, if we have a large number of variables, even a small proportion of missing values can affect a lot of records. Even with only 30 variables, if only 5% of the values are missing (spread randomly and independently among cases and variables), almost 80% of the records would have to be omitted from the analysis. (The chance that a given record would escape having a missing value is $0.95^{30} = 0.215$.)

An alternative to omitting records with missing values is to replace the missing value with an imputed value, based on the other values for that variable across all records. For example, if among 30 variables, household income is missing for a particular record, we might substitute the mean household income across all records. Doing so does not, of course, add any information about how household income affects the outcome variable. It merely allows us to proceed with the analysis and not lose the information contained in this record for the other 29 variables. Note that using such a technique will understate the variability in a dataset. However, we can assess variability and the performance of our data mining technique, using the validation data, and therefore this need not present a major problem.

Some datasets contain variables that have a very large number of missing values. In other words, a measurement is missing for a large number of records. In that case, dropping records with missing values will lead to a large loss of data. Imputing the missing values might also be useless, as the imputations are based on a small number of existing records. An alternative is to examine the importance of the predictor. If it is not very crucial, it can be dropped. If it is important, perhaps a proxy variable with fewer missing values can be used instead. When such a predictor is deemed central, the best solution is to invest in obtaining the missing data.

Significant time may be required to deal with missing data, as not all situations are susceptible to automated solution. In a messy dataset, for example, a 0 might mean two things: (1) the value is missing, or (2) the value is actually zero. In the credit industry, a 0 in the "past due" variable might mean a customer who is fully paid up, or a customer with no credit history at all—two very different situations. Human judgment may be required for individual cases or to determine a special rule to deal with the situation.

Normalizing (Standardizing) the Data   Some algorithms require that the data be normalized before the algorithm can be implemented effectively. To normalize the data, we subtract the mean from each value and divide by the standard deviation of the resulting deviations from the mean. In effect, we are expressing each value as the "number of standard deviations away from the mean," also called a *z-score*.

To consider why this might be necessary, consider the case of clustering. Clustering typically involves calculating a distance measure that reflects how far each record is from a cluster center or from other records. With multiple variables, different units will be used: days, dollars, counts, and so on. If the dollars are in the thousands and everything else is in the tens, the dollar variable will come to dominate the distance measure. Moreover, changing units from (say) days to hours or months could alter the outcome completely.

Data mining software, including XLMiner, typically has an option that normalizes the data in those algorithms where it may be required. It is an option rather than an automatic feature of such algorithms because there are situations where we want each variable to contribute to the distance measure in proportion to its scale.

Use and Creation of Partitions

In supervised learning, a key question presents itself: How well will our prediction or classification model perform when we apply it to new data? We are particularly interested in comparing the performance among various models so that we can choose the one we think will do the best when it is actually implemented.

At first glance, we might think it best to choose the model that did the best job of classifying or predicting the outcome variable of interest with the data at hand. However, when we use the same data both to develop the model and to assess its performance, we introduce bias. This is because when we pick the model that works best with the data, this model's superior performance comes from two sources:

- A superior model
- Chance aspects of the data that happen to match the chosen model better than they match other models

The latter is a particularly serious problem with techniques (such as trees and neural nets) that do not impose linear or other structure on the data, and thus end up overfitting it.

To address this problem, we simply divide (partition) our data and develop our model using only one of the partitions. After we have a model, we try it out on another partition and see how it performs, which we can measure in several ways. In a classification model, we can count the proportion of held-back records that were misclassified. In a prediction model, we can measure the residuals (errors) between the predicted values and the actual values. We typically deal with two or three partitions: a training set, a validation set, and sometimes an additional test set. Partitioning the data into training, validation, and test sets is done either randomly according to predetermined proportions or by specifying which records go into which partitioning according to some relevant variable (e.g., in time-series forecasting, the data are partitioned according to their chronological order). In most cases, the partitioning should be done randomly to avoid getting a biased partition. It is also possible (although cumbersome) to divide the data into more than three partitions by successive partitioning (e.g., divide the initial data into three partitions, then take one of those partitions and partition it further).

Training Partition   The training partition, typically the largest partition, contains the data used to build the various models we are examining. The same training partition is generally used to develop multiple models.

Validation Partition   This partition (sometimes called the *test partition*) is used to assess the performance of each model so that you can compare models and pick the best one. In some algorithms (e.g., classification and regression trees), the validation partition may be used in automated fashion to tune and improve the model.

Test Partition This partition (sometimes called the *holdout* or *evaluation partition*) is used if we need to assess the performance of the chosen model with new data.

Why have both a validation and a test partition? When we use the validation data to assess multiple models and then pick the model that does best with the validation data, we again encounter another (lesser) facet of the overfitting problem—chance aspects of the validation data that happen to match the chosen model better than they match other models.

The random features of the validation data that enhance the apparent performance of the chosen model will probably not be present in new data to which the model is applied. Therefore, we may have overestimated the accuracy of our model. The more models we test, the more likely it is that one of them will be particularly effective in explaining the noise in the validation data. Applying the model to the test data, which it has not seen before, will provide an unbiased estimate of how well it will do with new data. Figure 2.4 shows the three partitions and their use in the data mining process. When we are concerned mainly with finding the best model and less with exactly how well it will do, we might use only training and validation partitions.

Note that with some algorithms, such as nearest-neighbor algorithms, the training data itself is the model—records in the validation and test partitions, and in new data, are compared to records in the training data to find the nearest neighbor(s). As k-nearest neighbors is implemented in XLMiner and as discussed in this book, the use of two partitions is an essential part of the classification or prediction process, not merely a way to improve or assess it. Nonetheless, we

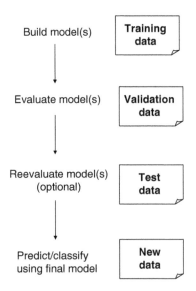

| FIGURE 2.4 | **THREE DATA PARTITIONS AND THEIR ROLE IN THE DATA MINING PROCESS** |

can still interpret the error in the validation data in the same way that we would interpret error from any other model.

> XLMiner has a facility for partitioning a dataset randomly or according to a user-specified variable. For user-specified partitioning, a variable should be created that contains the value *t* (training), *v* (validation), or *s* (test), according to the designation of that record.

2.6 BUILDING A MODEL: EXAMPLE WITH LINEAR REGRESSION

Let us go through the steps typical to many data mining tasks using a familiar procedure: multiple linear regression. This will help us understand the overall process before we begin tackling new algorithms. We illustrate the Excel procedure using XLMiner.

Boston Housing Data

The Boston housing data contain information on neighborhoods in Boston for which several measurements are taken (e.g., crime rate, pupil/teacher ratio). The outcome variable of interest is the median value of a housing unit in the neighborhood. This dataset has 14 variables, and a description of each variable is given in Table 2.2. A sample of the data is shown in Figure 2.5.

The first row in the data represents the first neighborhood, which had an average per capita crime rate of 0.006, had 18% of the residential land zoned for lots over 25,000 square feet (ft^2), 2.31% of the land devoted to nonretail business, no border on the Charles River, and so on.

TABLE 2.2   DESCRIPTION OF VARIABLES IN BOSTON HOUSING DATASET

CRIM	Crime rate
ZN	Percentage of residential land zoned for lots over 25,000 ft^2
INDUS	Percentage of land occupied by nonretail business
CHAS	Charles River dummy variable ($= 1$ if tract bounds river; $= 0$ otherwise)
NOX	Nitric oxide concentration (parts per 10 million)
RM	Average number of rooms per dwelling
AGE	Percentage of owner-occupied units built prior to 1940
DIS	Weighted distances to five Boston employment centers
RAD	Index of accessibility to radial highways
TAX	Full-value property tax rate per $10,000
PTRATIO	Pupil/teacher ratio by town
B	$1000(Bk \text{ minus } 0.63)^2$, where Bk is the proportion of blacks by town
LSTAT	% Lower status of the population
MEDV	Median value of owner-occupied homes in $1000s

CRIM	ZN	INDUS	CHAS	NOX	RM	AGE	DIS	RAD	TAX	PTRATIO	B	LSTAT	MEDV	CAT. MEDV
0.00632	18	2.31	0	0.538	6.58	65.2	4.09	1	296	15.3	396.9	4.98	24	0
0.02731	0	7.07	0	0.469	6.42	78.9	4.97	2	242	17.8	396.9	9.14	21.6	0
0.02729	0	7.07	0	0.469	7.19	61.1	4.97	2	242	17.8	392.83	4.03	34.7	1
0.03237	0	2.18	0	0.458	7	45.8	6.06	3	222	18.7	394.63	2.94	33.4	1
0.06905	0	2.18	0	0.458	7.15	54.2	6.06	3	222	18.7	396.9	5.33	36.2	1
0.02985	0	2.18	0	0.458	6.43	58.7	6.06	3	222	18.7	394.12	5.21	28.7	0
0.08829	13	7.87	0	0.524	6.01	66.6	5.56	5	311	15.2	395.6	12.43	22.9	0
0.14455	13	7.87	0	0.524	6.17	96.1	5.95	5	311	15.2	396.9	19.15	27.1	0
0.21124	13	7.87	0	0.524	5.63	100	6.08	5	311	15.2	386.63	29.93	16.5	0

FIGURE 2.5   FIRST NINE RECORDS IN THE BOSTON HOUSING DATA

Modeling Process

We now describe in detail the various model stages using the Boston housing example.

1. *Purpose.* Let us assume that the purpose of our data mining project is to predict the median house value in small Boston area neighborhoods.

2. *Obtain the Data.* We will use the Boston housing data. The dataset in question is small enough that we do not need to sample from it—we can use it in its entirety.

3. *Explore, Clean, and Preprocess the Data.* Let us look first at the description of the variables (e.g., crime rate, number of rooms per dwelling) to be sure that we understand them all. These descriptions are available on the "description" tab on the worksheet, as is a Web source for the dataset. They all seem fairly straightforward, but this is not always the case. Often, variable names are cryptic and their descriptions may be unclear or missing.

   It is useful to pause and think about what the variables mean and whether they should be included in the model. Consider the variable TAX. At first glance, we consider that the tax on a home is usually a function of its assessed value, so there is some circularity in the model—we want to predict a home's value using TAX as a predictor, yet TAX itself is determined by a home's value. TAX might be a very good predictor of home value in a numerical sense, but would it be useful if we wanted to apply our model to homes whose assessed value might not be known? Reflect, though, that the TAX variable, like all the variables, pertains to the average in a neighborhood, not to individual homes. Although the purpose of our inquiry has not been spelled out, it is possible that at some stage we might want to apply a model to individual homes, and in such a case, the neighborhood TAX value would be a useful predictor. So we will keep TAX in the analysis for now.

RM	AGE	DIS
79.29	96.2	2.04
8.78	82.9	1.90
8.75	83	2.89
8.70	88.8	1.00

FIGURE 2.6   OUTLIER IN BOSTON HOUSING DATA

In addition to these variables, the dataset also contains an additional variable, CAT.MEDV, which has been created by categorizing median value (MEDV) into two categories, high and low. (There are a couple of aspects of MEDV, the median house value, that bear noting. For one thing, it is quite low, since it dates from the 1970s. For another, there are a lot of 50s, the top value. It could be that median values above $50, 000$ were recorded as $50, 000$.) The variable CAT.MEDV is actually a categorical variable created from MEDV. If MEDV $\geq$ $30,000$, CAT.MEDV $= 1$. If MEDV $\leq$ $30,000$, CAT.MEDV $= 0$. If we were trying to categorize the cases into high and low median values, we would use CAT.MEDV instead of MEDV. As it is, we do not need CAT.MEDV, so we leave it out of the analysis. We are left with 13 independent (predictor) variables, which can all be used.

It is also useful to check for outliers that might be errors. For example, suppose that the RM (number of rooms) column looked like the one in Figure 2.6, after sorting the data in descending order based on rooms. We can tell right away that the 79.29 is in error—no neighborhood is going to have houses that have an average of 79 rooms. All other values are between 3 and 9. Probably, the decimal was misplaced and the value should be 7.929. (This hypothetical error is not present in the dataset supplied with XLMiner.)

4. *Reduce the Data and Partition Them into Training, Validation, and Test Partitions.* Our dataset has only 13 variables, so data reduction is not required. If we had many more variables, at this stage we might want to apply a variable reduction technique such as principal components analysis to consolidate multiple similar variables into a smaller number of variables. Our task is to predict the median house value and then assess how well that prediction does. We will partition the data into a training set to build the model and a validation set to see how well the model does. This technique is part of the "supervised learning" process in classification and prediction problems. These are problems in which we know the class or value of the outcome variable for some data, and we want to use those data in developing a model that can then be applied to other data where that value is unknown.

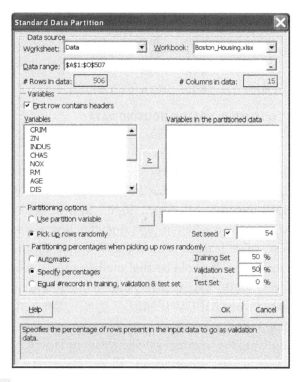

FIGURE 2.7 PARTITIONING THE DATA. THE DEFAULT IN XLMINER PARTITIONS THE DATA INTO 60% TRAINING DATA, 40% VALIDATION DATA, AND 0% TEST DATA. IN THIS EXAMPLE, A PARTITION OF 50% TRAINING AND 50% VALIDATION IS USED

In Excel, select *XLMiner > Partition* and the dialog box shown in Figure 2.7 appears. Here we specify which data range is to be partitioned and which variables are to be included in the partitioned dataset. The partitioning can be handled in one of two ways:

a. The dataset can have a partition variable that governs the division into training and validation partitions (e.g., 1 = training, 2 = validation).

b. The partitioning can be done randomly. If the partitioning is done randomly, we have the option of specifying a seed for randomization (which has the advantage of letting us duplicate the same random partition later should we need to). In this example, a seed of 54 is used.

In this case we divide the data into two partitions: training and validation. The training partition is used to build the model, and the validation partition is used to see how well the model does when applied to new data. We need to specify the percent of the data used in each partition.

Note: Although we are not using it here, a test partition might also be used.

Typically, a data mining endeavor involves testing multiple models, perhaps with multiple settings on each model. When we train just one model and try it out on the validation data, we can get an unbiased idea of how it might perform on more such data. However, when we train many models and use the validation data to see how each one does, then choose the best-performing model, the validation data no longer provide an unbiased estimate of how the model might do with more data. By playing a role in choosing the best model, the validation data have become part of the model itself. In fact, several algorithms (e.g., classification and regression trees) explicitly factor validation data into the model-building algorithm itself (e.g., in pruning trees). Models will almost always perform better with the data they were trained on than with fresh data. Hence, when validation data are used in the model itself, or when they are used to select the best model, the results achieved with the validation data, just as with the training data, will be overly optimistic.

The test data, which should not be used in either the model-building or model selection process, can give a better estimate of how well the chosen model will do with fresh data. Thus, once we have selected a final model, we apply it to the test data to get an estimate of how well it will actually perform.

5. *Determine the Data Mining Task*. In this case, as noted, the specific task is to predict the value of MEDV using the 13 predictor variables.

6. *Choose the Technique*. In this case, it is multiple linear regression. Having divided the data into training and validation partitions, we can use XLMiner to build a multiple linear regression model with the training data. We want to predict median house price on the basis of all the other values.

7. *Use the Algorithm to Perform the Task*. In XLMiner, we select *Prediction > Multiple Linear Regression*, as shown in Figure 2.8. The variable MEDV is selected as the output (dependent) variable, the variable CAT.MEDV is left unused, and the remaining variables are all selected as input (independent or predictor) variables. We ask XLMiner to show us the fitted values on the training data as well as the predicted values (scores) on the validation data, as shown in Figure 2.9. XLMiner produces standard regression output, but for now we defer that as well as the more advanced options displayed above. (See Chapter 6 or the user documentation for XLMiner for more information.) Rather, we review the predictions themselves. Figure 2.10 shows the predicted values for the first few records in the training data along with the actual values and the residual (prediction error). Note that the predicted values would often be called the *fitted values* since they are for the records to which the model was fit. The results for the validation

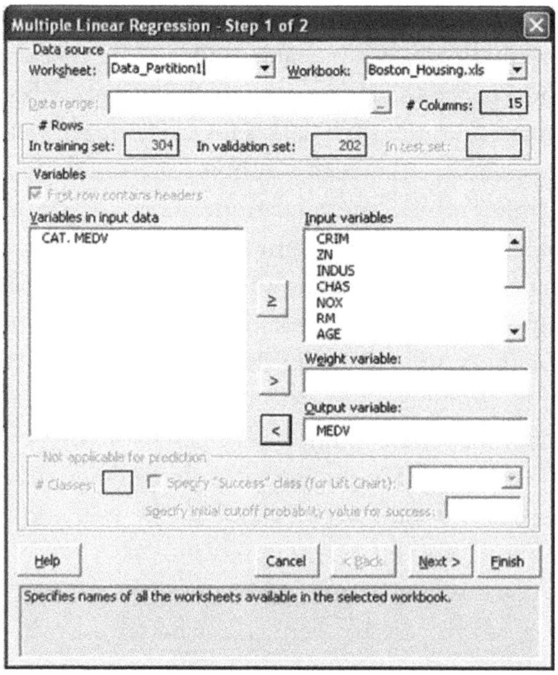

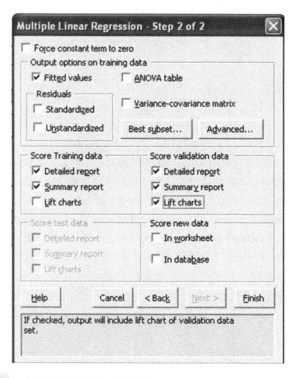

FIGURE 2.8 USING XLMINER FOR MULTIPLE LINEAR REGRESSION

FIGURE 2.9 SPECIFYING THE OUTPUT

XLMiner : Multiple Linear Regression - Prediction of Training Data

Data range | ['Boston_Housing']'Data_Partition2'!$C$19:$P$271 | Back to

Row Id.	Predicted Value	Actual Value	Residual	CRIM	ZN	INDUS	CHAS	NOX	RM	AGE	DIS	RAD	TAX	PTRATIO	B	LSTAT
1	30.02788078	24	-6.027880779	0.00632	18	2.31	0	0.538	6.575	65.2	4.09	1	296	15.3	396.9	4.98
2	24.90910941	21.6	-3.309109407	0.02731	0	7.07	0	0.469	6.421	78.9	4.9671	2	242	17.8	396.9	9.14
3	30.9549987	34.7	3.745001299	0.02729	0	7.07	0	0.469	7.185	61.1	4.9671	2	242	17.8	392.83	4.03
4	28.07549961	33.4	5.324500385	0.03237	0	2.18	0	0.458	6.998	45.8	6.0622	3	222	18.7	394.63	2.94
5	27.9091436	36.2	8.290856402	0.06905	0	2.18	0	0.458	7.147	54.2	6.0622	3	222	18.7	396.9	5.33
7	23.65328843	22.9	-0.753288432	0.08829	12.5	7.87	0	0.524	6.012	66.6	5.5605	5	311	15.2	395.6	12.43
8	21.11420949	27.1	5.98579051	0.14455	12.5	7.87	0	0.524	6.172	96.1	5.9505	5	311	15.2	396.9	19.15
11	21.09801414	15	-6.09801414	0.22489	12.5	7.87	0	0.524	6.377	94.3	6.3467	5	311	15.2	392.52	20.45
12	22.12589464	18.9	-3.225894637	0.11747	12.5	7.87	0	0.524	6.009	82.9	6.2267	5	311	15.2	396.9	13.27
14	17.61047844	20.4	2.789521563	0.62976	0	8.14	0	0.538	5.949	61.8	4.7075	4	307	21	396.9	8.26
21	11.62929772	13.6	1.970702281	1.25179	0	8.14	0	0.538	5.57	98.1	3.7979	4	307	21	376.57	21.02

FIGURE 2.10   **PREDICTIONS FOR THE TRAINING DATA**

data are shown in Figure 2.11. The prediction error for the training and validation data are compared in Figure 2.12.

Prediction error can be measured in several ways. Three measures produced by XLMiner are shown in Figure 2.12. On the right is the *average error*, simply the average of the residuals (errors). In both cases it is quite small relative to the units of MEDV, indicating that, on balance, predictions average about right—our predictions are "unbiased." Of course, this simply means that the positive and negative errors balance out. It tells us nothing about how large these errors are.

The *total sum of squared errors* on the left adds up the squared errors, so whether an error is positive or negative, it contributes just the same. However, this sum does not yield information about the size of the typical error.

The *RMS error* (root-mean-squared error) is perhaps the most useful term of all. It takes the square root of the average squared error; thus, it gives an idea of the typical error (whether positive or negative) in the same scale as that used for the original data. As we might expect, the RMS error for the validation data (5.66 thousand $), which the model is

XLMiner : Multiple Linear Regression - Prediction of Validation Data

Data range | ['Boston_Housing']'Data_Partition2'!$C$272:$P$524 | Back to

Row Id.	Predicted Value	Actual Value	Residual	CRIM	ZN	INDUS	CHAS	NOX	RM	AGE	DIS	RAD	TAX	PTRATIO	B	LSTAT
6	24.09753481	28.7	4.602465185	0.02985	0	2.18	0	0.458	6.43	58.7	6.0622	3	222	18.7	394.12	5.21
9	14.04293006	16.5	2.457069944	0.21124	12.5	7.87	0	0.524	5.631	100	6.0821	5	311	15.2	386.63	29.93
10	20.02731265	18.9	-1.127312654	0.17004	12.5	7.87	0	0.524	6.004	85.9	6.5921	5	311	15.2	386.71	17.1
13	22.31053056	21.7	-0.610530557	0.09378	12.5	7.87	0	0.524	5.889	39	5.4509	5	311	15.2	390.5	15.71
15	17.56538825	18.2	0.634611749	0.63796	0	8.14	0	0.538	6.096	84.5	4.4619	4	307	21	380.02	10.26
16	17.30239527	19.9	2.597604726	0.62739	0	8.14	0	0.538	5.834	56.5	4.4986	4	307	21	395.62	8.47
17	18.71477307	23.1	4.385226932	1.05393	0	8.14	0	0.538	5.935	29.3	4.4986	4	307	21	386.85	6.58
18	15.81653483	17.5	1.683465174	0.7842	0	8.14	0	0.538	5.99	81.7	4.2579	4	307	21	386.75	14.67
19	14.38144028	20.2	5.818559715	0.80271	0	8.14	0	0.538	5.456	36.6	3.7965	4	307	21	288.99	11.69
20	16.554546	18.2	1.645453996	0.7258	0	8.14	0	0.538	5.727	69.5	3.7965	4	307	21	390.95	11.28
22	16.30306818	19.6	3.296931816	0.85204	0	8.14	0	0.538	5.965	89.2	4.0123	4	307	21	392.53	13.83
25	14.58925181	15.6	1.010748192	0.75026	0	8.14	0	0.538	5.924	94.1	4.3996	4	307	21	394.33	16.3

FIGURE 2.11   **PREDICTIONS FOR THE VALIDATION DATA**

Training Data scoring - Summary Report

Total sum of squared errors	RMS Error	Average Error
4136.091425	4.043289187	-1.12501E-06

(a)

Validation Data scoring - Summary Report

Total sum of squared errors	RMS Error	Average Error
8117.85953	5.66448597	0.961240934

(b)

FIGURE 2.12   ERROR RATES FOR (*A*) TRAINING AND (*B*) VALIDATION DATA (ERROR FIGURES ARE IN THOUSANDS OF $)

seeing for the first time in making these predictions, is larger than for the training data (4.04 thousand $), which were used in training the model.

8. *Interpret the Results.* At this stage we would typically try other prediction algorithms (e.g., regression trees) and see how they do errorwise. We might also try different "settings" on the various models (e.g., we could use the *best subsets* option in multiple linear regression to chose a reduced set of variables that might perform better with the validation data). After choosing the best model (typically, the model with the lowest error on the validation data while also recognizing that "simpler is better"), we use that model to predict the output variable in fresh data. These steps are covered in more detail in the analysis of cases.

9. *Deploy the Model.* After the best model is chosen, it is applied to new data to predict MEDV for records where this value is unknown. This was, of course, the overall purpose.

2.7 USING EXCEL FOR DATA MINING

An important aspect of this process to note is that the heavy-duty analysis does not necessarily require huge numbers of records. The dataset to be analyzed may have millions of records, of course, but in doing multiple linear regression or applying a classification tree, the use of a sample of 20,000 is likely to yield as accurate an answer as that obtained when using the entire dataset. The principle involved is the same as the principle behind polling: If sampled judiciously, 2000 voters can give an estimate of the entire population's opinion within one or two percentage points. (See How Many Variables and How Much Data in Section 2.5 for further discussion.)

Therefore, in most cases, the number of records required in each partition (training, validation, and test) can be accommodated within the rows allowed by Excel. Of course, we need to get those records into Excel, and for this purpose the standard version of XLMiner provides an interface for random sampling of records from an external database.

Similarly, we need to apply the results of our analysis to a large database, and for this purpose the standard version of XLMiner has a facility for storing models and scoring them to an external database. For example, XLMiner would write an additional column (variable) to the database consisting of the predicted purchase amount for each record.

> XLMiner has a facility for drawing a sample from an external database. The sample can be drawn at random or it can be stratified. It also has a facility to score data in the external database using the model that was obtained from the training data.

DATA MINING SOFTWARE TOOLS: THE STATE OF THE MARKET By Herb Edelstein[1]

Data mining uses a variety of tools to discover patterns and relationships in data that can be used to explain the data or make meaningful predictions. The need for ever more powerful tools is driven by the increasing breadth and depth of analytical problems. In order to deal with tens of millions of cases (rows) and hundreds or even thousands of variables (columns), organizations need scalable tools. A carefully designed GUI (graphical user interface) also makes it easier to create, manage, and apply predictive models.

Data mining is a complete process, not just a particular technique or algorithm. Industrial-strength tools support all phases of this process, handle all sizes of databases, and manage even the most complex problems.

The software must first be able to pull all the data together. The data mining tool may need to access multiple databases across different database management systems. Consequently, the software should support joining and subsetting of data from a range of sources. Because some of the data may be a terabyte or more, the software also needs to support a variety of sampling methodologies.

Next, the software must facilitate exploring and manipulating the data to create understanding and suggest a starting point for model building. When a database has hundreds or thousands of variables, it becomes an enormous task to select the variables that best describe the data and lead to the most robust predictions. Visualization tools can make it easier to identify the most important variables and find meaningful patterns in very large databases. Certain algorithms are particularly suited to guiding the selection of the most relevant variables. However, often the best predictors are not the variables in the database themselves, but some mathematical combination of these variables. This not only increases the number of variables to be evaluated, but the more complex transformations require a scripting language. Frequently, the data access tools use the DBMS language itself to make transformations directly on the underlying database.

Because building and evaluating models is an iterative process, a dozen or more exploratory models may be built before settling on the best model. While any individual model may take only a modest amount of time for the software to construct, computer usage can really add up unless the tool is running on powerful hardware. Although some people consider this phase to be what data mining is all about, it usually represents a relatively small part of the total effort.

Finally, after building, testing, and selecting the desired model, it is necessary to deploy it. A model that was built using a small subset of the data may now be applied to millions of cases or integrated into a real-time application, processing hundreds of transactions each second. For example, the model may be integrated into credit scoring or fraud detection applications. Over time the model should be evaluated and refined as needed.

Data mining tools can be general purpose (either embedded in a DBMS or stand-alones) or they can be application specific.

All the major database management system vendors have incorporated data mining capabilities into their products. Leading products include IBM DB2 Intelligent Miner, Microsoft SQL Server 2005, Oracle Data Mining, and Teradata Warehouse Miner. The target user for embedded data mining is a database professional. Not surprisingly, these products take advantage of database functionality, including using the DBMS to transform variables, storing models in the database, and extending the data access language to include model building and scoring the database. A few products also supply a separate graphical interface for building data mining models. Where the DBMS has parallel processing capabilities, embedded data mining tools will generally take advantage of it, resulting in better performance. As with the data mining suites described below, these tools offer an assortment of algorithms.

Stand-alone data mining tools can be based on a single algorithm or a collection of algorithms called a suite. Target users include both statisticians and analysts. Well-known single-algorithm products include KXEN; RuleQuest Research C5.0; and Salford Systems CART, MARS, and Treenet. Most of the top single-algorithm tools have also been licensed to suite vendors. The leading suites include SAS Enterprise Miner, IBM Modeler (previously SPSS Clementine), and Spotfire Miner (previously Insightful Miner). Suites are characterized by providing a wide range of functionality and an interface designed to enhance model-building productivity. Many suites have outstanding visualization tools and links to statistical packages that extend the range of tasks they can perform, and most provide a procedural scripting language for more complex transformations. They use a graphical workflow interface to outline the entire data mining process. The suite vendors are working to link their tools more closely to underlying DBMSs; for example, data transformations might be handled by the DBMS. Data mining models can be exported to be incorporated into the DBMS either through generating SQL, procedural language code (e.g., C++ or Java), or a standardized data mining model language called Predictive Model Markup Language (PMML).

Application-specific tools, in contrast to the other types, are intended for particular analytic applications such as credit scoring, customer retention, or product marketing. Their focus may be further sharpened to address the needs of certain markets such as mortgage lending or financial services. The target user is an analyst with expertise in the applications domain. Therefore, the interfaces, the algorithms, and even the terminology are customized for that particular industry, application, or customer. While less flexible than general-purpose tools, they offer the advantage of already incorporating domain knowledge into the product design and can provide very good solutions with less effort. Data mining companies including SAS and SPSS offer vertical market tools, as do industry specialists such as Fair Isaac.

The tool used in this book, XLMiner, is a suite with both sampling and scoring capabilities. While Excel itself is not a suitable environment for dealing with thousands of columns and millions of rows, it is a familiar workspace to business analysts and can be used as a work platform to support other tools. An Excel add-in such as XLMiner (which uses non-Excel computational engines) is user friendly and can be used in conjunction with sampling techniques for prototyping, small-scale, and educational applications of data mining.

---

[1]Herb Edelstein is president of Two Crows Consulting (www.twocrows.com), a leading data mining consulting firm near Washington, D.C. He is an internationally recognized expert in data mining and data warehousing, a widely published author on these topics, and a popular speaker. 2006 Herb Edelstein.

SAS and *Enterprise Miner* are trademarks of SAS Institute, Inc. *CART, MARS*, and *TreeNet* are trademarks of Salford Systems. *XLMiner* is a trademark of Cytel Inc. *SPSS* and *Clementine* are trademarks of SPSS, Inc.

PROBLEMS

2.1   Assuming that data mining techniques are to be used in the following cases, identify whether the task required is supervised or unsupervised learning.

   a. Deciding whether to issue a loan to an applicant based on demographic and financial data (with reference to a database of similar data on prior customers).

   b. In an online bookstore, making recommendations to customers concerning additional items to buy based on the buying patterns in prior transactions.

   c. Identifying a network data packet as dangerous (virus, hacker attack) based on comparison to other packets whose threat status is known.

   d. Identifying segments of similar customers.

   e. Predicting whether a company will go bankrupt based on comparing its financial data to those of similar bankrupt and nonbankrupt firms.

   f. Estimating the repair time required for an aircraft based on a trouble ticket.

   g. Automated sorting of mail by zip code scanning.

   h. Printing of custom discount coupons at the conclusion of a grocery store checkout based on what you just bought and what others have bought previously.

2.2   Describe the difference in roles assumed by the validation partition and the test partition.

2.3   Consider the sample from a database of credit applicants in Figure 2.13. Comment on the likelihood that it was sampled randomly, and whether it is likely to be a useful sample.

2.4   Consider the sample from a bank database shown in Figure 2.14; it was selected randomly from a larger database to be the training set. Personal Loan indicates

OBS#	CHK_ACCT	DURATION	HISTORY	NEW_CAR	USED_CAR	FURNITURE	RADIO/TV	EDUCATION	RETRAINING	AMOUNT	SAV_ACCT	RESPONSE
1	0	6	4	0	0	0	1	0	0	1169	4	1
8	1	36	2	0	1	0	0	0	0	6948	0	1
16	0	24	2	0	0	0	1	0	0	1282	1	0
24	1	12	4	0	1	0	0	0	0	1804	1	1
32	0	24	2	0	0	1	0	0	0	4020	0	1
40	1	9	2	0	0	0	1	0	0	458	0	1
48	0	6	2	0	1	0	0	0	0	1352	2	1
56	3	6	1	1	0	0	0	0	0	783	4	1
64	1	48	0	0	0	0	0	0	1	14421	0	0
72	3	7	4	0	0	0	1	0	0	730	4	1
80	1	30	2	0	0	1	0	0	0	3832	0	1
88	1	36	2	0	0	0	0	1	0	12612	1	0
96	1	54	0	0	0	0	0	0	1	15945	0	0
104	1	9	4	0	0	1	0	0	0	1919	0	1
112	2	15	2	0	0	0	0	1	0	392	0	1

FIGURE 2.13   SAMPLE FROM A DATABASE OF CREDIT APPLICANTS

ID	Age	Experience	Income	ZIP Code	Family	CCAvg	Educ.	Mortgage	Personal Loan	Securities Account
1	25	1	49	91107	4	1.60	1	0	0	1
4	35	9	100	94112	1	2.70	2	0	0	0
5	35	8	45	91330	4	1.00	2	0	0	0
6	37	13	29	92121	4	0.40	2	155	0	0
9	35	10	81	90089	3	0.60	2	104	0	0
11	65	39	105	94710	4	2.40	3	0	0	0
12	29	5	45	90277	3	0.10	2	0	0	0
18	42	18	81	94305	4	2.40	1	0	0	0
20	55	28	21	94720	1	0.50	2	0	0	1
23	29	5	62	90277	1	1.20	1	260	0	0
26	43	19	29	94305	3	0.50	1	97	0	0
27	40	16	83	95064	4	0.20	3	0	0	0
29	56	30	48	94539	1	2.20	3	0	0	0
31	59	35	35	93106	1	1.20	3	122	0	0
32	40	16	29	94117	1	2.00	2	0	0	0
35	31	5	50	94035	4	1.80	3	0	0	0
36	48	24	81	92647	3	0.70	1	0	0	0
37	59	35	121	94720	1	2.90	1	0	0	0
38	51	25	71	95814	1	1.40	3	198	0	0
40	38	13	80	94115	4	0.70	3	285	0	0
41	57	32	84	92672	3	1.60	3	0	0	1

FIGURE 2.14   **SAMPLE FROM A BANK DATABASE**

whether a solicitation for a personal loan was accepted and is the response variable. A campaign is planned for a similar solicitation in the future, and the bank is looking for a model that will identify likely responders. Examine the data carefully and indicate what your next step would be.

2.5   Using the concept of overfitting, explain why when a model is fit to training data, zero error with those data is not necessarily good.

2.6   In fitting a model to classify prospects as purchasers or nonpurchasers, a certain company drew the training data from internal data that include demographic and purchase information. Future data to be classified will be lists purchased from other sources, with demographic (but not purchase) data included. It was found that "refund issued" was a useful predictor in the training data. Why is this not an appropriate variable to include in the model?

2.7   A dataset has 1000 records and 50 variables with 5% of the values missing, spread randomly throughout the records and variables. An analyst decides to remove records that have missing values. About how many records would you expect would be removed?

2.8   Normalize the data in Table 2.3, showing calculations.

TABLE 2.3

Age	Income ($)
25	49,000
56	156,000
65	99,000
32	192,000
41	39,000
49	57,000

2.9 Statistical distance between records can be measured in several ways. Consider Euclidean distance, measured as the square root of the sum of the squared differences. For the first two records in Table 2.3, it is

$$\sqrt{(25 - 56)^2 + (49,000 - 156,000)^2}.$$

Does normalizing the data change which two records are farthest from each other in terms of Euclidean distance?

2.10 Two models are applied to a dataset that has been partitioned. Model A is considerably more accurate than model B on the training data but slightly less accurate than model B on the validation data. Which model are you more likely to consider for final deployment?

2.11 The dataset ToyotaCorolla.xls contains data on used cars on sale during the late summer of 2004 in The Netherlands. It has 1436 records containing details on 38 attributes, including Price, Age, Kilometers, HP, and other specifications.

a. Explore the data using the data visualization (matrix plot) capabilities of XLMiner. Which of the pairs among the variables seem to be correlated?

b. We plan to analyze the data using various data mining techniques described in future chapters. Prepare the data for use as follows:

    i. The dataset has two categorical attributes, Fuel Type and Metallic.

      (a) Describe how you would convert these to binary variables.

      (b) Confirm this using XLMiner's utility to transform categorical data into dummies.

      (c) How would you work with these new variables to avoid including redundant information in models?

    ii. Prepare the dataset (as factored into dummies) for data mining techniques of supervised learning by creating partitions using XLMiner's data partitioning utility. Select all the variables and use default values for the random seed and partitioning percentages for training (50%), validation (30%), and test (20%) sets. Describe the roles that these partitions will play in modeling.

## PART TWO

Data Exploration and Dimension Reduction

## CHAPTER 3

Data Visualization

In this chapter we describe a set of plots that can be used to explore the multidimensional nature of a dataset. We present basic plots (bar charts, line graphs, and scatterplots), distribution plots (boxplots and histograms), and different enhancements that expand the capabilities of these plots to visualize more information. We focus on how the different visualizations and operations can support data mining tasks, from supervised (prediction, classification, and time series forecasting) to unsupervised tasks, and provide some guidelines on specific visualizations to use with each data mining task. We also describe the advantages of interactive visualization over static plots. The chapter concludes with a presentation of specialized plots that are suitable for data with special structure (hierarchical, network, and geographical).

3.1 USES OF DATA VISUALIZATION

The popular saying "a picture is worth a thousand words" refers to the ability to condense diffused verbal information into a compact and quickly understood graphical image. In the case of numbers, data visualization and numerical summarization provide us with both a powerful tool to explore data and an effective way to present results.

Where do visualization techniques fit into the data mining process, as described so far? Their use is primarily in the preprocessing portion of the data mining process. Visualization supports data cleaning by finding incorrect values (e.g., patients whose age is 999 or -1), missing values, duplicate rows, columns

Data Mining for Business Intelligence, By Galit Shmueli, Nitin R. Patel, and Peter C. Bruce
Copyright © 2010 Statistics.com and Galit Shmueli

with all the same value, and the like. Visualization techniques are also useful for variable derivation and selection: they can help determine which variables to include in the analysis and which might be redundant. They can also help with determining appropriate bin sizes, should binning of numerical variables be needed (e.g., a numerical outcome variable might need to be converted to a binary variable, as was done in the Boston housing data, if a yes/no decision is required). They can also play a role in combining categories as part of the data reduction process. Finally, if the data have yet to be collected and collection is expensive (as with the Pandora project at its outset, see Chapter 7), visualization methods can help determine, using a sample, which variables and metrics are useful.

In this chapter we focus on the use of graphical presentations for the purpose of *data exploration*, in particular with relation to predictive analytics. Although our focus is not on visualization for the purpose of data reporting, this chapter offers ideas as to the effectiveness of various graphical displays for the purpose of data presentation. These offer a wealth of useful presentations beyond tabular summaries and basic bar charts, currently the most popular form of data presentation in the business environment. For an excellent discussion of using graphs to report business data, see Few (2004). In terms of reporting data mining results graphically, we describe common graphical displays elsewhere in the book, some of which are technique specific [e.g., dendrograms for hierarchical clustering (Chapter 14), and tree charts for classification and regression trees (Chapter 9)] while others are more general [e.g., receiver operating characteristic (ROC) curves and lift charts for classification (Chapter 5) and profile plots for clustering (Chapter 14)].

Data exploration is a mandatory initial step whether or not more formal analysis follows. Graphical exploration can support free-form exploration for the purpose of understanding the data structure, cleaning the data (e.g., identifying unexpected gaps or "illegal" values), identifying outliers, discovering initial patterns (e.g., correlations among variables and surprising clusters), and generating interesting questions. Graphical exploration can also be more focused, geared toward specific questions of interest. In the data mining context a combination is needed: free-form exploration performed with the purpose of supporting a specific goal.

Graphical exploration can range from generating very basic plots to using operations such as filtering and zooming interactively to explore a set of interconnected plots that include advanced features such as color and multiple panels. This chapter is not meant to be an exhaustive guidebook on visualization techniques but instead discusses main principles and features that support data exploration in a data mining context. We start by describing varying levels of sophistication in terms of visualization and show the advantages of different features and operations. Our discussion is from the perspective of how visualization

supports the subsequent data mining goal. In particular, we distinguish between supervised and unsupervised learning; within supervised learning, we
also further distinguish between classification (categorical Y) and prediction
(numerical Y).

3.2   DATA EXAMPLES

To illustrate data visualization we use two datasets that are used in additional
chapters in the book. This allows the reader to compare some of the basic Excel
plots used in other chapters to the improved plots and easily see the merit of
advanced visualization.

Example 1: Boston Housing Data

The Boston housing data contain information on census tracts in Boston for
which several measurements are taken (e.g., crime rate, pupil/teacher ratio). It
has 14 variables (a description of each variable and the data are given in Chapter 2,
in Table 2.2 and Figure 2.5). We consider three possible tasks:

1. A supervised predictive task, where the outcome variable of interest is the
   median value of a home in the tract (MEDV)

2. A supervised classification task, where the outcome variable of interest
   is the binary variable CAT.MEDV that equals 1 for tracts with median
   home value above $30,000 and equals 0 otherwise

3. An unsupervised task, where the goal is to cluster census tracts

(MEDV and CAT.MEDV are not used together in any of the three cases.)

Example 2: Ridership on Amtrak Trains

Amtrak, a U.S. railway company, routinely collects data on ridership. Here we
focus on forecasting future ridership using the series of monthly ridership between January 1991 and March 2004. The data and their source are described in
Chapter 15. Hence our task here is (numerical) time series forecasting.

3.3   BASIC CHARTS: BAR CHARTS, LINE GRAPHS, AND SCATTERPLOTS

The three most effective basic plots are bar charts, line graphs, and scatterplots.
These plots are easy to create in Microsoft Excel and are the most commonly
used in the current business world, in both data exploration and presentation

(unfortunately, pie charts are also popular, although usually ineffective visualizations). Basic charts support data exploration by displaying one or two columns of data (variables) at a time. This is useful in the early stages of getting familiar with the data structure, the amount and types of variables, the volume and type of missing values, and the like.

The nature of the data mining task and domain knowledge about the data will affect the use of basic charts in terms of the amount of time and effort allocated to different variables. In supervised learning, there will be more focus on the outcome variable. In scatterplots, the outcome variable is typically associated with the y axis. In unsupervised learning (for the purpose of data reduction or clustering), basic plots that convey relationships (such as scatterplots) are preferred.

The top left panel in Figure 3.1 displays a line chart for the time series of monthly railway passengers on Amtrak. Line graphs are used primarily for showing time series. The choice of time frame to plot, as well as the temporal scale, should depend on the horizon of the forecasting task and on the nature of the data.

Bar charts are useful for comparing a single statistic (e.g., average, count, percentage) across groups. The height of the bar (or length, in a horizontal display) represents the value of the statistic, and different bars correspond to

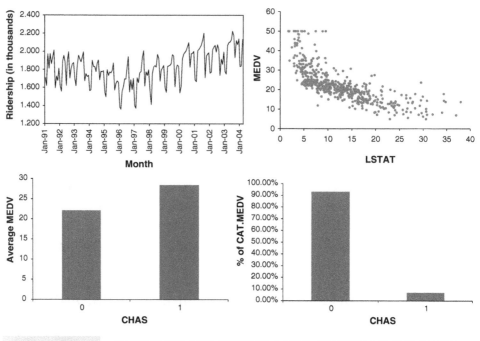

FIGURE 3.1   BASIC PLOTS: LINE GRAPH (TOP LEFT), SCATTERPLOT (TOP RIGHT), BAR CHART FOR NUMERICAL VARIABLE (BOTTOM LEFT), AND BAR CHART FOR CATEGORICAL VARIABLE (BOTTOM RIGHT)

different groups. Two examples are shown in the bottom panels in Figure 3.1. The left panel shows a bar chart for a numerical variable (MEDV) and the right panel shows a bar chart for a categorical variable (CAT.MEDV). In each, separate bars are used to denote homes in Boston that are near the Charles River versus those that are not (thereby comparing the two categories of CHAS). The chart with the numerical output MEDV (bottom left) uses the average MEDV on the y axis. This supports the predictive task: The numerical outcome is on the y axis and the x axis is used for a potential categorical predictor.[1] (Note that the x axis on a bar chart must be used only for categorical variables because the order of bars in a bar chart should be interchangeable.) For the classification task, CAT.MEDV is on the y axis (bottom right), but its aggregation is a percentage (the alternative would be a count). This graph shows us that the vast majority (over 90%) of the tracts do not border the Charles River (CHAS=0). Note that the labeling of the y axis can be confusing in this case: the value of CAT.MEDV plays no role and the y axis is simply a percentage of all records.

The top right panel in Figure 3.1 displays a scatterplot of MEDV versus LSTAT. This is an important plot in the prediction task. Note that the output MEDV is again on the y axis (and LSTAT on the x axis is a potential predictor). Because both variables in a basic scatterplot must be numerical, it cannot be used to display the relation between CAT.MEDV and potential predictors for the classification task (but we can enhance it to do so—see Section 3.4). For unsupervised learning, this particular scatterplot helps study the association between two numerical variables in terms of information overlap as well as identifying clusters of observations.

All three basic plots highlight global information such as the overall level of ridership or MEDV, as well as changes over time (line chart), differences between subgroups (bar chart), and relationships between numerical variables (scatterplot).

Distribution Plots: Boxplots and Histograms

Before moving on to more sophisticated visualizations that enable multidimensional investigation, we note two important plots that are usually not considered "basic charts" but are very useful in statistical and data mining contexts. The *boxplot* and the *histogram* are two plots that display the entire distribution of a numerical variable. Although averages are very popular and useful summary statistics, there is usually much to be gained by looking at additional statistics such as the median and standard deviation of a variable, and even more so by examining the entire distribution. Whereas bar charts can only use a single

---

[1] We refer here to a bar chart with vertical bars. The same principles apply if using a bar chart with horizontal bars, except that the x axis is now associated with the numerical variable and the y axis with the categorical variable.

aggregation, boxplots and histograms display the entire distribution of a numerical variable. Boxplots are also effective for comparing subgroups by generating side-by-side boxplots, or for looking at distributions over time by creating a series of boxplots.

Distribution plots are useful in supervised learning for determining potential data mining methods and variable transformations. For example, skewed numerical variables might warrant transformation (e.g., moving to a logarithmic scale) if used in methods that assume normality (e.g., linear regression, discriminant analysis).

A histogram represents the frequencies of all x values with a series of vertical connected bars. For example, in the top left panel of Figure 3.2, there are about 20 tracts where the median value (MEDV) is between $7500 and $12,500.

A boxplot represents the variable being plotted on the y axis (although the plot can potentially be turned in a 90° angle, so that the boxes are parallel to the x axis). In the top right panel of Figure 3.2 there are two boxplots (called a side-by-side boxplot). The box encloses 50% of the data—for example, in the right-hand box half of the tracts have median values (MEDV) between $20,000 and $33,000. The horizontal line inside the box represents the median (50th percentile). The top and bottom of the box represent the 75th and 25th percentiles, respectively. Lines extending above and below the box cover the rest of the data range; outliers may be depicted as points or circles. Sometimes the average is marked by a + (or similar) sign, as in the top right panel of Figure 3.2. Comparing the average and the median helps in assessing how skewed the data are. Boxplots are often arranged in a series with a different plot for each of the various values of a second variable, shown on the x axis.

Because histograms and boxplots are geared toward numerical variables, in their basic form they are useful for prediction tasks. Boxplots can also support unsupervised learning by displaying relationships between a numerical variable (y axis) and a categorical variable (x axis). To illustrate these points, see Figure 3.2. The top panel shows a histogram of MEDV, revealing a skewed distribution. Transforming the output variable to log(MEDV) would likely improve results of a linear regression predictor.

The right panel in Figure 3.2 shows side-by-side boxplots comparing the distribution of MEDV for homes that border the Charles River (1) or not (0), (similar to Figure 3.1). We see that not only is the average MEDV for river-bounding homes higher than the non-river-bounding homes, the entire distribution is higher (median, quartiles, min, and max). We also see that all river-bounding homes have MEDV above $10,000, unlike non-river-bounding homes. This information is useful for identifying the potential importance of this predictor (CHAS) and for choosing data mining methods that can capture the nonoverlapping area between the two distributions (e.g., trees). Boxplots and histograms applied to numerical variables can also provide directions for

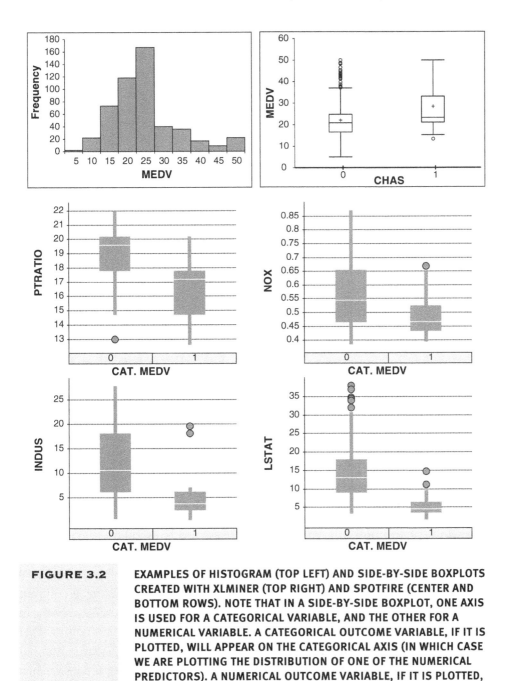

FIGURE 3.2   EXAMPLES OF HISTOGRAM (TOP LEFT) AND SIDE-BY-SIDE BOXPLOTS CREATED WITH XLMINER (TOP RIGHT) AND SPOTFIRE (CENTER AND BOTTOM ROWS). NOTE THAT IN A SIDE-BY-SIDE BOXPLOT, ONE AXIS IS USED FOR A CATEGORICAL VARIABLE, AND THE OTHER FOR A NUMERICAL VARIABLE. A CATEGORICAL OUTCOME VARIABLE, IF IT IS PLOTTED, WILL APPEAR ON THE CATEGORICAL AXIS (IN WHICH CASE WE ARE PLOTTING THE DISTRIBUTION OF ONE OF THE NUMERICAL PREDICTORS). A NUMERICAL OUTCOME VARIABLE, IF IT IS PLOTTED, WILL APPEAR ON THE NUMERICAL AXIS (IN WHICH CASE WE ARE PLOTTING THE DISTRIBUTION OF THE OUTCOME VARIABLE ITSELF, WITH A CATEGORICAL PREDICTOR ON THE CATEGORICAL AXIS)

deriving new variables, for example, they can indicate how to bin a numerical variable (e.g., binning a numerical outcome in order to use a naive Bayes classifier, or in the Boston housing example, choosing the cutoff to convert MEDV to CAT.MEDV).

Finally, side-by-side boxplots are useful in classification tasks for evaluating the potential of numerical predictors. This is done by using the x axis for the categorical outcome and the y axis for a numerical predictor. An example is shown in the center and bottom rows of Figure 3.2, where we can see the effects of four numerical predictors on CAT.MEDV. The pairs that are most separated (e.g., PTRATIO and INDUS) indicate potentially useful predictors.

Boxplots and histograms are not readily available in Microsoft Excel (although they can be constructed through a tedious manual process). They are available in a wide range of statistical software packages. In XLMiner they can be generated through the *Charts* menu (we note the current limitation of five categories for side-by-side boxplots).

The main weakness of basic charts and distribution plots, in their basic form (i.e., using position in relation to the axes to encode values), is that they can only display two variables and therefore cannot reveal high-dimensional information. Each of the basic charts has two dimensions, where each dimension is dedicated to a single variable. In data mining, the data are usually multivariate by nature, and the analytics are designed to capture and measure multivariate information. Visual exploration should therefore also incorporate this important aspect. In the next section we describe how to extend basic charts (and distribution charts) to multidimensional data visualization by adding features, employing manipulations, and incorporating interactivity. We then present several specialized charts that are geared toward displaying special data structures (Section 3.5).

Heatmaps: Visualizing Correlations and Missing Values

A *heatmap* is a graphical display of numerical data where color is used to denote values. In a data mining context, heatmaps are especially useful for two purposes: for visualizing correlation tables and for visualizing missing values in the data. In both cases the information is conveyed in a two-dimensional table. A correlation table for p variables has p rows and p columns. A data table contains p columns (variables) and n rows (records). If the number of rows is huge, then a subset can be used. In both cases it is much easier and faster to scan the color coding rather than the values. Note that heatmaps are useful when examining a large number of values, but they are not a replacement for more precise graphical display, such as bar charts, because color differences cannot be perceived accurately.

An example of a correlation table heatmap is shown in Figure 3.3, showing all the pairwise correlations between 14 variables (MEDV and 13 predictors). Darker shades correspond to stronger (positive or negative) correlation. It is easy to quickly spot the high and low correlations. This heatmap was produced using Excel's *Conditional Formatting*.

In a missing value heatmap rows correspond to records and columns to variables. We use a binary coding of the original dataset where 1 denotes a

	CRIM	ZN	INDUS	CHAS	NOX	RM	AGE	DIS	RAD	TAX	PTRATIO	B	LSTAT	MEDV
CRIM	1.00													
ZN	-0.20	1.00												
INDUS	0.41	-0.53	1.00											
CHAS	-0.06	-0.04	0.06	1.00										
NOX	0.42	-0.52	0.76	0.09	1.00									
RM	-0.22	0.31	-0.39	0.09	-0.30	1.00								
AGE	0.35	-0.57	0.64	0.09	0.73	-0.24	1.00							
DIS	-0.38	0.66	-0.71	-0.10	-0.77	0.21	-0.75	1.00						
RAD	0.63	-0.31	0.60	-0.01	0.61	-0.21	0.46	-0.49	1.00					
TAX	0.58	-0.31	0.72	-0.04	0.67	-0.29	0.51	-0.53	0.91	1.00				
PTRATIO	0.29	-0.39	0.38	-0.12	0.19	-0.36	0.26	-0.23	0.46	0.46	1.00			
B	-0.39	0.18	-0.36	0.05	-0.38	0.13	-0.27	0.29	-0.44	-0.44	-0.18	1.00		
LSTAT	0.46	-0.41	0.60	-0.05	0.59	-0.61	0.60	-0.50	0.49	0.54	0.37	-0.37	1.00	
MEDV	-0.39	0.36	-0.48	0.18	-0.43	0.70	-0.38	0.25	-0.38	-0.47	-0.51	0.33	-0.74	1.00

FIGURE 3.3   HEATMAP OF A CORRELATION TABLE. DARKER VALUES DENOTE STRONGER CORRELATION

missing value and 0 otherwise. This new binary table is then colored such that only missing value cells (with value 1) are colored. Figure 3.4 shows an example of a missing value heatmap for a dataset with over 1000 columns. The data include economic, social, political and "well-being" information on different countries around the world (each row is a country). The variables were merged from multiple sources, and for each source information was not always available on every country. The missing data heatmap helps visualize the level and amount of "missingness" in the merged data file. Some patterns of "missingness" easily emerge: variables that are missing for nearly all observations, as well as clusters of rows (countries) that are missing many values. Variables with little missingness are

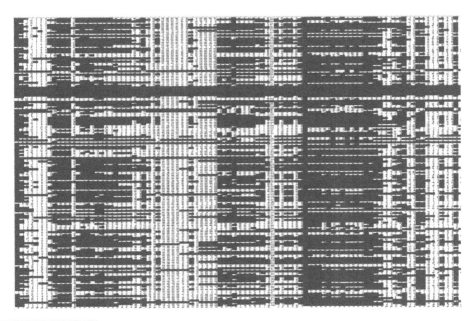

FIGURE 3.4   HEATMAP OF MISSING VALUES IN A DATASET. BLACK DENOTES MISSING VALUE

also visible. This information can then be used for determining how to handle the missingness (e.g., dropping some variables, dropping some records, imputing, or via other techniques).

3.4 MULTIDIMENSIONAL VISUALIZATION

Basic plots can convey richer information with features such as color, size, and multiple panels, and by enabling operations such as rescaling, aggregation, and interactivity. These additions allow looking at more than one or two variables at a time. The beauty of these additions is their effectiveness in displaying complex information in an easily understandable way. Effective features are based on understanding how visual perception works [see Few (2009) for a discussion]. The purpose is to make the information more understandable, not just represent the data in higher dimensions (such as three-dimensional plots that are usually ineffective visualizations).

Adding Variables: Color, Size, Shape, Multiple Panels, and Animation

In order to include more variables in a plot, we must consider the type of variable to include. To represent additional categorical information, the best way is to use hue, shape, or multiple panels. For additional numerical information we can use color intensity or size. Temporal information can be added via animation.

Incorporating additional categorical and/or numerical variables into the basic (and distribution) plots means that we can now use all of them for both prediction and classification tasks! For example, we mentioned earlier that a basic scatterplot cannot be used for studying the relationship between a categorical outcome and predictors (in the context of classification). However, a very effective plot for classification is a scatterplot of two numerical predictors color coded by the categorical outcome variable. An example is shown in the left panel of Figure 3.5, with color denoting CAT.MEDV.

In the context of prediction, color coding supports the exploration of the conditional relationship between the numerical outcome (on the y axis) and a numerical predictor. Color-coded scatterplots then help assess the need for creating interaction terms (e.g., is the relationship between MEDV and LSTAT different for homes near versus away from the river?).

Color can also be used to include further categorical variables into a bar chart, as long as the number of categories is small. When the number of categories is large, a better alternative is to use multiple panels. Creating multiple panels (also called "trellising") is done by splitting the observations according to a categorical variable and creating a separate plot (of the same type) for each category. An example is shown in the right panel of Figure 3.5, where a bar chart

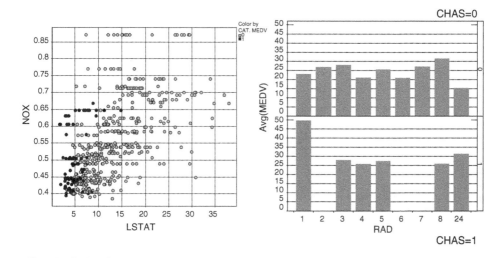

FIGURE 3.5   ADDING CATEGORICAL VARIABLES BY COLOR CODING AND MULTIPLE PANELS. (LEFT) SCATTERPLOT OF TWO NUMERICAL PREDICTORS, COLOR CODED BY THE CATEGORICAL OUTCOME (CAT.MEDV). (RIGHT) BAR CHART OF MEDV BY TWO CATEGORICAL PREDICTORS (CHAS AND RAD), USING MULTIPLE PANELS FOR CHAS.
(CHAS = 0 FOR UPPER PANEL, CHAS = 1 FOR LOWER PANEL)

of average MEDV by RAD is broken down into two panels by CHAS. We see that the average MEDV for different highway accessibility levels (RAD) behaves differently for homes near the river (lower panel) compared to homes away from the river (upper panel). This is especially salient for RAD=1. We also see that there are no near-river homes in RAD levels 2, 6, and 7. Such information might lead us to create an interaction term between RAD and CHAS and to consider condensing some of the bins in RAD. All these explorations are useful for prediction and classification.

A special plot that uses scatterplots with multiple panels is the *scatterplot matrix*. In it, all pairwise scatterplots are shown in a single display. The panels in a scatterplot matrix are organized in a special way, such that each column corresponds to a variable and each row corresponds to a variable, thereby the intersections create all the possible pairwise scatterplots. The scatterplot matrix plot is useful in unsupervised learning for studying the associations between numerical variables, detecting outliers and identifying clusters. For supervised learning, it can be used for examining pairwise relationships (and their nature) between predictors to support variable transformations and variable selection (see Section 4.4). For prediction it can also be used to depict the relationship of the outcome with the numerical predictors.

An example of a scatterplot matrix is shown in Figure 3.6, with MEDV and three predictors. To identify which pair is plotted, variable names are shown along the diagonal cells; plots in the row corresponding to a variable show the

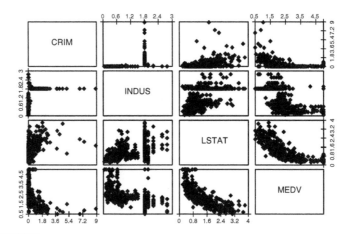

FIGURE 3.6   **SCATTERPLOT MATRIX FOR MEDV AND THREE NUMERICAL PREDICTORS**

variable's values along the y axis while plots in the corresponding column show the variable's values along the x axis. For example, the plots in the bottom row all have MEDV on the y axis (which allows studying the individual outcome–predictor relations). We can see different types of relationships from the different shapes (e.g., an exponential relationship between MEDV and LSTAT and a highly skewed relationship between CRIM and INDUS), which can indicate needed transformations. Note that the plots above and to the right of the diagonal are mirror images of those below and to the left.

Once hue is used, further categorical variables can be added via shape and multiple panels. However, one must proceed cautiously in adding multiple variables, as the display can become overcluttered and then visual perception is lost.

Adding a numerical variable via size is useful especially in scatterplots (thereby creating "bubble plots") because in a scatterplot, points represent individual observations. In plots that aggregate across observations (e.g., boxplots, histograms, bar charts) size and hue are not normally incorporated.

Finally, adding a temporal dimension to a plot to show how the information changes over time can be achieved via animation. A famous example is Rosling's animated scatterplots showing how world demographics changed over the years (www.gapminder.org). However, while animations of this type work for "statistical storytelling," they are not very effective for data exploration.

Manipulations: Rescaling, Aggregation and Hierarchies, Zooming, and Panning, and Filtering

Most of the time spent in data mining projects is spent in preprocessing. Typically, considerable effort is expended getting all the data in a format that can actually be used in the data mining software. Additional time is spent processing the data in ways that improve the performance of the data mining procedures. This pre-processing step in data mining includes variable transformation and derivation of new variables to help models perform more effectively. Transformations include changing the numeric scale of a variable, binning numerical variables, condensing categories in categorical variables, and the like. The following manipulations support the preprocessing step as well as the choice of adequate data mining methods. They do so by revealing patterns and their nature.

Rescaling Changing the scale in a display can enhance the plot and illuminate relationships. For example, in Figure 3.7 we see the effect of changing both axes of the scatterplot (top) and the y axis of a boxplot (bottom) to logarithmic (log) scale. Whereas the original plots (left) are hard to understand, the patterns become visible in log scale (right). In the scatterplot, the nature of the relationship between MEDV and CRIM is hard to determine in the original scale because too many of the points are "crowded" near the y axis. The rescaling removes this crowding and allows a better view of the linear relationship between the two log-scaled variables (indicating a log–log relationship). In the boxplot displays the crowding toward the x axis in the original units does not allow us to compare the two box sizes, their locations, lower outliers, and most of the distribution information. Rescaling removes the "crowding to the x axis" effect, thereby allowing a comparison of the two boxplots.

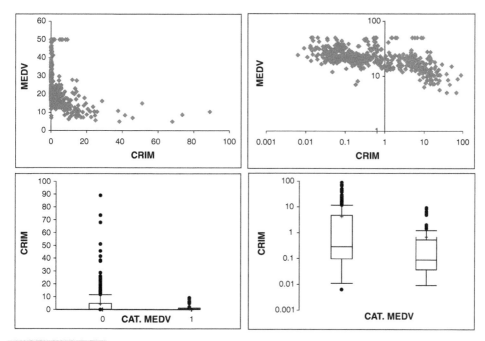

FIGURE 3.7   **RESCALING CAN ENHANCE PLOTS AND REVEAL PATTERNS. (LEFT) ORIGINAL SCALE. (RIGHT) LOG SCALE**

Aggregation and Hierarchies   Another useful manipulation of scaling is changing the level of aggregation. For a temporal scale, we can aggregate by different granularity (e.g., monthly, daily, hourly) or even by a "seasonal" factor of interest such as month of year or day of week. A popular aggregation for time series is a moving average, where the average of neighboring values within a given window size is plotted. Moving-average plots enhance global trend visualization (see Chapter 15).

Nontemporal variables can be aggregated if some meaningful hierarchy exists: geographical (tracts within a zip code in the Boston housing example), organizational (people within departments within units), and so on. Figure 3.8 illustrates two types of aggregation for the railway ridership time series. The original monthly series is shown in the top left panel. Seasonal aggregation (by month of year) is shown in the top right panel, where it is easy to see the peak in ridership in July–Aug and the dip in Jan–Feb. The bottom right panel shows temporal aggregation, where the series is now displayed in yearly aggregates. This plot reveals the global long-term trend in ridership and the generally increasing trend from 1996 on.

Examining different scales, aggregations, or hierarchies supports both supervised and unsupervised tasks in that it can reveal patterns and relationships at various levels and can suggest new sets of variables with which to work.

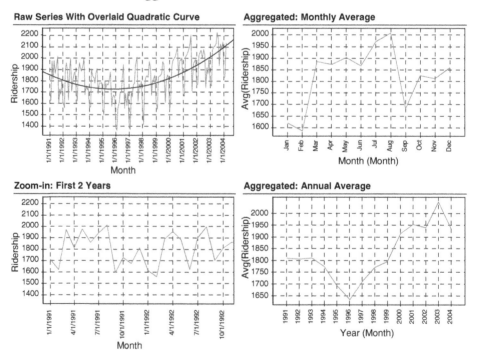

FIGURE 3.8   TIME SERIES LINE GRAPHS USING DIFFERENT AGGREGATIONS (RIGHT PANELS), ADDING CURVES (TOP LEFT PANEL), AND ZOOMING IN (BOTTOM LEFT PANEL). CREATED WITH SPOTFIRE

Zooming and Panning   The ability to zoom in and out of certain areas of the data on a plot is important for revealing patterns and outliers. We are often interested in more detail on areas of dense information or of special interest. Panning refers to the operation of moving the zoom window to other areas (popular in mapping applications such as Google Maps). An example of zooming is shown in the bottom left panel of Figure 3.8, where the ridership series is zoomed in to the first 2 years of the series.

Zooming and panning support supervised and unsupervised methods by detecting areas of different behavior, which may lead to creating new inter-action terms, new variables, or even separate models for data subsets. In addition, zooming and panning can help choose between methods that assume global behavior (e.g., regression models) and data-driven methods (e.g., exponential smoothing forecasters and k-nearest neighbors classifiers) and indicate the level of global–local behavior (as manifested by parameters such as k in k-nearest neighbors, the size of a tree, or the smoothing parameters in exponential smoothing).

Filtering   Filtering means removing some of the observations from the plot. The purpose of filtering is to focus the attention on certain data while eliminating "noise" created by other data. Filtering supports supervised and unsupervised learning in a similar way to zooming and panning: It assists in identifying different or unusual local behavior.

Reference: Trend Lines and Labels

Trend lines and using in-plot labels also help to detect patterns and outliers. Trend lines serve as a reference and allow us to more easily assess the shape of a pattern. Although linearity is easy to visually perceive, more elaborate relationships such as exponential and polynomial trends are harder to assess by eye. Trend lines are useful in line graphs as well as in scatterplots. An example is shown in the top left panel of Figure 3.8, where a polynomial curve is overlaid on the original line graph (see also Chapter 15).

In displays that are not overcrowded, the use of in-plot labels can be useful for better exploration of outliers and clusters. An example is shown in Figure 3.9 (a reproduction of Figure 14.1 with the addition of labels). The figure shows different utilities on a scatterplot that compares fuel cost with total sales. We might be interested in clustering the data, and using clustering algorithms to identify clusters that differ markedly with respect to fuel cost and sales. Figure 14.1, with the labels, helps visualize these clusters and their members (e.g., Nevada and Puget are part of a clear cluster with low fuel costs and high sales). For more on clustering, and this example, see Chapter 14.

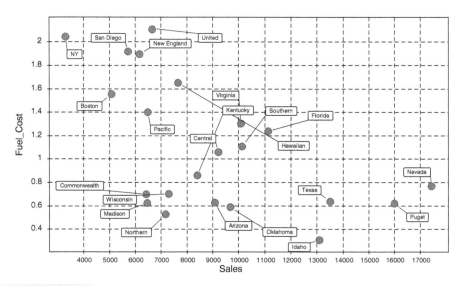

FIGURE 3.9 SCATTERPLOT WITH LABELED POINTS (CREATED WITH SPOTFIRE). COMPARE TO FIGURE 14.1

Scaling up: Large Datasets

When the number of observations (rows) is large, plots that display each individual observation (e.g., scatterplots) can become ineffective. Aside from applying aggregated charts such as boxplots, some alternatives are:

1. Sampling: drawing a random sample and using it for plotting (XLMiner has a sampling utility)
2. Reducing marker size
3. Using more transparent marker colors and removing fill
4. Breaking down the data into subsets (e.g., by creating multiple panels)
5. Using aggregation (e.g., bubble plots where size corresponds to number of observations in a certain range)
6. Using jittering (slightly moving each marker by adding a small amount of noise)

An example of the advantage of plotting a sample over the large dataset is shown in Figure 12.2 in Chapter 12, where a scatterplot of 5000 records is plotted alongside a scatterplot of a sample. Those plots were generated in Excel. We illustrate (Figure 3.10) an improved plot of the full dataset by applying smaller markers, using jittering to uncover overlaid points, and more transparent colors. We can see that larger areas of the plot are dominated by the gray class, the black class is mainly on the right, while there is a lot of overlap in the top right area.

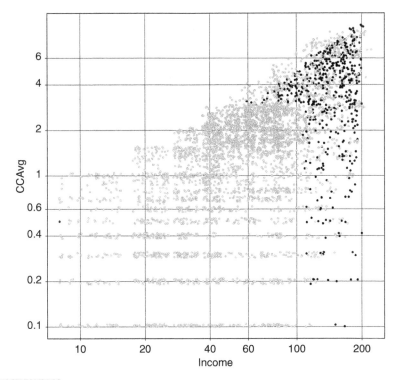

FIGURE 3.10   **REPRODUCTION OF FIGURE 12.2 WITH REDUCED MARKER SIZE, JITTERING, AND MORE TRANSPARENT COLORING**

Multivariate Plot: Parallel Coordinates Plot

Another approach toward presenting multidimensional information in a two-dimensional plot is via specialized plots such as the *parallel coordinates plot*. In this plot a vertical axis is drawn for each variable. Then each observation is represented by drawing a line that connects its values on the different axes, thereby creating a "multivariate profile." An example is shown in Figure 3.11 for the Boston housing data. In this display separate panels are used for the two values of CAT.MEDV in order to compare the profiles of homes in the two classes (for a classification task). We see that the more expensive homes (bottom panel) consistently have low CRIM, low LSAT, high B, and high RM compared to cheaper homes (top panel), which are more mixed on CRIM, LSAT, and B, and have a medium level of RM. This observation gives an indication of useful predictors and suggests possible binning for some numerical predictors.

Parallel coordinate plots are also useful in unsupervised tasks. They can reveal clusters, outliers, and information overlap across variables. A useful manipulation is to reorder the columns to better reveal observation clusterings. Parallel coordinate plots are not implemented in Excel. However, a free Excel add-in is currently available at http://ibmi.mf.uni-lj.si/ibmi-english/biostat-center/programje/excel/ParallelCoordinates.xls.

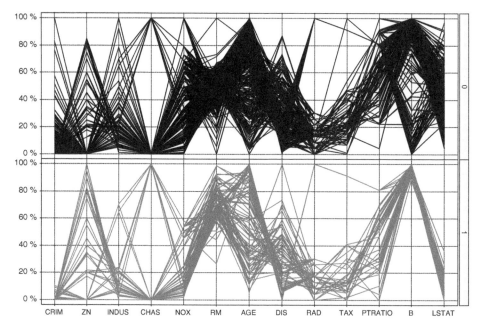

FIGURE 3.11   PARALLEL COORDINATES PLOT FOR BOSTON HOUSING DATA. EACH
OF THE VARIABLES (SHOWN ON THE HORIZONTAL AXIS) IS SCALED TO
0--100%. PANELS ARE USED TO DISTINGUISH CAT.MEDV (TOP PANEL
= HOMES BELOW $30,000). CREATED USING SPOTFIRE

Interactive Visualization

Similar to the interactive nature of the data mining process, interactivity is key
to enhancing our ability to gain information from graphical visualization. In the
words of Stephen Few (Few, 2009, p. 55), an expert in data visualization,

> We can only learn so much when staring at a static visualization such as a
> printed graph ... If we can't interact with the data ... we hit the wall.

By interactive visualization we mean an interface that supports the following
principles:

1. Making changes to a plot is *easy, rapid, and reversible.*
2. Multiple concurrent plots can be easily combined and displayed on a single
   screen.
3. A set of visualizations can be linked such that operations in one display
   are reflected in the other displays.

Let us consider a few examples where we contrast a static plot generator
(e.g., Excel) with an interactive visualization interface.

Histogram rebinning   Consider the need to bin a numerical variable using a histogram. A static histogram would require replotting for each new binning choice (in Excel it would require creating the new bins manually). If the user generates multiple plots, then the screen becomes cluttered. If the same plot is recreated, then it is hard to compare to other binning choices. In contrast, an interactive visualization would provide an easy way to change bin width interactively (see, e.g., the slider below the histogram in Figure 3.12), and then the histogram would automatically and rapidly replot as the user changes the bin width.

Aggregation and Zooming   Consider a time series forecasting task, given a long series of data. Temporal aggregation at multiple levels is needed for determining short- and long-term patterns. Zooming and panning are used to identify unusual periods. A static plotting software requires the user to create new data columns for each temporal aggregation (e.g., aggregate daily data to obtain weekly aggregates). Zooming and panning in Excel requires manually changing the min and max values on the axis scale of interest (thereby losing the ability to quickly move between different areas without creating multiple charts). An interactive visualization would provide immediate temporal hierarchies between which the

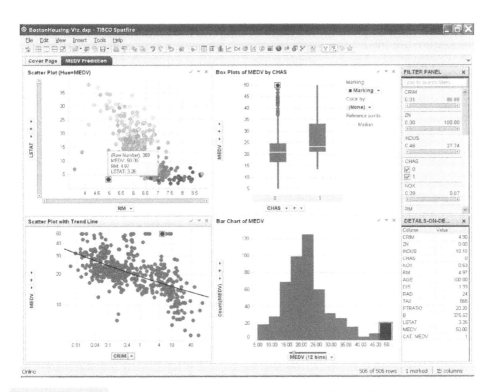

FIGURE 3.12   **MULTIPLE INTERLINKED PLOTS IN A SINGLE VIEW (IN SPOTFIRE). NOTE THE MARKED OBSERVATION IN THE TOP LEFT PANEL, WHICH IS ALSO HIGHLIGHTED IN ALL OTHER PLOTS**

user can easily switch. Zooming would be enabled as a slider near the axis (see, e.g., the sliders on the top left panel in Figure 3.12), thereby allowing direct manipulation and rapid reaction.

Combining Multiple Linked Plots That Fit in a Single Screen   To support a classification task, multiple plots are created of the outcome variable versus potential categorical and numerical predictors. These can include side-by-side boxplots, color-coded scatterplots, multipanel bar charts, and the like. The user wants to detect multidimensional relationships (and identify outliers) by selecting a certain subset of the data (e.g., a single category of some variable) and locating the observations on the other plots. In a static interface, the user would have to manually organize the plots of interest and resize them in order to fit within a single screen. A static interface would usually not support interplot linkage, and even if so, the entire set of plots would have to be regenerated each time that a selection is made. In contrast, an interactive visualization would provide an easy way to automatically organize and resize the set of plots to fit within a screen. Linking the set of plots would be easy, and in response to the users selection on one plot, the appropriate selection would be automatically highlighted in the other plots (see example in Figure 3.12).

In earlier sections we used plots to illustrate the advantages of visualizations because "a picture is worth a thousand words." The advantages of an interactive visualization are even greater. As Ben Shneiderman, a well-known researcher in information visualization and interfaces, notes:

> A picture is worth a thousand words. An interface is worth a thousand pictures.

Interactive Visualization Software   Some added features such as color, shape, and size are often available in software that produces static plots, while others (multiple panels, hierarchies, labels) are only available in more advanced visualization tools. Even when a feature is available (e.g., color), the ease of applying it to a plot can widely vary. For example, incorporating color into an Excel scatterplot is a daunting task.[2] Plot manipulation possibilities (e.g., zooming, filtering, and aggregation) and ease of implementation are also quite limited in standard "static plot" software.

Although we do not intend to provide a market survey of interactive visualization tools, we do mention a few prominent packages. Spotfire (http://spotfire.tibco.com) and Tableau (www.tableausoftware.com) are two dedicated data visualization tools (several of the plots in this chapter were created using Spotfire). They both provide a high level of interactivity, can support large datasets, and produce high-quality plots that are also easy to export. JMP

---

[2] See http://blog.bzst.com/2009/08/creating-color-coded-scatterplots-in.html.

by SAS (www.jmp.com) is a "statistical discovery" software that also has strong interactive visualization capabilities. All three offer free trial versions. Finally, we mention Many Eyes by IBM (http://manyeyes.alphaworks.ibm.com/manyeyes) that allows uploading your data and visualizing it via different interactive visualizations.

3.5 SPECIALIZED VISUALIZATIONS

In this section we mention a few specialized visualizations that are able to capture data structures beyond the standard time series and cross-sectional structures—special types of relationships that are usually hard to capture with ordinary plots. In particular, we address hierarchical data, network data, and geographical data—three types of data that are becoming more available.

Visualizing Networked Data

With the explosion of social and product network data, network analysis has become a hot topic. Examples of social networks include networks of sellers and buyers on eBay and networks of people on Facebook. An example of a product network is the network of products on Amazon (linked through the recommendation system). Network data visualization is available in various network-specialized software, and also in general-purpose software.

A network diagram consists of actors and relations between them. "Nodes" are the actors (e.g., people in a social network or products in a product network) and represented by circles. "Edges" are the relations between nodes and are represented by lines connecting nodes. For example, in a social network such as Facebook, we can construct a list of users (nodes) and all the pairwise relations (edges) between users who are "Friends." Alternatively, we can define edges as a posting that one user posts on another user's Facebook page. In this setup we might have more than a single edge between two nodes. Networks can also have nodes of multiple types. A common structure is networks with two types of nodes. An example of a two-type node network is shown in Figure 3.13, where we see a set of transactions between a network of sellers and buyers on the online auction site www.eBay.com [the data are for auctions selling Swarowski beads and took place during a period of several months; from Jank and Yahav (2010)]. The circles on the left side represent sellers and on the right side buyers. Circle size represents the number of transactions that the node (seller or buyer) was involved in within this network. Line width represents the number of auctions that the bidder–seller pair interacted in (in this case we use arrows to denote the directional relationship from buyer to seller). We can see that this marketplace is dominated by three or four high-volume sellers. We can also see that many

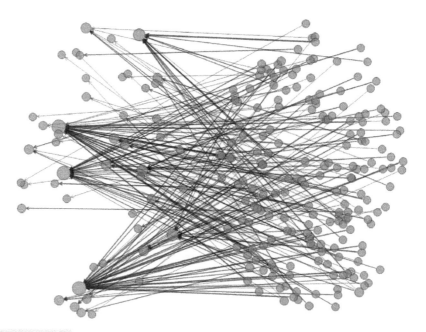

FIGURE 3.13   NETWORK GRAPH OF EBAY SELLERS (LEFT SIDE) AND BUYERS (RIGHT SIDE) OF SWAROSKI BEADS. CIRCLE SIZE REPRESENTS THE NODE'S NUMBER OF TRANSACTIONS. LINE WIDTH REPRESENTS THE NUMBER OF TRANSACTIONS BETWEEN THAT PAIR OF SELLER--BUYER (CREATED WITH SPOTFIRE)

buyers interact with a single seller. The market structures for many individual products could be reviewed quickly in this way. Network providers could use the information, for example, to identify possible partnerships to explore with sellers.

Figure 3.13 was produced using Spotfire's network visualization. An Excel-based tool is NodeXL (http://nodexl.codeplex.com), which is a template for Excel 2007 that allows entering a network edge list. The graph's appearance can be customized and various interactive features are available such as zooming, scaling and panning the graph, dynamically filtering nodes and edges, altering the graph's layout, finding clusters of related nodes, and calculating graph metrics. Networks can be imported from and exported to a variety of data formats, and built-in connections for getting networks from Twitter, Flickr, and your local e-mail are provided.

Network graphs can be potentially useful in the context of association rules (see Chapter 13). For example, consider a case of mining a dataset of consumers' grocery purchases to learn which items are purchased together ("what goes with what"). A network can be constructed with items as nodes and edges connecting items that were purchased together. After a set of rules is generated by the data mining algorithm (which often contains an excessive number of rules, many of which are unimportant), the network graph can help visualize different rules for the purpose of choosing the interesting ones. For example, a popular "beer and diapers" combination would appear in the network graph as a pair of nodes with very high connectivity. An item that is almost always purchased regardless of other items (e.g., milk) would appear as a very large node with high connectivity to all other nodes.

Visualizing Hierarchical Data: Treemaps

We discussed hierarchical data and the exploration of data at different hierarchy levels in the context of plot manipulations. *Treemaps* are useful visualizations for exploring large data sets that are hierarchically structured (tree structured). They enable exploration of various dimensions of the data while maintaining the data's hierarchical nature. An example is shown in Figure 3.14, which displays a large set of eBay auctions, hierarchically ordered by item category, subcategory, and brand. The levels in the hierarchy of the treemap are visualized as rectangles containing subrectangles. Categorical variables can be included in the display by using hue. Numerical variables can be included via rectangle size and color intensity (ordering of the rectangles is sometimes used to reinforce size). In Figure 3.14 size is used to represent the average closing price (which reflects item value), and color intensity represents the percent of sellers with negative feedback

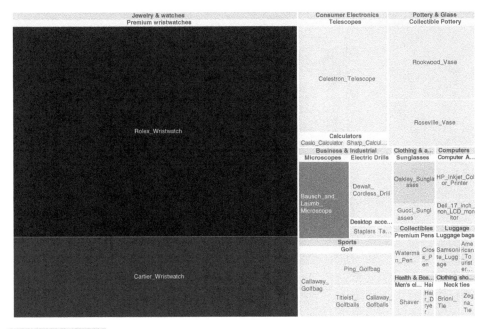

FIGURE 3.14   TREEMAP SHOWING NEARLY 11,000 EBAY AUCTIONS, ORGANIZED BY ITEM CATEGORY, SUBCATEGORY, AND BRAND. RECTANGLE SIZE REPRESENTS AVERAGE CLOSING PRICE (REFLECTING ITEM VALUE). SHADE REPRESENTS % OF SELLERS WITH NEGATIVE FEEDBACK (DARKER = HIGHER %)

(a negative seller feedback indicates buyer dissatisfaction in past transactions and often indicative of fraudulent seller behavior). Consider the task of classifying ongoing auctions in terms of a fraudulent outcome. From the treemap we see that the highest proportion of sellers with negative ratings (black) is concentrated in expensive item auctions (Rolex and Cartier wristwatches).

Ideally, treemaps should be explored interactively, zooming to different levels of the hierarchy. An interactive online application of treemaps is "Map of the Market" by Smart-Money (www.smartmoney.com/map-of-the-market), which displays stock market information in an interactive treemap display.

A free treemap add-in for Excel was developed by Microsoft research and is available at http://research.microsoft.com/apps/dp/dl/downloads.aspx (search for "Treemapper").

Visualizing Geographical Data: Map Charts

With the growing availability of location data, many datasets used for data mining now include geographical information. Zip codes are one example of a categorical variable with many categories, where creating meaningful variables for analysis is not straightforward. Plotting the data on a geographical map can often reveal patterns that are otherwise harder to identify. A map chart uses a

FIGURE 3.15   WORLD MAPS COMPARING "WELL-BEING" TO GDP. (TOP) SHADING BY AVERAGE "GLOBAL WELL-BEING" SCORE OF COUNTRY (DARKER CORRESPONDS TO HIGHER SCORE OR LEVEL). (BOTTOM) SHADING ACCORDING TO GDP. DATA FROM VEENHOVEN'S WORLD DATABASE OF HAPPINESS

geographical map as its background, and then color, hue, and other features can be used to include categorical or numerical variables. Besides specialized mapping software, maps are now becoming part of general-purpose software. Figure 3.15 shows two world maps (created with Spotfire), comparing countries' "well-being" (according to a 2006 Gallup survey) in the top map to gross domestic product (GDP) in the bottom map. A darker shade means higher value (white areas are missing data).

3.6 SUMMARY OF MAJOR VISUALIZATIONS AND OPERATIONS, ACCORDING TO DATA MINING GOAL

Prediction

- Plot outcome on the y axis of vertical boxplots, vertical bar charts, and scatterplots.
- Study relation of outcome to categorical predictors via side-by-side boxplots, bar charts, and multiple panels.
- Study relation of outcome to numerical predictors via scatterplots.
- Use distribution plots (boxplot, histogram) for determining needed transformations of the outcome variable (and/or numerical predictors).
- Examine scatterplots with added color/panels/size to determine the need for interaction terms.
- Use various aggregation levels and zooming to determine areas of the data with different behavior, and to evaluate the level of global versus local patterns.

Classification

- Study relation of outcome to categorical predictors using bar charts with the outcome on the y axis.
- Study relation of outcome to pairs of numerical predictors via color-coded scatterplots (color denotes the outcome).
- Study relation of outcome to numerical predictors via side-by-side boxplots: Plot boxplots of a numerical variable by outcome. Create similar displays for each numerical predictor. The most separable boxes indicate potentially useful predictors.
- Use color to represent the outcome variable on a parallel coordinate plot.
- Use distribution plots (boxplot, histogram) for determining needed transformations of the outcome variable.

- Examine scatterplots with added color/panels/size to determine the need for interaction terms.
- Use various aggregation levels and zooming to determine areas of the data with different behavior, and to evaluate the level of global versus local patterns.

Time Series Forecasting

- Create line graphs at different temporal aggregations to determine types of patterns.
- Use zooming and panning to examine various shorter periods of the series to determine areas of the data with different behavior.
- Use various aggregation levels to identify global and local patterns.
- Identify missing values in the series (that require handling).
- Overlay trend lines of different types to determine adequate modeling choices.

Unsupervised Learning

- Create scatterplot matrices to identify pairwise relationships and clustering of observations.
- Use heatmaps to examine the correlation table.
- Use various aggregation levels and zooming to determine areas of the data with different behavior.
- Generate a parallel coordinate plot to identify clusters of observations.

PROBLEMS

3.1   Shipments of Household Appliances: Line Graphs. The file ApplianceShipments.xls contains the series of quarterly shipments (in million $) of U.S. household appliances between 1985 and 1989 (data courtesy of Ken Black).

  a. Create a well-formatted time plot of the data using Excel.

  b. Does there appear to be a quarterly pattern? For a closer view of the patterns, zoom in to the range of 3500–5000 on the y axis.

  c. Create four separate lines for Q1, Q2, Q3, and Q4, using Excel. In each, plot a line graph. In Excel, order the data by Q1, Q2, Q3, Q4 (alphabetical sorting will work), and plot them as separate series on the line graph. Zoom in to the range of 3500–5000 on the y axis. Does there appear to be a difference between quarters?

  d. Using Excel, create a line graph of the series at a yearly aggregated level (i.e., the total shipments in each year).

  e. Re-create the above plots using an interactive visualization tool. Make sure to enter the quarter information in a format that is recognized by the software as a date.

  f. Compare the two processes of generating the line graphs in terms of the effort as well as the quality of the resulting plots. What are the advantages of each?

3.2   Sales of Riding Mowers: Scatterplots. A company that manufactures riding mowers wants to identify the best sales prospects for an intensive sales campaign. In particular, the manufacturer is interested in classifying households as prospective owners or nonowners on the basis of Income (in $1000s) and Lot Size (in 1000 ft^2). The marketing expert looked at a random sample of 24 households, included in the file RidingMowers.xls.

  a. Using Excel, create a scatterplot of Lot Size vs. Income, color coded by the outcome variable owner/nonowner. Make sure to obtain a well-formatted plot (remove excessive background and gridlines; create legible labels and a legend, etc.). The result should be similar to Figure 9.2. *Hint:* First sort the data by the outcome variable, and then plot the data for each category as separate series.

  b. Create the same plot, this time using an interactive visualization tool.

  c. Compare the two processes of generating the plot in terms of the effort as well as the quality of the resulting plots. What are the advantages of each?

3.3   Laptop Sales at a London Computer Chain: Bar Charts and Boxplots. The file LaptopSalesJanuary2008.xls contains data for all sales of laptops at a computer chain in London in January 2008. This is a subset of the full dataset that includes data for the entire year.

  a. Create a bar chart, showing the average retail price by store. Which store has the highest average? Which has the lowest?

  b. To better compare retail prices across stores, create side-by-side boxplots of retail price by store. Now compare the prices in the two stores above. Do you see a difference between their price distributions? Explain.

3.4   Laptop Sales at a London Computer Chain: Interactive Visualization. *The next exercises are designed for use with an interactive visualization tool. The file LaptopSales.txt is a comma-separated file with nearly 300,000 rows. ENBIS (the European Network for Business and Industrial Statistics) provided these data as part of a contest organized in the fall of 2009.*

Scenario: You are a new analyst for Acell, a company selling laptops. You have been provided with data about products and sales. Your task is to help the company to plan product strategy and pricing policies that will maximize Acell's projected revenues in 2009. Using an interactive visualization tool, answer the following questions.

a. Price Questions

  i. At what prices are the laptops actually selling?

  ii. Does price change with time? (*Hint*: Make sure that the date column is recognized as such. The software should then enable different temporal aggregation choices, e.g., plotting the data by weekly or monthly aggregates, or even by day of week.)

  iii. Are prices consistent across retail outlets?

  iv. How does price change with configuration?

b. Location Questions

  i. Where are the stores and customers located?

  ii. Which stores are selling the most?

  iii. How far would customers travel to buy a laptop?

   ○ *Hint 1*: you should be able to aggregate the data, for example, plot the sum or average of the prices.

   ○ *Hint 2*: Use the coordinated highlighting between multiple visualizations in the same page, for example, select a store in one view to see the matching customers in another visualization.

   ○ *Hint 3*: Explore the use of filters to see differences. Make sure to filter in the zoomed out view. For example, try to use a "store location" slider as an alternative way to dynamically compare store locations. This is especially useful for spotting outlier patterns if there are many store locations to compare.

  iv. Try an alternative way of looking at how far customers traveled. Do this by creating a new data column that computes the distance between customer and store.

c. Revenue Questions

  i. How do the sales volume in each store relate to Acell's revenues?

  ii. How does this depend on the configuration?

d. Configuration Questions

  i. What are the details of each configuration? How does this relate to price?

  ii. Do all stores sell all configurations?

## CHAPTER 4

Dimension Reduction

In this chapter we describe the important step of dimension reduction. The dimension of a dataset, which is the number of variables, must be reduced for the data mining algorithms to operate efficiently. We present and discuss several dimension reduction approaches: (1) Incorporating domain knowledge to remove or combine categories, (2) using data summaries to detect information overlap between variables (and remove or combine redundant variables or categories), (3) using data conversion techniques such as converting categorical variables into numerical variables, and (4) employing automated reduction techniques, such as principal components analysis (PCA), where a new set of variables (which are weighted averages of the original variables) is created. These new variables are uncorrelated and a small subset of them usually contains most of their combined information (hence, we can reduce dimension by using only a subset of the new variables). Finally, we mention data mining methods such as regression models and regression and classification trees, which can be used for removing redundant variables and for combining "similar" categories of categorical variables.

4.1 INTRODUCTION

In data mining one often encounters situations where there are a large number of variables in the database. In such situations it is very likely that subsets of variables are highly correlated with each other. Included in a classification or prediction model, highly correlated variables, or variables that are unrelated to the outcome of interest, can lead to overfitting, and accuracy and reliability can suffer. Large numbers of variables also pose computational problems for some models (aside from questions of correlation). In model deployment, superfluous variables can increase costs due to the collection and processing of these variables. The *dimensionality* of a model is the number of independent or input variables used by the model. One of the key steps in data mining, therefore, is finding ways to reduce dimensionality without sacrificing accuracy. In the artificial intelligence literature, dimension reduction is often referred to as *factor selection* or *feature extraction*.

4.2   PRACTICAL CONSIDERATIONS

Although data mining prefers automated methods over domain knowledge, it is important at the first step of data exploration to make sure that the variables measured are reasonable for the task at hand. The integration of expert knowledge through a discussion with the data provider (or user) will probably lead to better results. Practical considerations include: Which variables are most important for the task at hand, and which are most likely to be useless? Which variables are likely to contain much error? Which variables will be available for measurement (and what will it cost to measure them) in the future if the analysis is repeated? Which variables can actually be measured before the outcome occurs? (For example, if we want to predict the closing price of an ongoing online auction, we cannot use the number of bids as a predictor because this will not be known until the auction closes.)

Example 1: House Prices in Boston

We return to the Boston housing example introduced in Chapter 2. For each neighborhood, a number of variables are given, such as the crime rate, the student/teacher ratio, and the median value of a housing unit in the neighborhood. A description of all 14 variables is given in Table 4.1. The first 10 records of

TABLE 4.1	DESCRIPTION OF VARIABLES IN THE BOSTON HOUSING DATASET
CRIM	Crime rate
ZN	Percentage of residential land zoned for lots over 25,000 ft^2
INDUS	Percentage of land occupied by nonretail business
CHAS	Charles River dummy variable ($= 1$ if tract bounds river; $=0$ otherwise)
NOX	Nitric oxide concentration (parts per 10 million)
RM	Average number of rooms per dwelling
AGE	Percentage of owner-occupied units built prior to 1940
DIS	Weighted distances to five Boston employment centers
RAD	Index of accessibility to radial highways
TAX	Full-value property tax rate per $10,000
PTRATIO	Pupil/teacher ratio by town
B	1000(Bk minus 0.63)^2, where Bk is the proportion of blacks by town
LSTAT	% Lower status of the population
MEDV	Median value of owner-occupied homes in $1000s

CRIM	ZN	INDUS	CHAS	NOX	RM	AGE	DIS	RAD	TAX	PTRATIO	B	LSTAT	MEDV	CAT. MEDV
0.00632	18	2.31	0	0.538	6.58	65.2	4.09	1	296	15.3	396.9	4.98	24	0
0.02731	0	7.07	0	0.469	6.42	78.9	4.97	2	242	17.8	396.9	9.14	21.6	0
0.02729	0	7.07	0	0.469	7.19	61.1	4.97	2	242	17.8	392.83	4.03	34.7	1
0.03237	0	2.18	0	0.458	7	45.8	6.06	3	222	18.7	394.63	2.94	33.4	1
0.06905	0	2.18	0	0.458	7.15	54.2	6.06	3	222	18.7	396.9	5.33	36.2	1
0.02985	0	2.18	0	0.458	6.43	58.7	6.06	3	222	18.7	394.12	5.21	28.7	0
0.08829	13	7.87	0	0.524	6.01	66.6	5.56	5	311	15.2	395.6	12.43	22.9	0
0.14455	13	7.87	0	0.524	6.17	96.1	5.95	5	311	15.2	396.9	19.15	27.1	0
0.21124	13	7.87	0	0.524	5.63	100	6.08	5	311	15.2	386.63	29.93	16.5	0

FIGURE 4.1   FIRST NINE RECORDS IN THE BOSTON HOUSING DATASET

the data are shown in Figure 4.1. The first row in this figure represents the first neighborhood, which had an average per capita crime rate of 0.006, 18% of the residential land zoned for lots over 25,000 ft^2, 2.31% of the land devoted to nonretail business, no border on the Charles River, and so on.

4.3   DATA SUMMARIES

As we have seen in Chapter 3 on data visualization, an important initial step of data exploration is getting familiar with the data and their characteristics through summaries and graphs. The importance of this step cannot be overstated. The better you understand the data, the better the results from the modeling or mining process will be.

Numerical summaries and graphs of the data are very helpful for data reduction. The information that they convey can assist in combining categories of a categorical variable, in choosing variables to remove, in assessing the level of information overlap between variables, and more. Before discussing such strategies for reducing the dimension of a data set, let us consider useful summaries and tools.

Summary Statistics

Excel has several functions and facilities that assist in summarizing data. The functions *average, stdev, min, max, median,* and *count* are very helpful for learning about the characteristics of each variable. First, they give us information about the scale and type of values that the variable takes. The min and max functions can be used to detect extreme values that might be errors. The average and median give a sense of the central values of that variable, and a large deviation between the two also indicates skew. The standard deviation gives a sense of how dispersed the data are (relative to the mean). Other functions, such as *countblank,* which gives the number of empty cells, can tell us about missing values. It is also possible to use Excel's *Descriptive Statistics* facility in the *Data > Data Analysis* menu (in Excel

	Average	Median	Min	Max	Std	Count	Countblank
CRIM	3.61	0.26	0.01	88.98	8.60	506	0
ZN	11.36	0.00	0.00	100.00	23.32	506	0
INDUS	11.14	9.69	0.46	27.74	6.86	506	0
CHAS	0.07	0.00	0.00	1.00	0.25	506	0
NOX	0.55	0.54	0.39	0.87	0.12	506	0
RM	6.28	6.21	3.56	8.78	0.70	506	0
AGE	68.57	77.50	2.90	100.00	28.15	506	0
DIS	3.80	3.21	1.13	12.13	2.11	506	0
RAD	9.55	5.00	1.00	24.00	8.71	506	0
TAX	408.24	330.00	187.00	711.00	168.54	506	0
PTRATIO	18.46	19.05	12.60	22.00	2.16	506	0
B	356.67	391.44	0.32	396.90	91.29	506	0
LSTAT	12.65	11.36	1.73	37.97	7.14	506	0
MEDV	22.53	21.20	5.00	50.00	9.20	506	0

FIGURE 4.2   SUMMARY STATISTICS FOR THE BOSTON HOUSING DATA

2003: *Tools > Data Analysis*). This will generate a set of 13 summary statistics for each of the variables.

Figure 4.2 shows six summary statistics for the Boston housing example. We see immediately that the different variables have very different ranges of values. We will see soon how variation in scale across variables can distort analyses if not treated properly. Another observation that can be made is that the average of the first variable, CRIM (as well as several others), is much larger than the median, indicating right skew. None of the variables have empty cells. There also do not appear to be indications of extreme values that might result from typing errors.

Next, we summarize relationships between two or more variables. For numerical variables, we can compute pairwise correlations (using the Excel function *correl*). We can also obtain a complete matrix of correlations between each pair of variables in the data using Excel's *Correlation* facility in the *Data > Data Analysis* menu (in Excel 2003, *Tools > Data Analysis*). Figure 4.3 shows the correlation matrix for a subset of the Boston housing variables. We see that most are low and that many are negative. Recall also the visual display of a correlation matrix via a heatmap (see Figure 3.3 for the heatmap corresponding to this correlation

	CRIM	ZN	INDUS	CHAS	NOX	RM	AGE	DIS	RAD	TAX	PTRATIO	B	LSTAT	MEDV
CRIM	1.00													
ZN	-0.20	1.00												
INDUS	0.41	-0.53	1.00											
CHAS	-0.06	-0.04	0.06	1.00										
NOX	0.42	-0.52	0.76	0.09	1.00									
RM	-0.22	0.31	-0.39	0.09	-0.30	1.00								
AGE	0.35	-0.57	0.64	0.09	0.73	-0.24	1.00							
DIS	-0.38	0.66	-0.71	-0.10	-0.77	0.21	-0.75	1.00						
RAD	0.63	-0.31	0.60	-0.01	0.61	-0.21	0.46	-0.49	1.00					
TAX	0.58	-0.31	0.72	-0.04	0.67	-0.29	0.51	-0.53	0.91	1.00				
PTRATIO	0.29	-0.39	0.38	-0.12	0.19	-0.36	0.26	-0.23	0.46	0.46	1.00			
B	-0.39	0.18	-0.36	0.05	-0.38	0.13	-0.27	0.29	-0.44	-0.44	-0.18	1.00		
LSTAT	0.46	-0.41	0.60	-0.05	0.59	-0.61	0.60	-0.50	0.49	0.54	0.37	-0.37	1.00	
MEDV	-0.39	0.36	-0.48	0.18	-0.43	0.70	-0.38	0.25	-0.38	-0.47	-0.51	0.33	-0.74	1.00

FIGURE 4.3   CORRELATION TABLE FOR BOSTON HOUSING DATA, GENERATED USING EXCEL'S DATA ANALYSIS MENU

table). We will return to the importance of the correlation matrix soon, in the context of correlation analysis.

Pivot Tables

Another very useful tool is Excel's *pivot tables*, in the *Insert > Data* menu (in Excel 2003, in the *Data* menu). These are interactive tables that can combine information from multiple variables and compute a range of summary statistics (count, average, percentage, etc.). A simple example is the average MEDV for neighborhoods that bound the Charles River versus those that do not. First, we get a count of neighborhoods bordering the river. The Excel pivot table in Figure 4.4 (top panel) was obtained by selecting CHAS as a "row labels" field and MEDV or any other variable as a "values" field, using the "count" summary. It appears that the majority of neighborhoods (471 of 506) do not bound the river. By double-clicking on a certain cell, the complete data for records in that cell are shown on a new worksheet. For instance, double-clicking on the cell containing 471 will display the complete records of neighborhoods that do not bound the river.

Pivot tables can be used for multiple variables. For categorical variables we obtain a breakdown of the records by the combination of categories. For instance, the bottom panel of Figure 4.4 shows the average MEDV by CHAS (column) and RM (row). Note that the numerical variable RM (the average number of rooms per dwelling in the neighborhood) is grouped into bins of 3–4, 5–6, and so on. Note also the empty cells, denoting that there are no neighborhoods in the dataset with those combinations (e.g., bounding the river and having on average three or four rooms). There are many more possibilities and options for using Excel's pivot tables. We leave it to the reader to explore these using Excel's documentation.

Count of MEDV	
CHAS	Total
0	471
1	35
Grand Total	506

Average of MEDV	CHAS		
RM	0	1	Grand Total
3-4	25.3		25.3
4-5	16.023077		16.02307692
5-6	17.133333	22.21818182	17.48734177
6-7	21.76917	25.91875	22.01598513
7-8	35.964444	44.06666667	36.91764706
8-9	45.7	35.95	44.2
Grand Total	22.093843	28.44	22.53280632

FIGURE 4.4   PIVOT TABLES FOR THE BOSTON HOUSING DATA

In classification tasks, where the goal is to find predictor variables that do a good job of distinguishing between two classes, a good exploratory step is to produce summaries for each class. This can assist in detecting useful predictors that display some separation between the two classes. Data summaries are useful for almost any data mining task and are therefore an important preliminary step for cleaning and understanding the data before carrying out further analyses.

4.4 CORRELATION ANALYSIS

In datasets with a large number of variables (which are likely to serve as predictors), there is usually much overlap in the information covered by the set of variables. One simple way to find redundancies is to look at a correlation matrix. This shows all the pairwise correlations between variables. Pairs that have a very strong (positive or negative) correlation contain a lot of overlap in information and are good candidates for data reduction by removing one of the variables. Removing variables that are strongly correlated to others is useful for avoiding multicollinearity problems that can arise in various models. (*Multicollinearity* is the presence of two or more predictors sharing the same linear relationship with the outcome variable.)

Correlation analysis is also a good method for detecting duplications of variables in the data. Sometimes, the same variable appears accidentally more than once in the dataset (under a different name) because the dataset was merged from multiple sources, the same phenomenon is measured in different units, and so on. Using correlation table heatmaps, as shown in Chapter 3, can make the task of identifying strong correlations easier.

4.5 REDUCING THE NUMBER OF CATEGORIES IN CATEGORICAL VARIABLES

When a categorical variable has many categories, and this variable is destined to be a predictor, many data mining methods will require converting it into many dummy variables. In particular, a variable with m categories will be transformed into $m - 1$ dummy variables. This means that even if we have very few original categorical variables, they can greatly inflate the dimension of the dataset. One way to handle this is to reduce the number of categories by combining close or similar categories. To combine categories requires incorporating expert knowledge and common sense. Pivot tables are useful for this task: We can examine the sizes of the various categories and how the response behaves at each category. Generally, categories that contain very few observations are good candidates for combining with other categories. Use only the categories that are

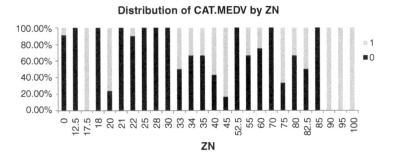

FIGURE 4.5   DISTRIBUTION OF CAT.MEDV (BLACK DENOTES CAT.MEDV=0) BY ZN. SIMILAR BARS INDICATE LOW SEPARATION BETWEEN CLASSES AND CAN BE COMBINED

most relevant to the analysis, and label the rest as "other." In classification tasks (with a categorical output), a pivot table broken down by the output classes can help identify categories that do not separate the classes. Those categories too are candidates for inclusion in the "other" category. An example is shown in Figure 4.5, where the distribution of output variable CAT.MEDV is broken down by ZN (treated here as a categorical variable). We can see that the distribution of CAT.MEDV is identical for ZN=17.5, 90, 95, and 100 (where all neighborhoods have CAT.MEDV=1). These four categories can then be combined into a single category. Similarly categories ZN=12.5, 25, 28, 30, and 70 can be combined. Further combination is also possible based on similar bars.

In a time series context where we might have a categorical variable denoting season (such as month, or hour of day) that will serve as a predictor, reducing categories can be done by examining the time series plot and identifying similar periods. For example, the time plot in Figure 4.6 shows the quarterly revenues of Toys "R" Us between 1992 and 1995. Only quarter 4 periods appear different, and therefore we can combine quarters 1–3 into a single category.

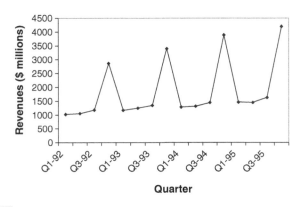

FIGURE 4.6   QUARTERLY REVENUES OF TOYS "R" US, 1992–1995

4.6 CONVERTING A CATEGORICAL VARIABLE TO A NUMERICAL VARIABLE

Sometimes the categories in a categorical variable represent intervals. Common examples are age group or income group. If the interval values are known (e.g., category 2 is the age interval 20–30), we can replace the categorical value ("2" in the example) with the midinterval value (here "25"). The result will be a numerical variable that no longer requires multiple dummy variables.

4.7 PRINCIPAL COMPONENTS ANALYSIS

Principal components analysis (PCA) is a useful procedure for reducing the number of predictors in the model by analyzing the input variables. It is especially valuable when we have subsets of measurements that are highly correlated. In that case it provides a few variables (often as few as three) that are weighted linear combinations of the original variables that retain the explanatory power of the full original set. PCA is intended for use with quantitative variables. For categorical variables, other methods, such as correspondence analysis, are more suitable.

Example 2: Breakfast Cereals

Data were collected on the nutritional information and consumer rating of 77 breakfast cereals.[1] For each cereal the data include 13 numerical variables, and we are interested in reducing this dimension. For each cereal the information is based on a bowl of cereal rather than a serving size because most people simply fill a cereal bowl (resulting in constant volume, but not weight). A snapshot of these data is given in Figure 4.7, and the description of the different variables is given in Table 4.2.

We focus first on two variables: *calories* and *consumer rating*. These are given in Table 4.3. The average calories across the 75 cereals is 106.88 and the average consumer rating is 42.67. The estimated covariance matrix between the two variables is

$$S = \begin{bmatrix} 379.63 & -188.68 \\ -188.68 & 197.32 \end{bmatrix}.$$

It can be seen that the two variables are strongly correlated with a negative correlation of

$$-0.69 = \frac{-188.68}{\sqrt{(379.63)(197.32)}}.$$

---

[1] The data are available at http://lib.stat.cmu.edu/DASL/Stories/HealthyBreakfast.html.

Cereal Name	mfr	type	calories	protein	fat	sodium	fiber	carbo	sugars	potass	vitamins
100% Bran	N	C	70	4	1	130	10	5	6	280	25
100% Natural Bran	Q	C	120	3	5	15	2	8	8	135	0
All-Bran	K	C	70	4	1	260	9	7	5	320	25
All-Bran with Extra Fiber	K	C	50	4	0	140	14	8	0	330	25
Almond Delight	R	C	110	2	2	200	1	14	8		25
Apple Cinnamon Cheerios	G	C	110	2	2	180	1.5	10.5	10	70	25
Apple Jacks	K	C	110	2	0	125	1	11	14	30	25
Basic 4	G	C	130	3	2	210	2	18	8	100	25
Bran Chex	R	C	90	2	1	200	4	15	6	125	25
Bran Flakes	P	C	90	3	0	210	5	13	5	190	25
Cap'n'Crunch	Q	C	120	1	2	220	0	12	12	35	25
Cheerios	G	C	110	6	2	290	2	17	1	105	25
Cinnamon Toast Crunch	G	C	120	1	3	210	0	13	9	45	25
Clusters	G	C	110	3	2	140	2	13	7	105	25
Cocoa Puffs	G	C	110	1	1	180	0	12	13	55	25
Corn Chex	R	C	110	2	0	280	0	22	3	25	25
Corn Flakes	K	C	100	2	0	290	1	21	2	35	25
Corn Pops	K	C	110	1	0	90	1	13	12	20	25
Count Chocula	G	C	110	1	1	180	0	12	13	65	25
Cracklin' Oat Bran	K	C	110	3	3	140	4	10	7	160	25

FIGURE 4.7   SAMPLE FROM THE 77 BREAKFAST CEREALS DATASET

Roughly speaking, 69% of the total variation in both variables is actually "co-variation," or variation in one variable that is duplicated by similar variation in the other variable. Can we use this fact to reduce the number of variables, while making maximum use of their unique contributions to the overall variation? Since there is redundancy in the information that the two variables contain, it might be possible to reduce the two variables to a single variable without losing too much information. The idea in PCA is to find a linear combination of the two variables that contains most, even if not all, of the information, so

TABLE 4.2   DESCRIPTION OF THE VARIABLES IN THE BREAKFAST CEREAL DATASET

Variable	Description
mfr	Manufacturer of cereal (American Home Food Products, General Mills, Kellogg, etc.)
type	Cold or hot
calories	Calories per serving
protein	Grams of protein
fat	Grams of fat
sodium	Milligrams of sodium
fiber	Grams of dietary fiber
carbo	Grams of complex carbohydrates
sugars	Grams of sugars
potass	Milligrams of potassium
vitamins	Vitamins and minerals: 0, 25, or 100, indicating the typical percentage of FDA recommended
shelf	Display shelf (1, 2, or 3, counting from the floor)
weight	Weight in ounces of one serving
cups	Number of cups in one serving
rating	Rating of the cereal calculated by *Consumer Reports*

TABLE 4.3   CEREAL CALORIES AND RATINGS

Cereal	Calories	Rating	Cereal	Calories	Rating
100% Bran	70	68.40297	Just Right Crunchy	110	36.52368
100% Natural Bran	120	33.98368	Nuggets		
All-Bran	70	59.42551	Just Right Fruit & Nut	140	36.471512
All-Bran with Extra	50	93.70491	Kix	110	39.241114
Fiber			Life	100	45.328074
Almond Delight	110	34.38484	Lucky Charms	110	26.734515
Apple Cinnamon	110	29.50954	Maypo	100	54.850917
Cheerios			Muesli Raisins, Dates	150	37.136863
Apple Jacks	110	33.17409	& Almonds		
Basic 4	130	37.03856	Muesli Raisins,	150	34.139765
Bran Chex	90	49.12025	Peaches & Pecans		
Bran Flakes	90	53.31381	Mueslix Crispy Blend	160	30.313351
Cap'n'Crunch	120	18.04285	Multi-Grain Cheerios	100	40.105965
Cheerios	110	50.765	Nut&Honey Crunch	120	29.924285
Cinnamon Toast	120	19.82357	Nutri-Grain Almond-	140	40.69232
Crunch			Raisin		
Clusters	110	40.40021	Nutri-grain Wheat	90	59.642837
Cocoa Puffs	110	22.73645	Oatmeal Raisin Crisp	130	30.450843
Corn Chex	110	41.44502	Post Nat. Raisin Bran	120	37.840594
Corn Flakes	100	45.86332	Product 19	100	41.50354
Corn Pops	110	35.78279	Puffed Rice	50	60.756112
Count Chocula	110	22.39651	Puffed Wheat	50	63.005645
Cracklin' Oat Bran	110	40.44877	Quaker Oat Squares	100	49.511874
Cream of Wheat	100	64.53382	Quaker Oatmeal	100	50.828392
(Quick)			Raisin Bran	120	39.259197
Crispix	110	46.89564	Raisin Nut Bran	100	39.7034
Crispy Wheat &	100	36.1762	Raisin Squares	90	55.333142
Raisins			Rice Chex	110	41.998933
Double Chex	100	44.33086	Rice Krispies	110	40.560159
Froot Loops	110	32.20758	Shredded Wheat	80	68.235885
Frosted Flakes	110	31.43597	Shredded Wheat	90	74.472949
Frosted	100	58.34514	'n'Bran		
Mini-Wheats			Shredded Wheat	90	72.801787
Fruit & Fibre Dates,	120	40.91705	spoon size		
Walnuts & Oats			Smacks	110	31.230054
Fruitful Bran	120	41.01549	Special K	110	53.131324
Fruity Pebbles	110	28.02577	Strawberry Fruit	90	59.363993
Golden Crisp	100	35.25244	Wheats		
Golden Grahams	110	23.80404	Total Corn Flakes	110	38.839746
Grape Nuts Flakes	100	52.0769	Total Raisin Bran	140	28.592785
Grape-Nuts	110	53.37101	Total Whole Grain	100	46.658844
Great Grains Pecan	120	45.81172	Triples	110	39.106174
Honey Graham Ohs	120	21.87129	Trix	110	27.753301
Honey Nut Cheerios	110	31.07222	Wheat Chex	100	49.787445
Honey-comb	110	28.74241	Wheaties	100	51.592193
			Wheaties Honey Gold	110	36.187559

that this new variable can replace the two original variables. Information here is in the sense of variability: What can explain the most variability *among* the 77 cereals? The total variability here is the sum of the variances of the two variables, which in this case is $379.63 + 197.32 = 577$. This means that *calories* accounts for $66\% = 379.63/577$ of the total variability, and *rating* for the remaining 34%. If we drop one of the variables for the sake of dimension reduction, we lose at least 34% of the total variability. Can we redistribute the total variability between two new variables in a more polarized way? If so, it might be possible to keep only the one new variable that (hopefully) accounts for a large portion of the total variation.

Figure 4.8 shows a scatterplot of *rating* versus *calories*. The line z_1 is the direction in which the variability of the points is largest. It is the line that captures the most variation in the data if we decide to reduce the dimensionality of the data from two to one. Among all possible lines, it is the line for which, if we project the points in the dataset orthogonally to get a set of 77 (one-dimensional) values, the variance of the z_1 values will be maximum. This is called the *first principal component*. It is also the line that minimizes the sum-of-squared perpendicular distances from the line. The z_2 axis is chosen to be perpendicular to the z_1-axis. In the case of two variables, there is only one line that is perpendicular to z_1, and it has the second largest variability, but its information is uncorrelated with z_1. This is called the *second principal component*. In general, when we have more than two variables, once we find the direction z_1 with the largest variability, we search among all the orthogonal directions to z_1 for the one with the next highest variability. That is z_2. The idea is then to find the coordinates of these lines and to see how they redistribute the variability.

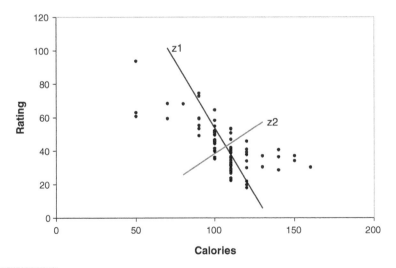

FIGURE 4.8   SCATTERPLOT OF *RATING* VS. *CALORIES* FOR 77 BREAKFAST CEREALS, WITH THE TWO PRINCIPAL COMPONENT DIRECTIONS

Principal Components

	Components	
Variable	**1**	**2**
calories	-0.84705347	0.53150767
rating	0.53150767	0.84705347

Variance	498.0244751	78.932724
Variance%	86.31913757	13.68086338
Cum%	86.31913757	100
P-value	0	1

FIGURE 4.9   OUTPUT FROM PRINCIPAL COMPONENTS ANALYSIS OF *CALORIES* AND *RATING*

Figure 4.9 shows the XLMiner output from running PCA on these two variables. The principal components table gives the weights that are used to project the original points onto the two new directions. The weights for z_1 are given by $(-0.847, 0.532)$, and for z_2 they are given by $(0.532, 0.847)$. Figure 4.9 gives the reallocated variance: z_1 accounts for 86% of the total variability and z_2 for the remaining 14%. Therefore, if we drop z_2, we still maintain 86% of the total variability.

The weights are used to compute principal component scores, which are the projected values of *calories* and *rating* onto the new axes (after subtracting the means). Figure 4.10 shows the scores for the two dimensions. The first column is the projection onto z_1 using the weights $(-0.847, 0.532)$. The second column is the projection onto z_2 using the weights $(0.532, 0.847)$. For example, the

Row Id.	1	2
100% Bran	44.92152786	2.19717932
100% Natural Bran	-15.7252636	-0.38241446
All-Bran	40.14993668	-5.40721178
All-Bran with Extra Fiber	75.31076813	12.99912071
Almond Delight	-7.04150867	-5.35768652
Apple Cinnamon Cheerios	-9.63276863	-9.48732758
Apple Jacks	-7.68502998	-6.38325357
Basic 4	-22.57210541	7.52030993
Bran Chex	17.7315464	-3.50615811
Bran Flakes	19.96045494	0.04600986
Cap'n'Crunch	-24.19793701	-13.88514996
Cheerios	1.66467071	8.5171833
Cinnamon Toast Crunch	-23.25147057	-12.37678337
Clusters	-3.84429598	-0.26235023
Cocoa Puffs	-13.23272038	-15.2244997
Corn Chex	-3.28897071	0.62266076
Corn Flakes	7.5299263	-0.94987571

FIGURE 4.10   PRINCIPAL SCORES FROM PRINCIPAL COMPONENTS ANALYSIS OF *CALORIES* AND *RATING* FOR THE FIRST 17 CEREALS

first score for the 100% Bran cereal (with 70 calories and a rating of 68.4) is $(-0.847)(70 - 106.88) + (0.532)(68.4 - 42.67) = 44.92$.

Note that the means of the new variables z_1 and z_2 are zero (because we have subtracted the mean of each variable). The sum of the variances $\text{var}(z_1) + \text{var}(z_2)$ is equal to the sum of the variances of the original variables, *calories* and *rating*. Furthermore, the variances of z_1 and z_2 are 498 and 79, respectively, so the first principal component, z_1, accounts for 86% of the total variance. Since it captures most of the variability in the data, it seems reasonable to use one variable, the first principal score, to represent the two variables in the original data. Next, we generalize these ideas to more than two variables.

Principal Components

Let us formalize the procedure described above so that it can easily be generalized to $p > 2$ variables. Denote by $X_1, X_2, \ldots, X_p$ the original p variables. In PCA we are looking for a set of new variables $Z_1, Z_2, \ldots, Z_p$ that are weighted averages of the original variables (after subtracting their mean):

$$Z_i = a_{i,1}(X_1 - \bar{X}_1) - +a_{i,2}(X_2 - \bar{X}_2) + \cdots + a_{i,p}(X_p - \bar{X}_p) \qquad i = 1, \ldots, p$$

where each pair of Z's has correlation $= 0$. We then order the resulting Z's by their variance, with Z_1 having the largest variance and Z_p having the smallest variance. The software computes the weights $a_{i,j}$, which are then used in computing the principal component scores.

A further advantage of the principal components compared to the original data is that they are uncorrelated (correlation coefficient $= 0$). If we construct regression models using these principal components as independent variables, we will not encounter problems of multicollinearity.

Let us return to the breakfast cereal dataset with all 15 variables, and apply PCA to the 13 numerical variables. The resulting output is shown in Figure 4.11. For simplicity, we removed three cereals that contained missing values. Note that the first three components account for more than 96% of the total variation associated with all 13 of the original variables. This suggests that we can capture most of the variability in the data with less than 25% of the number of original dimensions in the data. In fact, the first two principal components alone capture 92.6% of the total variation. However, these results are influenced by the scales of the variables, as we describe next.

Normalizing the Data

A further use of PCA is to understand the structure of the data. This is done by examining the weights to see how the original variables contribute to the different principal components. In our example it is clear that the first principal component is dominated by the sodium content of the cereal: it has the

Variable	1	2	3	4	5	6	7
calories	0.07798425	-0.00931156	0.62920582	-0.60102159	0.45495847	0.11884782	0.09385654
protein	-0.00075678	0.00880103	0.00102611	0.00319992	0.05617596	0.11274506	0.25810272
fat	-0.00010178	0.00269915	0.01619579	-0.02526222	-0.01609845	-0.13181572	0.37258437
sodium	0.98021454	0.14089581	-0.13590187	-0.00096808	0.01394816	0.02279307	0.00450823
fiber	-0.00541276	0.03068075	-0.01819105	0.0204722	0.01360502	0.2628414	0.0431139
carbo	0.01724625	-0.0167833	0.01736996	0.02594825	0.34926692	-0.53783643	-0.67243195
sugars	0.00298888	-0.00025348	0.09770504	-0.11548097	-0.29906642	0.64792335	-0.5669753
potass	-0.13490002	0.98656207	0.03678251	-0.0421758	-0.04715054	-0.04999856	-0.01795866
vitamins	0.09429332	0.01672884	0.69197786	0.714118	-0.03700861	0.01575723	0.01210225
shelf	-0.00154142	0.0043604	0.01248884	0.00564718	-0.00787646	-0.0599014	0.09221537
weight	0.000512	0.00099922	0.00380597	-0.00254643	0.00302211	0.00905157	-0.02361298
cups	0.00051012	-0.00159098	0.00069433	0.00098539	0.00214846	-0.01030537	-0.01959434
rating	-0.07529629	0.07174215	-0.30794701	0.33453393	0.75770795	0.41302064	0.01832427
Variance	7016.42041	5028.831543	512.7391968	367.9292603	70.95076752	4.3750844	2.8880403
Variance%	53.95025635	38.66740417	3.94252491	2.82906055	0.54555058	0.03364065	0.02220655
Cum%	53.95025635	92.61766052	96.56018829	99.38924408	99.93479919	99.96843719	99.99064636

FIGURE 4.11   PCA OUTPUT USING ALL 13 NUMERICAL VARIABLES IN THE BREAKFAST CEREALS DATASET. RESULTS ARE GIVEN FOR THE FIRST SEVEN PRINCIPAL COMPONENTS

highest (in this case, positive) weight. This means that the first principal component is measuring how much sodium is in the cereal. Similarly, the second principal component seems to be measuring the amount of potassium. Since both these variables are measured in milligrams, whereas the other nutrients are measured in grams, the scale is obviously leading to this result. The variances of potassium and sodium are much larger than the variances of the other variables, and thus the total variance is dominated by these two variances. A solution is to normalize the data before performing the PCA. Normalization (or standardization) means replacing each original variable by a standardized version of the variable that has unit variance. This is easily accomplished by dividing each variable by its standard deviation. The effect of this normalization (standardization) is to give all variables equal importance in terms of the variability.

When should we normalize the data like this? It depends on the nature of the data. When the units of measurement are common for the variables (e.g., dollars), and when their scale reflects their importance (sales of jet fuel, sales of heating oil), it is probably best not to normalize (i.e., not to rescale the data so that they have unit variance). If the variables are measured in quite differing units so that it is unclear how to compare the variability of different variables (e.g., dollars for some, parts per million for others) or if for variables measured in the same units, scale does not reflect importance (earnings per share, gross revenues), it is generally advisable to normalize. In this way, the changes in units of measurement do not change the principal components' weights. In the rare situations where we can give relative weights to variables, we multiply the normalized variables by these weights before doing the principal components analysis.

Variable	1	2	3	4	5	6	7
calories	0.2995424	0.39314792	0.11485746	0.20435865	0.20389892	-0.25590625	-0.02559552
protein	-0.30735639	0.16532333	0.27728197	0.30074316	0.319749	0.120752	0.28270504
fat	0.03991544	0.34572428	-0.20489009	0.18683317	0.58689332	0.34796733	-0.05115468
sodium	0.18339655	0.13722059	0.38943109	0.12033724	-0.33836424	0.66437215	-0.28370309
fiber	-0.45349041	0.17981192	0.06976604	0.03917367	-0.255119	0.0642436	0.11232537
carbo	0.19244903	-0.14944831	0.56245244	0.0878355	0.18274252	-0.32639283	-0.26046798
sugars	0.22806853	0.35143444	-0.35540518	-0.02270711	-0.31487244	-0.15208226	0.22798519
potass	-0.40196434	0.30054429	0.06762024	0.09087842	-0.14836049	0.02515389	0.14880823
vitamins	0.11598022	0.1729092	0.38785872	-0.6041106	-0.04928682	0.12948574	0.29427618
shelf	-0.17126338	0.26505029	-0.00153102	-0.63887852	0.32910112	-0.05204415	-0.17483434
weight	0.05029929	0.45030847	0.24713831	0.15342878	-0.22128329	-0.39877367	0.01392053
cups	0.29463556	-0.21224795	0.13999969	0.04748911	0.12081645	0.09946091	0.74856687
rating	-0.43837839	-0.25153893	0.1818424	0.0383162	0.05758421	-0.18614525	0.06344455
Variance	3.63360572	3.1480546	1.90934956	1.01947618	0.98935974	0.72206175	0.67151642
Variance%	27.95081329	24.21580505	14.6873045	7.84212446	7.61045933	5.55432129	5.16551113
Cum%	27.95081329	52.16661835	66.85391998	74.69604492	82.3065033	87.86082458	93.02633667

FIGURE 4.12 PCA OUTPUT USING ALL *NORMALIZED* 13 NUMERICAL VARIABLES IN THE BREAKFAST CEREALS DATASET. RESULTS ARE GIVEN FOR THE FIRST SEVEN PRINCIPAL COMPONENTS

Thus far, we have calculated principal components using the covariance matrix. An alternative to normalizing and then performing PCA is to perform PCA on the correlation matrix instead of the covariance matrix. Most software programs allow the user to choose between the two. Remember that using the correlation matrix means that you are operating on the normalized data.

Returning to the breakfast cereal data, we normalize the 13 variables due to the different scales of the variables and then perform PCA (or equivalently, we use PCA applied to the correlation matrix). The output is shown in Figure 4.12. Now we find that we need 7 principal components to account for more than 90% of the total variability. The first 2 principal components account for only 52% of the total variability, and thus reducing the number of variables to 2 would mean losing a lot of information. Examining the weights, we see that the first principal component measures the balance between 2 quantities: (1) calories and cups (large positive weights) versus (2) protein, fiber, potassium, and consumer rating (large negative weights). High scores on principal component 1 mean that the cereal is high in calories and the amount per bowl, and low in protein, fiber, and potassium. Unsurprisingly, this type of cereal is associated with a low consumer rating. The second principal component is most affected by the weight of a serving, and the third principal component by the carbohydrate content. We can continue labeling the next principal components in a similar fashion to learn about the structure of the data.

When the data can be reduced to two dimensions, a useful plot is a scatterplot of the first versus the second principal scores with labels for the observations (if the dataset is not too large). To illustrate this, Figure 4.13 displays the first two principal component scores for the breakfast cereals.

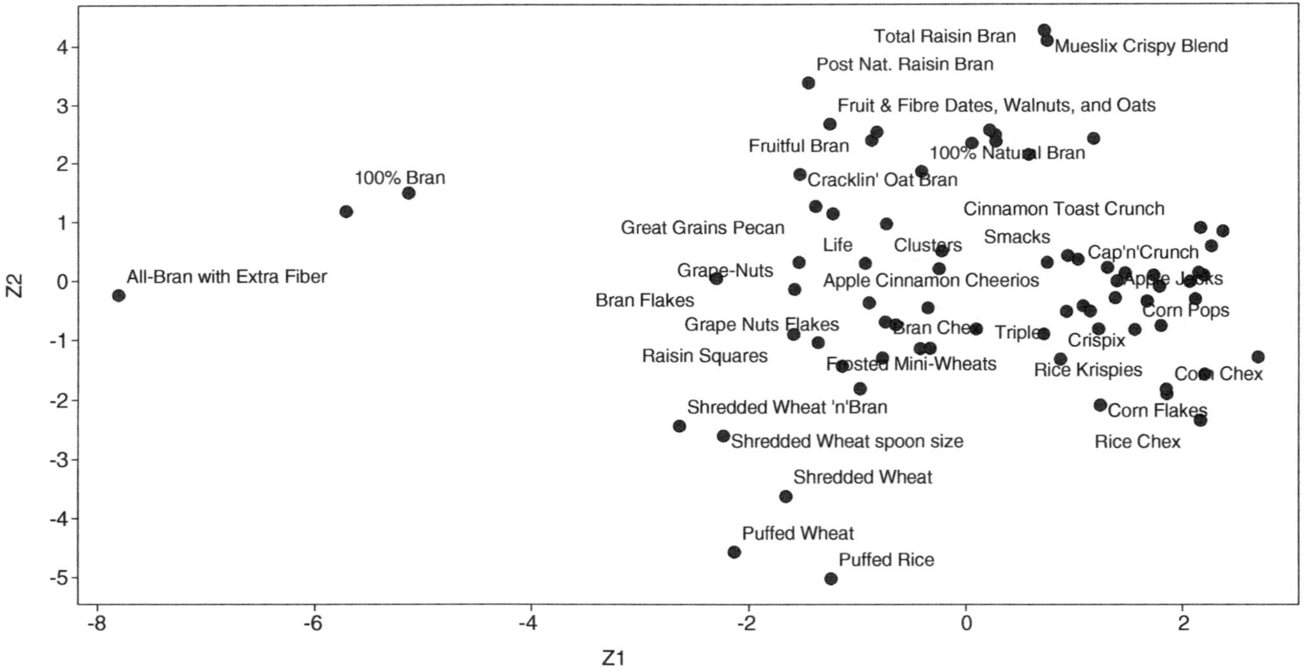

FIGURE 4.13 SCATTERPLOT OF THE SECOND VS. FIRST PRINCIPAL COMPONENTS SCORES FOR THE NORMALIZED BREAKFAST CEREAL OUTPUT

We can see that as we move from left (bran cereals) to right, the cereals are less "healthy" in the sense of high calories, low protein and fiber, and so on. Also, moving from bottom to top, we get heavier cereals (moving from puffed rice to raisin bran). These plots are especially useful if interesting clusterings of observations can be found. For instance, we see here that children's cereals are close together on the middle-right part of the plot.

Using Principal Components for Classification and Prediction

When the goal of the data reduction is to have a smaller set of variables that will serve as predictors, we can proceed as follows: Apply PCA to the training data. Use the output to determine the number of principal components to be retained. The predictors in the model now use the (reduced number of) principal scores columns. For the validation set we can use the weights computed from the training data to obtain a set of principal scores by applying the weights to the variables in the validation set. These new variables are then treated as the predictors.

4.8 DIMENSION REDUCTION USING REGRESSION MODELS

In this chapter we discussed methods for reducing the number of columns using summary statistics, plots, and principal components analysis. All these are considered exploratory methods. Some of them completely ignore the output variable (e.g., PCA), whereas in other methods we informally try to incorporate the relationship between the predictors and the output variable (e.g., combining similar categories, in terms of their behavior with y). Another approach to reducing the number of predictors, which directly considers the predictive or classification task, is by fitting a regression model. For prediction a linear regression model is used (see Chapter 6), and for classification a logistic regression model (see Chapter 10) is used. In both cases we can employ subset selection procedures that algorithmically choose a subset of variables among the larger set (see details in the relevant chapters).

Fitted regression models can also be used to further combine similar categories: categories that have coefficients that are not statistically significant (i.e., have a high p-value) can be combined with the reference category because their distinction from the reference category appears to have no significant effect on the output variable. Moreover, categories that have similar coefficient values (and the same sign) can often be combined because their effect on the output variable is similar. See the example in Chapter 10 on predicting delayed flights for an illustration of how regression models can be used for dimension reduction.

4.9 DIMENSION REDUCTION USING CLASSIFICATION AND REGRESSION TREES

Another method for reducing the number of columns and for combining categories of a categorical variable is by applying classification and regression trees (see Chapter 9). Classification trees are used for classification tasks and regression trees for prediction tasks. In both cases the algorithm creates binary splits on the predictors that best classify/predict the outcome (e.g., above/below age 30). Although we defer the detailed discussion to Chapter 9, we note here that the resulting tree diagram can be used for determining the important predictors. Predictors (numerical or categorical) that do not appear in the tree can be removed. Similarly, categories that do not appear in the tree can be combined.

PROBLEMS

4.1 **Breakfast Cereals.** Use the data for the breakfast cereal example in Section 4.7 to explore and summarize the data as follows: (Note that a few records contain missing values; since there are just a few, a simple solution is to remove them first. You can use the "Missing Data Handling" utility in XLMiner.)

    a. Which variables are quantitative/numerical? Which are ordinal? Which are nominal?

    b. Create a table with the average, median, min, max, and standard deviation for each of the quantitative variables. This can be done through Excel's functions or Excel's *Tools > DataAnalysis > DescriptiveStatistics* menu.

    c. Use XLMiner to plot a histogram for each of the quantitative variables. Based on the histograms and summary statistics, answer the following questions:

        i. Which variables have the largest variability?

        ii. Which variables seem skewed?

        iii. Are there any values that seem extreme?

    d. Use XLMiner to plot a side-by-side boxplot comparing the calories in hot versus cold cereals. What does this plot show us?

    e. Use XLMiner to plot a side-by-side boxplot of consumer rating as a function of the shelf height. If we were to predict consumer rating from shelf height, does it appear that we need to keep all three categories of shelf height?

    f. Compute the correlation table for the quantitative variable (use Excel's *Tools > Data-Analysis > Correlation* menu). In addition, use XLMiner to generate a matrix plot for these variables.

        i. Which pair of variables is most strongly correlated?

        ii. How can we reduce the number of variables based on these correlations?

        iii. How would the correlations change if we normalized the data first?

    g. Consider the first column on the left in Figure 4.11. Describe briefly what this column represents.

4.2 **Chemical Features of Wine.** Figure 4.14 shows the PCA output on data (nonnormalized) in which the variables represent chemical characteristics of wine, and each case is a different wine.

    a. The data are in the file Wine.xls. Consider the row near the bottom labeled "Variance." Explain why column 1's variance is so much greater than that of any other column.

    b. Comment on the use of normalization (standardization) in part (a).

4.3 **University Rankings.** The dataset on American college and university rankings (available from www.dataminingbook.com) contains information on 1302 American colleges and universities offering an undergraduate program. For each university there are 17 measurements that include continuous measurements (such as tuition and graduation rate) and categorical measurements (such as location by state and whether it is a private or a public school).

    a. Remove all categorical variables. Then remove all records with missing numerical measurements from the dataset (by creating a new worksheet).

    b. Conduct a principal components analysis on the cleaned data and comment on the results. Should the data be normalized? Discuss what characterizes the components you consider key.

	Components					
Variable	1	2	3	4	5	6
Alcohol	0.00165926	0.00120342	0.01687386	-0.14144674	0.02033708	0.19412018
Malic_Acid	-0.00068102	0.00215498	0.12200337	-0.16038956	-0.61288345	0.74247289
Ash	0.00019491	0.00459369	0.05198744	0.00977282	0.02017558	0.04175295
Ash_Alcalinity	-0.0046713	0.02645036	0.93859297	0.33096525	0.06435229	-0.02406531
Magnesium	0.01786801	0.99934423	-0.02978026	0.00539375	-0.00614938	-0.0019238
Total_Phenols	0.00098983	0.00087797	-0.04048461	0.07458466	0.31524512	0.2787168
Flavanoids	0.00156729	-0.00005184	-0.08544329	0.16908674	0.5247612	0.43359798
Nonflavanoid_Ph	-0.00012309	-0.00135448	0.01351078	-0.01080556	-0.02964753	-0.02195283
Proanthocyanins	0.00060061	0.0050044	-0.02465936	0.05012095	0.25118256	0.24188447
Color_Intensity	0.00232714	0.01510037	0.29139856	-0.87889373	0.33174714	0.00273963
Hue	0.00017138	-0.00076267	-0.02597765	0.06003497	0.05152407	-0.02377617
OD280_OD315	0.00070493	-0.00349536	-0.07032393	0.17820027	0.26063919	0.28891277
Proline	0.99982297	-0.01777381	0.00452868	0.00311292	-0.00229857	-0.00121226
Variance	99201.78906	172.5352631	9.43811321	4.99117851	1.22884524	0.84106386
Variance%	99.80912018	0.17359155	0.0094959	0.00502174	0.00123637	0.00084621
Cum%	99.80912018	99.98271179	99.99221039	99.99723053	99.99846649	99.99931335

FIGURE 4.14   PRINCIPAL COMPONENTS OF NONNORMALIZED WINE DATA

4.4   Sales of Toyota Corolla Cars. The file ToyotaCorolla.xls contains data on used cars (Toyota Corollas) on sale during late summer of 2004 in The Netherlands. It has 1436 records containing details on 38 attributes, including Price, Age, Kilometers, HP, and other specifications. The goal will be to predict the price of a used Toyota Corolla based on its specifications.

a. Identify the categorical variables.

b. Explain the relationship between a categorical variable and the series of binary dummy variables derived from it.

c. How many dummy binary variables are required to capture the information in a categorical variable with N categories?

d. Using XLMiner's data utilities, convert the categorical variables in this dataset into dummy binaries, and explain in words, for one record, the values in the derived binary dummies.

e. Use Excel's correlation command (*Tools > DataAnalysis > Correlation* menu) to produce a correlation matrix and XLMiner's matrix plot to obtain a matrix of all scatterplots. Comment on the relationships among variables.

## PART THREE

Performance Evaluation

## CHAPTER 5

Evaluating Classification and Predictive Performance

In this chapter we discuss how the predictive performance of data mining methods can be assessed. We point out the danger of overfitting to the training data and the need for testing model performance on data that were not used in the training step. We discuss popular performance metrics. For prediction, metrics include Average Error, MAPE, and RMSE (based on the validation data). For classification tasks, metrics include the classification matrix, specificity and sensitivity, and metrics that account for misclassification costs. We also show the relation between the choice of cutoff value and method performance, and present the receiver operating characteristic (ROC) curve, which is a popular plot for assessing method performance at different cutoff values. When the goal is to accurately classify the top tier of a new sample rather than accurately classify the entire sample (e.g., the 10% of customers most likely to respond to an offer), lift charts are used to assess performance. We also discuss the need for oversampling rare classes and how to adjust performance metrics for the oversampling. Finally, we mention the usefulness of comparing metrics based on the validation data to those based on the training data for the purpose of detecting overfitting. While some differences are expected, extreme differences can be indicative of overfitting.

5.1 INTRODUCTION

In supervised learning we are interested in predicting the class (classification) or continuous value (prediction) of an outcome variable. In Chapter 2 we worked

Data Mining for Business Intelligence, By Galit Shmueli, Nitin R. Patel, and Peter C. Bruce

through a simple example. Let us now examine the questions of how to judge the usefulness of a classifier or predictor and how to compare different ones.

5.2 JUDGING CLASSIFICATION PERFORMANCE

The need for performance measures arises from the wide choice of classifiers and predictive methods. Not only do we have several different methods, but even within a single method there are usually many options that can lead to completely different results. A simple example is the choice of predictors used within a particular predictive algorithm. Before we study these various algorithms in detail and face decisions on how to set these options, we need to know how we will measure success.

A natural criterion for judging the performance of a classifier is the probability of making a *misclassification error*. Misclassification means that the observation belongs to one class but the model classifies it as a member of a different class. A classifier that makes no errors would be perfect, but we do not expect to be able to construct such classifiers in the real world due to "noise" and to not having all the information needed to classify cases precisely. Is there a minimal probability of misclassification that we should require of a classifier?

Benchmark: The Naive Rule

A very simple rule for classifying a record into one of m classes, ignoring all predictor information $(X_1, X_2, ..., X_p)$ that we may have, is to classify the record as a member of the majority class. In other words, "classify as belonging to the most prevalent class." The *naive rule* is used mainly as a baseline or benchmark for evaluating the performance of more complicated classifiers. Clearly, a classifier that uses external predictor information (on top of the class membership allocation) should outperform the naive rule. There are various performance measures based on the naive rule that measure how much better than the naive rule a certain classifier performs. One example is the multiple R^2 reported by XLMiner, which measures the distance between the fit of the classifier to the data and the fit of the naive rule to the data (for further details, see Section 10.5.

The equivalent of the naive rule for classification when considering a quantitative response is to use $\hat{y}$, the sample mean, to predict the value of y for a new record. In both cases the predictions rely solely on the y information and exclude any additional predictor information.

Class Separation

If the classes are well separated by the predictor information, even a small dataset will suffice in finding a good classifier, whereas if the classes are not separated at all by the predictors, even a very large dataset will not help. Figure 5.1 illustrates this for a two-class case. Figure 5.1(*a*) includes a small dataset (*n* = 24 observations) where two predictors (income and lot size) are used for separating owners from nonowners [we thank Dean Wichern for this example, described in Johnson and Wichern (2002)]. Here, the predictor information seems useful in that it separates the two classes (owners/nonowners). Figure 5.1(*b*) shows a much larger dataset (*n* = 5000 observations) where the two predictors (income and average credit card spending) do not separate the two classes well in most of the higher ranges (loan acceptors/nonacceptors).

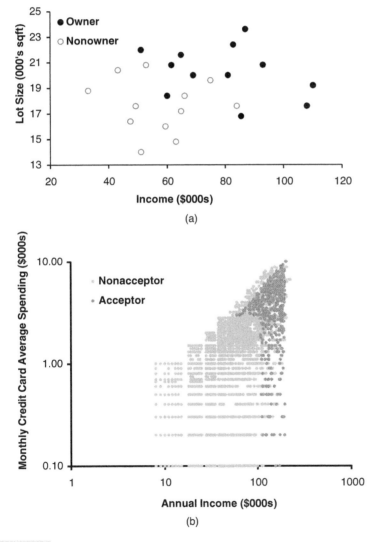

FIGURE 5.1   (*A*) HIGH AND (*B*) LOW LEVELS OF SEPARATION BETWEEN TWO CLASSES, USING TWO PREDICTORS

Classification Confusion Matrix		
	Predicted Class	
Actual Class	1	0
1	201	85
0	25	2689

FIGURE 5.2 **CLASSIFICATION MATRIX BASED ON 3000 OBSERVATIONS AND TWO CLASSES**

Classification Matrix

In practice, most accuracy measures are derived from the *classification matrix* (also called the *confusion matrix*). This matrix summarizes the correct and incorrect classifications that a classifier produced for a certain dataset. Rows and columns of the classification matrix correspond to the true and predicted classes, respectively. Figure 5.2 shows an example of a classification (confusion) matrix for a two-class (0/1) problem resulting from applying a certain classifier to 3000 observations. The two diagonal cells (upper left, lower right) give the number of correct classifications, where the predicted class coincides with the actual class of the observation. The off-diagonal cells give counts of misclassification. The top right cell gives the number of class 1 members that were misclassified as 0's (in this example, there were 85 such misclassifications). Similarly, the lower left cell gives the number of class 0 members that were misclassified as 1's (25 such observations).

The classification matrix gives estimates of the true classification and misclassification rates. Of course, these are estimates and they can be incorrect, but if we have a large enough dataset and neither class is very rare, our estimates will be reliable. Sometimes, we may be able to use public data such as U.S. Census data to estimate these proportions. However, in most practical business settings, we will not know them.

Using the Validation Data

To obtain an honest estimate of classification error, we use the classification matrix that is computed from the *validation data*. In other words, we first partition the data into training and validation sets by random selection of cases. We then construct a classifier using the training data and apply it to the validation data. This will yield the predicted classifications for observations in the validation set (see Figure 2.4). We then summarize these classifications in a classification matrix. Although we can summarize our results in a classification matrix for training data as well, the resulting classification matrix is not useful for getting an honest estimate of the misclassification rate for new data due to the danger of overfitting.

In addition to examining the validation data classification matrix to assess the classification performance on new data, we compare the training data classification matrix to the validation data classification matrix, in order to detect overfitting: although we expect inferior results on the validation data, a large discrepancy in training and validation performance might be indicative of overfitting.

Accuracy Measures

Different accuracy measures can be derived from the classification matrix. Consider a two-class case with classes C_0 and C_1 (e.g., buyer/nonbuyer). The schematic classification matrix in Table 5.1 uses the notation $n_{i,j}$ to denote the number of cases that are class C_i members and were classified as C_j members. Of course, if $i \neq j$, these are counts of misclassifications. The total number of observations is $n = n_{0,0} + n_{0,1} + n_{1,0} + n_{1,1}$.

A main accuracy measure is the *estimated misclassification rate*, also called the *overall error rate*. It is given by

$$\text{err} = \frac{n_{0,1} + n_{1,0}}{n},$$

where n is the total number of cases in the validation dataset. In the example in Figure 5.2, we get $\text{err} = (25 + 85)/3000 = 3.67\%$.

We can measure accuracy by looking at the correct classifications instead of the misclassifications. The *overall accuracy* of a classifier is estimated by

$$\text{Accuracy} = 1 - \text{err} = \frac{n_{0,0} + n_{1,1}}{n}.$$

In the example we have $(201 + 2689)/3000 = 96.33$.

Cutoff for Classification

The first step in most classification algorithms is to estimate the probability that a case belongs to each of the classes. If overall classification accuracy (involving all the classes) is of interest, the case can be assigned to the class with the highest

TABLE 5.1   **CLASSIFICATION MATRIX: MEANING OF EACH CELL**

Actual Class	Predicted Class	
	C_0	C_1
C_0	$n_{0,0} =$ number of C_0 cases classified correctly	$n_{0,1} =$ number of C_0 cases classified incorrectly as C_1
C_1	$n_{1,0} =$ number of C_1 cases classified incorrectly as C_0	$n_{1,1} =$ number of C_1 cases classified correctly

probability. In many cases, a single class is of special interest, so we will focus on that particular class and compare the estimated probability of belonging to that class to a *cutoff value*. This approach can be used with two classes or more than two classes, though it may make sense in such cases to consolidate classes so that you end up with two: the class of interest and all other classes. If the probability of belonging to the class of interest is above the cutoff, the case is assigned to that class.

The default cutoff value in two-class classifiers is 0.5. Thus, if the probability of a record being a class 1 member is greater than 0.5, that record is classified as a 1. Any record with an estimated probability of less than 0.5 would be classified as a 0. It is possible, however, to use a cutoff that is either higher or lower than 0.5. A cutoff greater than 0.5 will end up classifying fewer records as 1's, whereas a cutoff less than 0.5 will end up classifying more records as 1. Typically, the misclassification rate will rise in either case.

Consider the data in Table 5.2, showing the actual class for 24 records, sorted by the probability that the record is a 1 (as estimated by a data mining algorithm). If we adopt the standard 0.5 as the cutoff, our misclassification rate is 3/24, whereas if we instead adopt a cutoff of 0.25, we classify more records as 1's and the misclassification rate goes up (comprising more 0's misclassified as 1's) to 5/24. Conversely, if we adopt a cutoff of 0.75, we classify fewer records as 1's. The misclassification rate goes up (comprising more 1's misclassified as 0's) to 6/24. All this can be seen in the classification tables in Figure 5.3.

To see the entire range of cutoff values and how the accuracy or misclassification rates change as a function of the cutoff, we can use one-variable tables in Excel (see the accompanying box), and then plot the performance measure of interest versus the cutoff. The results for the data above are shown in Figure 5.4.

TABLE 5.2   **24 RECORDS WITH THEIR ACTUAL CLASS AND PROBABILITY OF BEING CLASS 1 MEMBERS, AS ESTIMATED BY A CLASSIFIER**

Actual Class	Probability of Class 1	Actual Class	Probability of Class 1
1	0.995976726	1	0.505506928
1	0.987533139	0	0.47134045
1	0.984456382	0	0.337117362
1	0.980439587	1	0.21796781
1	0.948110638	0	0.199240432
1	0.889297203	0	0.149482655
1	0.847631864	0	0.047962588
0	0.762806287	0	0.038341401
1	0.706991915	0	0.024850999
1	0.680754087	0	0.021806029
1	0.656343749	0	0.016129906
0	0.622419543	0	0.003559986

Cut off Prob.Val. for Success (Updatable)	0.5

Classification Confusion Matrix		
	Predicted Class	
Actual Class	Owner	Nonowner
Owner	11	1
Nonowner	2	10

Cut off Prob.Val. for Success (Updatable)	0.25

Classification Confusion Matrix		
	Predicted Class	
Actual Class	Owner	Nonowner
Owner	11	1
Nonowner	4	8

Cut off Prob.Val. for Success (Updatable)	0.75

Classification Confusion Matrix		
	Predicted Class	
Actual Class	Owner	Nonowner
Owner	7	5
Nonowner	1	11

FIGURE 5.3   CLASSIFICATION MATRICES BASED ON CUTOFFS OF 0.5, 0.25, AND 0.75

We can see that the accuracy level is pretty stable around 0.8 for cutoff values between 0.2 and 0.8.

Why would we want to use cutoffs different from 0.5 if they increase the misclassification rate? The answer is that it might be more important to classify 1's properly than 0's, and we would tolerate a greater misclassification of the latter. Or the reverse might be true; in other words, the costs of misclassification might be asymmetric. We can adjust the cutoff value in such a case to classify more records as the high-value class (in other words, accept more misclassifications where the misclassification cost is low). Keep in mind that we are doing so after the data mining model has already been selected—we are not changing that model. It is also possible to incorporate costs into the picture before deriving the model. These subjects are discussed in greater detail below.

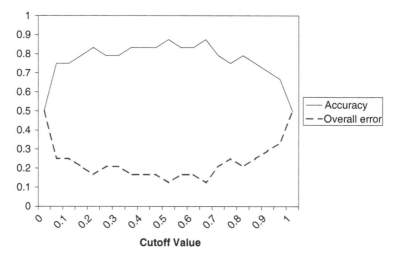

FIGURE 5.4   **PLOTTING RESULTS FROM ONE-WAY TABLE: ACCURACY AND OVERALL ERROR AS A FUNCTION OF THE CUTOFF VALUE**

ONE-VARIABLE TABLES IN EXCEL

Excel's one-variable data tables are very useful for studying how the cutoff affects different performance measures. It will change the cutoff values to values in a user-specified column and calculate different functions based on the corresponding classification matrix. To create a one-variable data table (see Figure 5.5):

1. In the top row, create column names for each of the measures you wish to compute. (We created "overall error" and "accuracy" in B11 and C11.) The leftmost column should be titled "cutoff" (A11).

2. In the row below, add formulas, using references to the relevant classification matrix cells. [The formula in B12 is $= (B6 + C7)/(B6 + C6 + B7 + C7)$.]

3. In the leftmost column, list the cutoff values that you want to evaluate. (We chose 0, 0.05, . . . , 1 in B13 to B33.)

4. Select the range excluding the first row (B12:C33). In Excel 2007 go to *Data › WhatifAnalysis › Data Table* (in Excel 2003 select *Table* from the *Data* menu).

5. In "column input cell," select the cell that changes (here, the cell with the cutoff value, D1). Click OK.

6. The table will now be automatically completed.

Performance in Unequal Importance of Classes

Suppose that it is more important to predict membership correctly in class 1 than in class 0. An example is predicting the financial status (bankrupt/solvent) of firms. It may be more important to predict correctly a firm that is going bankrupt than to predict correctly a firm that is going to stay solvent. The

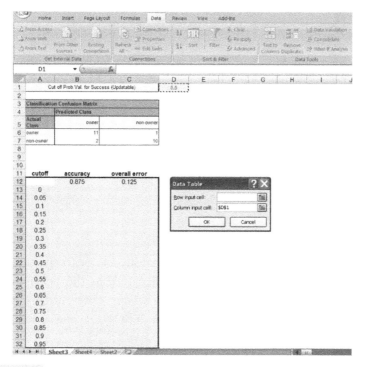

FIGURE 5.5   CREATING ONE-VARIABLE TABLES IN EXCEL. ACCURACY AND OVERALL
ERROR ARE COMPUTED FOR DIFFERENT VALUES OF THE CUTOFF

classifier is essentially used as a system for detecting or signaling bankruptcy. In
such a case, the overall accuracy is not a good measure for evaluating the classifier.
Suppose that the important class is C_1. The following pair of accuracy measures
are the most popular:

The sensitivity of a classifier is its ability to detect the important class
members correctly. This is measured by $n_{1,1}/(n_{1,0} + n_{1,1})$, the percentage of
C_1 members classified correctly.

The specificity of a classifier is its ability to rule out C_0 members correctly. This is measured by $n_{0,0}/(n_{0,0} + n_{0,1})$, the percentage of C_0 members
classified correctly.

It can be useful to plot these measures versus the cutoff value (using one-
variable tables in Excel, as described above) in order to find a cutoff value that
balances these measures.

ROC Curve   A more popular method for plotting the two measures is
through *ROC* (receiver operating characteristic) *curves*. The ROC curve plots
the pairs {sensitivity, 1-specificity} as the cutoff value increases from 0 and 1.

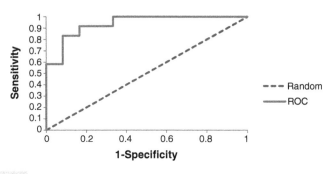

FIGURE 5.6   **ROC CURVE FOR THE EXAMPLE**

Better performance is reflected by curves that are closer to the top left corner. The comparison curve is the diagonal, which reflects the performance of the naive rule, using varying cutoff values (i.e., setting different thresholds on the level of majority used by the majority rule). The ROC curve for our 24–case example above is shown in Figure 5.6.

FALSE-POSITIVE AND FALSE-NEGATIVE RATES

Sensitivity and specificity measure the performance of a classifier from the point of view of the "classifying agency" (e.g., a company classifying customers or a hospital classifying patients). They answer the question: How well does the classifier segregate the important class members? It is also possible to measure accuracy from the perspective of the entity that is being predicted (e.g., the customer or the patient), who asks: What is my chance of belonging to the important class? This question, however, is usually less relevant in a data mining application. The terms *false-positive rate* and *false-negative rate*, which are sometimes used erroneously to describe 1-sensitivity and 1-specificity, are measures of performance from the perspective of the individual entity. They are defined as:

The *false-positive rate* is the proportion of C_1 predictions that are wrong: $n_{0,1}/(n_{0,1} + n_{1,1})$. Notice that this is a ratio within the column of C_1 predictions (i.e., it uses only records that were classified as C_1).

The *false-negative rate* is the proportion of C_0 predictions that are wrong: $n_{1,0}/(n_{0,0} + n_{1,0})$. Notice that this is a ratio within the column of C_0 predictions (i.e., it uses only records that were classified as C_0).

Lift Charts   Let us continue further with the case in which a particular class is relatively rare and of much more interest than the other class: tax cheats, debt defaulters, or responders to a mailing. We would like our classification model to sift through the records and sort them according to which ones are most likely to be tax cheats, responders to the mailing, and so on. We can then make more informed decisions. For example, we can decide how many, and which tax returns to examine, looking for tax cheats. The model will give us an estimate of the extent to which we will encounter more and more noncheaters as we proceed through the sorted data starting with the records most likely to be tax cheats. Or we can use the sorted data to decide to which potential customers a limited-budget mailing should be targeted. In other words, we are describing the case when our goal is to obtain a rank ordering among the records according to their estimated probabilities of class membership.

In such cases, when the classifier gives a probability of belonging to each class and not just a binary classification to C_1 or C_0, we can use a very useful device known as the *lift curve*, also called a *gains curve* or *gains chart*. The lift curve is a popular technique in direct marketing. One useful way to think of a lift curve is to consider a data mining model that attempts to identify the likely responders to a mailing by assigning each case a "probability of responding" score. The lift curve helps us determine how effectively we can "skim the cream" by selecting a relatively small number of cases and getting a relatively large portion of the responders. The input required to construct a lift curve is a validation dataset that has been "scored" by appending to each case the estimated probability that it will belong to a given class.

Let us return to the example in Table 5.2. We have shown that different choices of a cutoff value lead to different classification matrices (as in Figure 5.3). Instead of looking at a large number of classification matrices, it is much more convenient to look at the *cumulative lift curve* (sometimes called a *gains chart*), which summarizes all the information in these multiple classification matrices into a graph. The graph is constructed with the cumulative number of cases (in descending order of probability) on the x axis and the cumulative number of true positives on the y axis. Figure 5.7 gives the table of cumulative values of the class 1 classifications and the corresponding lift chart. The line joining the points (0,0) to (24,12) is a reference line. For any given number of cases (the x-axis value), it represents the expected number of C_1 predictions if we did not have a model but simply selected cases at random. It provides a benchmark against which we can see performance of the model. If we had to choose 10 cases as class 1 (the important class) members and used our model to pick the ones most likely to be 1's, the lift curve tells us that we would be right about 9 of them. If we simply select 10 cases at random, we expect to be right for $10 \times 12/24 = 5$ cases. The model gives us a "lift" in predicting class 1 of 9/5 = 1.8. The lift will vary with the number of cases on which we choose to act. A good classifier will give us a high lift when we act on only a few cases (i.e., use the prediction for those at the top). As we include more cases, the lift will decrease. The lift curve for the best possible classifier—a classifier that makes no errors—would overlap the existing curve at the start, continue with a slope of

Serial no.	Predicted prob of 1	Actual Class	Cumulative Actual class
1	0.995976726	1	1
2	0.987533139	1	2
3	0.984456382	1	3
4	0.980439587	1	4
5	0.948110638	1	5
6	0.889297203	1	6
7	0.847631864	1	7
8	0.762806287	0	7
9	0.706991915	1	8
10	0.680754087	1	9
11	0.656343749	1	10
12	0.622419543	0	10
13	0.505506928	1	11
14	0.47134045	0	11
15	0.337117362	0	11
16	0.21796781	1	12
17	0.199240432	0	12
18	0.149482655	0	12
19	0.047962588	0	12
20	0.038341401	0	12
21	0.024850999	0	12
22	0.021806029	0	12
23	0.016129906	0	12
24	0.003559986	0	12

FIGURE 5.7   TABLE AND LIFT CHART FOR THE EXAMPLE

1 until it reached 12 successes (all the successes), then continue horizontally to the right.

The same information can be portrayed as a *decile chart*, shown in Figure 5.8, which is widely used in direct marketing predictive modeling. The bars show the factor by which our model outperforms a random assignment of 0's and 1's, taking one decile at a time. Reading the first bar on the left, we see that taking

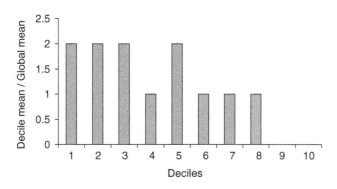

FIGURE 5.8   DECILE LIFT CHART

the 10% of the records that are ranked by the model as "the most probable 1's" yields twice as many 1's as would a random selection of 10% of the records.

> XLMiner automatically creates lift (and decile) charts from probabilities predicted by classifiers for both training and validation data. Of course, the lift curve based on the validation data is a better estimator of performance for new cases.

Asymmetric Misclassification Costs

Implicit in our discussion of the lift curve, which measures how effective we are in identifying the members of one particular class, is the assumption that the error of misclassifying a case belonging to one class is more serious than for the other class. For example, misclassifying a household as unlikely to respond to a sales offer when it belongs to the class that would respond incurs a greater cost (the opportunity cost of the foregone sale) than the converse error. In the former case, you are missing out on a sale worth perhaps tens or hundreds of dollars. In the latter, you are incurring the costs of mailing a letter to someone who will not purchase. In such a scenario, using the misclassification rate as a criterion can be misleading.

Note that we are assuming that the cost (or benefit) of making correct classifications is zero. At first glance, this may seem incomplete. After all, the benefit (negative cost) of classifying a buyer correctly as a buyer would seem substantial. And in other circumstances (e.g., scoring our classification algorithm to fresh data to implement our decisions), it will be appropriate to consider the actual net dollar impact of each possible classification (or misclassification). Here, however, we are attempting to assess the value of a classifier in terms of classification error, so it greatly simplifies matters if we can capture all cost–benefit information in the misclassification cells. So, instead of recording the benefit of classifying a respondent household correctly, we record the cost of failing to classify it as a respondent household. It amounts to the same thing and our goal becomes the minimization of costs, whether the costs are actual costs or missed benefits (opportunity costs).

Consider the situation where the sales offer is mailed to a random sample of people for the purpose of constructing a good classifier. Suppose that the offer is accepted by 1% of those households. For these data, if a classifier simply classifies every household as a nonresponder, it will have an error rate of only 1% but it will be useless in practice. A classifier that misclassifies 2% of buying households as nonbuyers and 20% of the nonbuyers as buyers would have a higher error rate but would be better if the profit from a sale is substantially higher than the cost of sending out an offer. In these situations, if we have estimates of the cost of both types of misclassification, we can use the classification matrix to compute the expected cost of misclassification for each case in the validation data. This enables us to compare different classifiers using overall expected costs (or profits) as the criterion.

Suppose that we are considering sending an offer to 1000 more people, 1% of whom respond (1), on average. Naively classifying everyone as a 0 has an error rate of only 1%. Using a data mining routine, suppose that we can produce these classifications:

	Predict Class 0	Predict Class 1
Actual 0	970	20
Actual 1	2	8

These classifications have an error rate of $100 \times (20 + 2)/1000 = 2.2\%$—higher than the naive rate.

Now suppose that the profit from a 1 is $10 and the cost of sending the offer is $1. Classifying everyone as a 0 still has a misclassification rate of only 1% but yields a profit of $0. Using the data mining routine, despite the higher misclassification rate, yields a profit of $60.

The matrix of profit is as follows (nothing is sent to the predicted 0's so there are no costs or sales in that column):

Profit	Predict Class 0	Predict Class 1
Actual 0	0	− $20
Actual 1	0	$80

Looked at purely in terms of costs, when everyone is classified as a 0, there are no costs of sending the offer; the only costs are the opportunity costs of failing to make sales to the ten 1's = \$100. The cost (actual costs of sending the offer, plus the opportunity costs of missed sales) of using the data mining routine to select people to send the offer to is only \$48, as follows:

Costs	Predict Class 0	Predict Class 1
Actual 0	0	\$20
Actual 1	\$20	\$8

However, this does not improve the actual classifications themselves. A better method is to change the classification rules (and hence the misclassification rates), as discussed in the preceding section, to reflect the asymmetric costs.

A popular performance measure that includes costs is the *average misclassification cost*, which measures the average cost of misclassification per classified observation. Denote by q_0 the cost of misclassifying a class 0 observation (as belonging to class 1) and by q_1 the cost of misclassifying a class 1 observation (as belonging to class 0). The average misclassification cost is

$$\frac{q_0 n_{0,1} + q_1 n_{1,0}}{n}.$$

Thus, we are looking for a classifier that minimizes this quantity. This can be computed, for instance, for different cutoff values.

It turns out that the optimal parameters are affected by the misclassification costs only through the ratio of these costs. This can be seen if we write the foregoing measure slightly differently:

$$\frac{q_0 n_{0,1} + q_1 n_{1,0}}{n} = \frac{n_{0,1}}{n_{0,0} + n_{0,1}} \frac{n_{0,0} + n_{0,1}}{n} q_0 + \frac{n_{1,0}}{n_{1,0} + n_{1,1}} \frac{n_{1,0} + n_{1,1}}{n} q_1.$$

Minimizing this expression is equivalent to minimizing the same expression divided by a constant. If we divide by q_0, it can be seen clearly that the minimization depends only on q_1/q_0 and not on their individual values. This is very practical because in many cases it is difficult to assess the cost associated with misclassifying a 0 member and that associated with misclassifying a 1 member, but estimating the ratio is easier.

This expression is a reasonable estimate of future misclassification cost if the proportions of classes 0 and 1 in the sample data are similar to the proportions of classes 0 and 1 that are expected in the future. If instead of a random sample, we draw a sample such that one class is oversampled (as described in the next section), then the sample proportions of 0's and 1's will be distorted compared to the future or population. We can then correct the average misclassification cost measure for the distorted sample proportions by incorporating estimates of the true proportions (from external data or domain knowledge), denoted by $p(C_0)$ and $p(C_1)$, into the formula:

$$\frac{n_{0,1}}{n_{0,0} + n_{0,1}} p(C_0) \, q_0 + \frac{n_{1,0}}{n_{1,0} + n_{1,1}} p(C_1) \, q_1.$$

Using the same logic as above, it can be shown that optimizing this quantity depends on the costs only through their ratio (q_1/q_0) and on the prior probabilities only through their ratio $[p(C_0)/p(C_1)]$. This is why software packages that incorporate costs and prior probabilities might prompt the user for ratios rather than actual costs and probabilities.

Generalization to More Than Two Classes   All the comments made above about two-class classifiers extend readily to classification into more than two classes. Let us suppose that we have m classes $C_0, C_1, C_2, \ldots, C_{m-1}$. The classification matrix has m rows and m columns. The misclassification cost associated with the diagonal cells is, of course, always zero. Incorporating prior probabilities of the various classes (where now we have m such numbers) is still done in the same manner. However, evaluating misclassification costs becomes much more complicated: For an m-class case we have $m(m-1)$ types of misclassifications. Constructing a matrix of misclassification costs thus becomes prohibitively complicated.

A lift chart cannot be used with a multiclass classifier, unless a single "important class" is defined, and the classifications are reduced to "important" and "unimportant" classes.

Lift Charts Incorporating Costs and Benefits   When the benefits and costs of correct and incorrect classification are known or can be estimated, the lift chart is still a useful presentation and decision tool. As before, a classifier is needed that assigns to each record a probability that it belongs to a particular class. The procedure is then as follows:

1. Sort the records in order of predicted probability of success (where *success* = belonging to the class of interest).

2. For each record, record the cost (benefit) associated with the actual outcome.

3. For the highest probability (i.e., first) record, the value in step 2 is the y coordinate of the first point on the lift chart. The x coordinate is index number 1.

4. For the next record, again calculate the cost (benefit) associated with the actual outcome. Add this to the cost (benefit) for the previous record.

This sum is the y coordinate of the second point on the lift curve. The x coordinate is index number 2.

5. Repeat step 4 until all records have been examined. Connect all the points, and this is the lift curve.

6. The reference line is a straight line from the origin to the point y = total net benefit and $x = N(N$ = number of records).

Note: It is entirely possible for a reference line that incorporates costs and benefits to have a negative slope if the net value for the entire dataset is negative. For example, if the cost of mailing to a person is $0.65, the value of a responder is $25, and the overall response rate is 2%, the expected net value of mailing to a list of 10,000 is $(0.02 \times \$25 \times 10,000) - (\$0.65 \times 10,000) = \$5000 - \$6500 = -\$1500$. Hence, the y value at the far right of the lift curve ($x = 10,000$) is -1500, and the slope of the reference line from the origin will be negative. The optimal point will be where the lift curve is at a maximum (i.e., mailing to about 3000 people) in Figure 5.9.

Lift as Function of Cutoff   We could also plot the lift as a function of the cutoff value. The only difference is the scale on the x axis. When the goal is to select the top records based on a certain budget, the lift versus number of records is preferable. In contrast, when the goal is to find a cutoff that distinguishes well between the two classes, the lift versus cutoff value is more useful.

Oversampling and Asymmetric Costs

As we saw briefly in Chapter 2, when classes are present in very unequal proportions, simple random sampling may produce too few of the rare class to yield useful information about what distinguishes them from the dominant class. In such cases, stratified sampling is often used to oversample the cases from the more rare class and improve the performance of classifiers. It is often the case that the

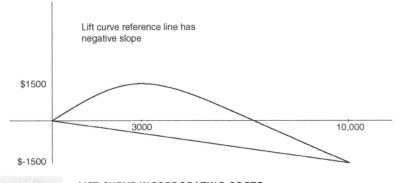

FIGURE 5.9   **LIFT CURVE INCORPORATING COSTS**

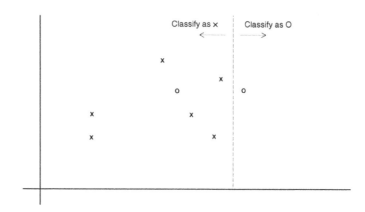

FIGURE 5.10 **CLASSIFICATION ASSUMING EQUAL COSTS OF MISCLASSIFICATION**

more rare events are the more interesting or important ones: responders to a mailing, those who commit fraud, defaulters on debt, and the like.

> In all discussions of *oversampling* (also called *weighted sampling*), we assume the common situation in which there are two classes, one of much greater interest than the other. Data with more than two classes do not lend themselves to this procedure.

Consider the data in Figure 5.10, where × represents nonresponders, and O, responders. The two axes correspond to two predictors. The dashed vertical line does the best job of classification under the assumption of equal costs: It results in just one misclassification (one O is misclassified as an ×). If we incorporate more realistic misclassification costs—let us say that failing to catch an O is five times as costly as failing to catch a ×—the costs of misclassification jump to 5. In such a case, a horizontal line as shown in Figure 5.11, does a better job: It results in misclassification costs of just 2.

Oversampling is one way of incorporating these costs into the training process. In Figure 5.12, we can see that classification algorithms would automatically determine the appropriate classification line if four additional O's were present at each existing O. We can achieve appropriate results either by taking five times as many o's as we would get from simple random sampling (by sampling with replacement if necessary), or by replicating the existing o's fourfold.

Oversampling without replacement in accord with the ratio of costs (the first option above) is the optimal solution but may not always be practical. There may not be an adequate number of responders to assure that there will be enough nonresponders to fit a model if the latter constitutes only a small proportion of the former. Also, it is often the case that our interest in discovering responders is known to be much greater than our interest in discovering nonresponders, but the exact ratio of costs is difficult to determine. When faced with very low response

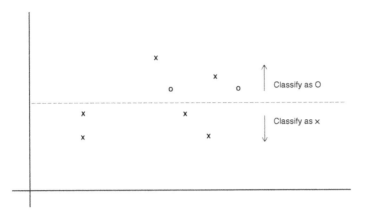

FIGURE 5.11   **CLASSIFICATION ASSUMING UNEQUAL COSTS OF MISCLASSIFICATION**

rates in a classification problem, practitioners often sample equal numbers of responders and nonresponders as a relatively effective and convenient approach. Whatever approach is used, when it comes time to assess and predict model performance, we will need to adjust for the oversampling in one of two ways:

1. Score the model to a validation set that has been selected without oversampling (i.e., via simple random sampling).
2. Score the model to an oversampled validation set, and reweight the results to remove the effects of oversampling.

The first method is more straightforward and easier to implement. We describe how to oversample and how to evaluate performance for each of the two methods.

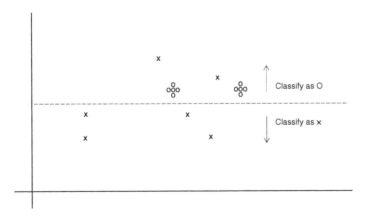

FIGURE 5.12   **CLASSIFICATION USING OVERSAMPLING TO ACCOUNT FOR UNEQUAL COSTS**

When classifying data with very low response rates, practitioners typically:

- Train models on data that are 50% responder, 50% nonresponder.
- Validate the models with an unweighted (simple random) sample from the original data.

Oversampling the Training Set   How is weighted sampling done? One common procedure, where responders are sufficiently scarce that you will want to use all of them, follows:

1. First, the response and nonresponse data are separated into two distinct sets, or *strata*.
2. Records are then randomly selected for the training set from each stratum. Typically, one might select half the (scarce) responders for the training set, then an equal number of nonresponders.
3. The remaining responders are put in the validation set.
4. Nonresponders are randomly selected for the validation set in sufficient numbers to maintain the original ratio of responders to nonresponders.
5. If a test set is required, it can be taken randomly from the validation set.

XLMiner has a utility for this purpose.

Evaluating Model Performance Using a Nonoversampled Validation Set   Although the oversampled data can be used to train models, they are often not suitable for predicting model performance because the number of responders will (of course) be exaggerated. The most straightforward way of gaining an unbiased estimate of model performance is to apply the model to regular data (i.e., data not oversampled). To recap: Train the model on oversampled data, but validate it with regular data.

Evaluating Model Performance If Only Oversampled Validation Set Exists   In some cases, very low response rates may make it more practical to use oversampled data not only for the training data, but also for the validation data. This might happen, for example, if an analyst is given a data sample for exploration and prototyping, and it is more convenient to transfer and work with a smaller dataset in which a sizable proportion of cases are those with the rare response (typically the response of interest). In such cases it is still possible to assess how well the model will do with real data, but this requires the oversampled validation set to be reweighted, in order to restore the class of observations that were underrepresented in the sampling process. This adjustment should be made to the classification matrix and to the lift chart in order to derive good accuracy measures. These adjustments are described next.

I. Adjusting the Confusion Matrix for Oversampling   Let us say that the response rate in the data as a whole is 2%, and that the data were oversampled, yielding a sample in which the response rate is 25 times as great $= 50\%$. Assume that the validation classification matrix looks like this:

CLASSIFICATION MATRIX, OVERSAMPLED DATA (VALIDATION)

	Predicted 0	Predicted 1	Total
Actual 0	390	110	500
Actual 1	80	420	500
Total	470	530	1000

At this point, the (inaccurate) misclassification rate appears to be $(80 + 110)/1000 = 19\%$, and the model ends up classifying 53% of the records as 1's.

There were 500 (actual) 1's in the sample and 500 (actual) 0's. If we had not oversampled, there would have been far fewer 1's. Put another way, there would be many more 0's for each 1. So we can either take away 1's or add 0's to reweight the sample. The calculations for the latter are shown: We need to add enough 0's so that the 1's constitute only 2% of the total, and the 0's, 98% (where X is the total):

$$500 + 0.98X = X.$$

Solving for X, we find that $X = 25,000$.

The total is 25,000, so the number of 0's is $(0.98)(25,000) = 24,500$. We can now redraw the classification matrix by augmenting the number of (actual) nonresponders, assigning them to the appropriate cells in the same ratio in which they appear in the classification table above (3.545 predicted 0's for every predicted 1):

CLASSIFICATION MATRIX, REWEIGHTED

	Predicted 0	Predicted 1	Total
Actual 0	19,110	5,390	24,500
Actual 1	80	420	500
Total	19,190	5,810	25,000

The adjusted misclassification rate is $(80 + 5390)/25,000 = 21.9\%$, and the model ends up classifying 5810/25,000 of the records as 1's, or 21.4%.

II. Adjusting the Lift Curve for Oversampling   The lift curve is likely to be a more useful measure in low-response situations, where our interest lies not so much in classifying all the records correctly as in finding a model that guides us toward those records most likely to contain the response of interest (under the assumption that scarce resources preclude examining or contacting all the records). Typically, our interest in such a case is in maximizing value or minimizing cost, so we will show the adjustment process incorporating the cost-benefit element. The following procedure can be used (and easily implemented in Excel):

1. Sort the validation records in order of the predicted probability of success (where success = belonging to the class of interest).

2. For each record, record the cost (benefit) associated with the actual outcome.

3. Multiply that value by the proportion of the original data having this outcome; this is the adjusted value.

4. For the highest probability (i.e., first) record, the value above is the y coordinate of the first point on the lift chart. The x coordinate is index number 1.

5. For the next record, again calculate the adjusted value associated with the actual outcome. Add this to the adjusted cost (benefit) for the previous record. This sum is the y coordinate of the second point on the lift curve. The x coordinate is index number 2.

6. Repeat step 5 until all records have been examined. Connect all the points, and this is the lift curve.

7. The reference line is a straight line from the origin to the point $y =$ total net benefit and $x = N (N =$ number of records).

Classification Using a Triage Strategy

In some cases it is useful to have a "cannot say" option for the classifier. In a two-class situation, this means that for a case, we can make one of three predictions: The case belongs to C_0, or the case belongs to C_1, or we cannot make a prediction because there is not enough information to pick C_0 or C_1 confidently. Cases that the classifier cannot classify are subjected to closer scrutiny either by using expert judgment or by enriching the set of predictor variables by gathering additional information that is perhaps more difficult or expensive to obtain. This is analogous to the strategy of triage, which is often employed during retreat in battle. The wounded are classified into those who are well enough to retreat, those who are too ill to retreat even if treated medically under the prevailing conditions, and those who are likely to become well enough to retreat if given medical attention. An example is in processing credit card transactions, where a classifier may be used to identify clearly legitimate cases and obviously fraudulent ones while referring the remaining cases to a human decision maker who may look up a database to form a judgment. Since the vast majority of transactions are legitimate, such a classifier would substantially reduce the burden on human experts.

5.3   EVALUATING PREDICTIVE PERFORMANCE

When the response variable is continuous, the evaluation of model performance is slightly different from the categorical response case. First, let us emphasize that predictive accuracy is not the same as goodness of fit. Classical measures of performance are aimed at finding a model that fits the data well, whereas in data mining we are interested in models that have high predictive accuracy. Measures such as R^2 and standard error of estimate are very popular strength of fit measures in classical regression modeling, and residual analysis is used to gauge goodness of fit where the goal is to find the best fit for the data. However, these measures do not tell us much about the ability of the model to predict new cases.

For prediction performance, there are several measures that are used to assess the predictive accuracy of a regression model. In all cases, the measures are based on the validation set, which serves as a more objective ground than the training set to assess predictive accuracy. This is because records in the validation set are not used to select predictors or to estimate the model coefficients. Measures of accuracy use the prediction error that results from predicting the validation data with the model (that was trained on the training data).

Benchmark: The Average

Recall that the benchmark criterion in prediction is using the average outcome (thereby ignoring all predictor information). In other words, the prediction for a new record is simply the average outcome of the records in the training set. A good predictive model should outperform the benchmark criterion in terms of predictive accuracy.

Prediction Accuracy Measures

The prediction error for record i is defined as the difference between its actual y value and its predicted y value: $e_i = y_i - \hat{y}_i$. A few popular numerical measures of predictive accuracy are:

- *MAE* or *MAD* (mean absolute error/deviation) $= 1/n \sum_{i=1}^{n} |e_i|$. This gives the magnitude of the average absolute error.
- *Average error* $= 1/n \sum_{i=1}^{n} e_i$. This measure is similar to MAD except that it retains the sign of the errors, so that negative errors cancel out positive errors of the same magnitude. It therefore gives an indication of whether the predictions are on average over- or underpredicting the response.
- *MAPE* (mean absolute percentage error) $= 100\% \times 1/n \sum_{i=1}^{n} |e_i/y_i|$. This measure gives a percentage score of how predictions deviate (on average) from the actual values.
- *RMSE* (root-mean-squared error) $= \sqrt{1/n \sum_{i=1}^{n} e_i^2}$. This is similar to the standard error of estimate, except that it is computed on the validation data rather than on the training data. It has the same units as the variable predicted.
- Total *SSE* (total sum of squared errors) $= \sum_{i=1}^{n} e_i^2$.

Such measures can be used to compare models and to assess their degree of prediction accuracy. Notice that all these measures are influenced by outliers. To check outlier influence, we can compute median-based measures (and compare to the mean-based measures) or simply plot a histogram or boxplot of the errors. It is important to note that a model with high predictive accuracy might not coincide with a model that fits the training data best.

Finally, a graphical way to assess predictive performance is through a lift chart. This compares the model's predictive performance to a baseline model that has no predictors. Predictions from the baseline model are simply the average $\bar{y}$. A lift chart for a continuous response is relevant only when we are searching for a set of records that gives the highest cumulative predicted values.

To illustrate this, consider a car rental firm that renews its fleet regularly so that customers drive late-model cars. This entails disposing of a large quantity of used vehicles on a continuing basis. Since the firm is not primarily in the used car sales business, it tries to dispose of as much of its fleet as possible through volume sales to used car dealers. However, it is profitable to sell a limited number of cars through its own channels. Its volume deals with the used car dealers leave it flexibility to pick and choose which cars to sell in this fashion, so it would like to have a model for selecting cars for resale through its own channels. Since all cars were purchased some time ago and the deals with the used car dealers are for fixed prices (specifying a given number of cars of a certain make and model class), the cars' costs are now irrelevant and the dealer is interested only in maximizing revenue. This is done by selecting for its own resale the cars likely to generate the most revenue. The lift chart in this case gives the predicted lift for revenue.

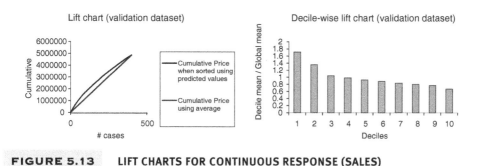

FIGURE 5.13   LIFT CHARTS FOR CONTINUOUS RESPONSE (SALES)

Figure 5.13 shows a lift chart based on fitting a linear regression model to a dataset that includes the car prices (y) and a set of predictor variables that describe a car's features (mileage, color, etc.) The lift chart is based on the validation data of 400 cars. It can be seen that the model's predictive performance is better than the baseline model since its lift curve is higher than that of the baseline model. The lift (and decile-wise) charts in Figure 5.13 would be useful in the following scenario: Choosing the top 10% of the cars that gave the highest predicted sales, for example, we would gain 1.7 times the amount compared to choosing 10% of the cars at random. This can be seen from the decile chart (Figure 5.13). This number can also be computed from the lift chart by comparing the sales predicted for 40 random cars (the value of the baseline curve at $x = 40$), which is \$486,871 (= the sum of the predictions of the 400 validation set cars divided by 10) with the sales of the 40 cars that have the highest predicted values by the model (the value of the lift curve at $x = 40$), \$835,883. The ratio between these numbers is 1.7.

PROBLEMS

5.1 A data mining routine has been applied to a transaction dataset and has classified 88 records as fraudulent (30 correctly so) and 952 as nonfraudulent (920 correctly so). Construct the classification matrix and calculate the error rate.

5.2 Suppose that this routine has an adjustable cutoff (threshold) mechanism by which you can alter the proportion of records classified as fraudulent. Describe how moving the cutoff up or down would affect the following:

  a. The classification error rate for records that are truly fraudulent

  b. The classification error rate for records that are truly nonfraudulent

5.3 Consider Figure 5.14, the decile-wise lift chart for the transaction data model, applied to new data

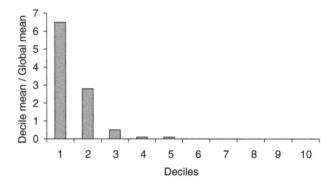

FIGURE 5.14 DECILE-WISE LIFT CHART FOR TRANSACTION DATA

  a. Interpret the meaning of the first and second bars from the left.

  b. Explain how you might use this information in practice.

  c. Another analyst comments that you could improve the accuracy of the model by classifying everything as nonfraudulent. If you do that, what is the error rate?

  d. Comment on the usefulness, in this situation, of these two metrics of model performance (error rate and lift).

5.4 A large number of insurance records are to be examined to develop a model for predicting fraudulent claims. Of the claims in the historical database, 1% were judged to be fraudulent. A sample is taken to develop a model, and oversampling is used to provide a balanced sample in light of the very low response rate. When applied to this sample ($N = 800$), the model ends up correctly classifying 310 frauds, and 270 nonfrauds. It missed 90 frauds, and classified 130 records incorrectly as frauds when they were not.

  a. Produce the classification matrix for the sample as it stands.

  b. Find the adjusted misclassification rate (adjusting for the oversampling).

  c. What percentage of new records would you expect to be classified as fraudulent?

## PART FOUR

Prediction and Classification Methods

## CHAPTER 6

Multiple Linear Regression

In this chapter we introduce linear regression models for the purpose of prediction. We discuss the differences between fitting and using regression models for the purpose of inference (as in classical statistics) and for prediction. A predictive goal calls for evaluating model performance on a validation set and for using predictive metrics. We then raise the challenges of using many predictors and describe variable selection algorithms that are often implemented in linear regression procedures.

6.1 INTRODUCTION

The most popular model for making predictions is the *multiple linear regression model* encountered in most introductory statistics classes and textbooks. This model is used to fit a linear relationship between a quantitative *dependent variable* Y (also called the *outcome* or *response variable*) and a set of *predictors* $X_1, X_2, \ldots, X_p$ (also referred to as *independent variables*, *input variables*, *regressors*, or *covariates*). The assumption is that in the population of interest, the following relationship holds:

$$Y = \beta_0 + \beta_1 x_1 + \beta_2 x_2 + \cdots + \beta_p x_p + \epsilon, \tag{6.1}$$

where $\beta_0, \ldots, \beta_p$ are *coefficients* and ϵ is the *noise* or *unexplained* part. The data, which are a sample from this population, are then used to estimate the coefficients and the variability of the noise.

The two popular objectives behind fitting a model that relates a quantitative outcome with predictors are for understanding the relationship between these factors and for predicting the outcomes of new cases. The classical statistical

Data Mining for Business Intelligence, By Galit Shmueli, Nitin R. Patel, and Peter C. Bruce

approach has focused on the first objective: fitting the best model to the data in an attempt to learn about the underlying relationship in the population. In data mining, however, the focus is typically on the second goal: predicting new observations. Important differences between the approaches stem from the fact that in the classical statistical world we are interested in drawing conclusions from a limited supply of data and in learning how reliable those conclusions might be. In data mining, by contrast, data are typically plentiful, so the performance and reliability of our model can easily be established by applying it to fresh data.

Multiple linear regression is applicable to numerous data mining situations. Examples are predicting customer activity on credit cards from their demographics and historical activity patterns, predicting the time to failure of equipment based on utilization and environment conditions, predicting expenditures on vacation travel based on historical frequent flyer data, predicting staffing requirements at help desks based on historical data and product and sales information, predicting sales from cross selling of products from historical information, and predicting the impact of discounts on sales in retail outlets. Although a linear regression model is used for both goals, the modeling step and performance assessment differ depending on the goal. Therefore, the choice of model is closely tied to whether the goal is explanatory or predictive.

6.2 EXPLANATORY VERSUS PREDICTIVE MODELING

Both explanatory and predictive modeling involve using a dataset to fit a model (i.e., to estimate coefficients), checking model validity, assessing its performance, and comparing to other models. However, there are several major differences between the two:

1. A good explanatory model is one that fits the data closely, whereas a good predictive model is one that predicts new cases accurately.

2. In explanatory models (classical statistical world, scarce data) the entire dataset is used for estimating the best-fit model, to maximize the amount of information that we have about the hypothesized relationship in the population. When the goal is to predict outcomes of new cases (data mining, plentiful data), the data are typically split into a training set and a validation set. The training set is used to estimate the model, and the validation, or *holdout*, set is used to assess this model's performance on new, unobserved data.

3. Performance measures for explanatory models measure how close the data fit the model (how well the model approximates the data), whereas in predictive models performance is measured by predictive accuracy (how well the model predicts new cases).

For these reasons it is extremely important to know the goal of the analysis before beginning the modeling process. A good predictive model can have a looser fit to the data on which it is based, and a good explanatory model can have low prediction accuracy. In the remainder of this chapter we focus on predictive models because these are more popular in data mining and because most textbooks focus on explanatory modeling.

6.3 ESTIMATING THE REGRESSION EQUATION AND PREDICTION

The coefficients $\beta_0, \ldots, \beta_p$ and the standard deviation of the noise (σ) determine the relationship in the population of interest. Since we only have a sample from that population, these coefficients are unknown. We therefore estimate them from the data using a method called *ordinary least squares* (OLS). This method finds values $\hat{\beta}_0, \hat{\beta}_1, \hat{\beta}_2, \ldots, \hat{\beta}_p$ that minimize the sum of squared deviations between the actual values (Y) and their predicted values based on that model ($\hat{Y}$).

To predict the value of the dependent value from known values of the predictors, $x_1, x_2, \ldots, x_p$, we use sample estimates for $\beta_0, \ldots, \beta_p$ in the linear regression model (6.1) since $\beta_0, \ldots, \beta_p$ cannot be observed directly unless we have available the entire population of interest. The predicted value, $\hat{Y}$, is computed from the equation $\hat{Y} = \hat{\beta}_0 + \hat{\beta}_1 x_1 + \hat{\beta}_2 x_2 + \cdots + \hat{\beta}_p x_p$. Predictions based on this equation are the best predictions possible in the sense that they will be unbiased (equal to the true values on average) and will have the smallest average squared error compared to any unbiased estimates *if* we make the following assumptions:

1. The noise ϵ (or equivalently, the dependent variable) follows a normal distribution.
2. The linear relationship is correct.
3. The cases are independent of each other.
4. The variability in Y values for a given set of predictors is the same regardless of the values of the predictors (*homoskedasticity*).

An important and interesting fact for the predictive goal is that *even if we drop the first assumption and allow the noise to follow an arbitrary distribution, these estimates are very good for prediction,* in the sense that among all linear models, as defined by equation (6.1), the model using the least-squares estimates, $\hat{\beta}_0, \hat{\beta}_1, \hat{\beta}_2, \ldots, \hat{\beta}_p$, will have the smallest average squared errors. An assumption of a normal distribution is required in the classical implementation of multiple linear regression to derive confidence intervals for predictions. In this classical world, data are scarce and the same data are used to fit the regression model and to assess its reliability (with confidence limits). In data mining applications we have two distinct sets of data: The training dataset and the validation dataset are both representative of the relationship between the dependent and independent variables. The training data is used to fit the model and estimate the regression coefficients $\beta_0, \beta_1, \ldots, \beta_p$. The validation dataset constitutes a holdout sample and is not used in computing the coefficient estimates. The estimates are then used to make predictions for each case in the validation data. This enables us to estimate the error in our predictions by using the validation set without having to assume that the noise follows a normal distribution. The prediction for each case is then compared to the value of the dependent variable that was actually observed in the validation data. The average of the square of this error enables us to compare different models and to assess the prediction accuracy of the model.

Example: Predicting the Price of Used Toyota Corolla Automobiles

A large Toyota car dealership offers purchasers of new Toyota cars the option to buy their used car as part of a trade-in. In particular, a new promotion promises to pay high prices for used Toyota Corolla cars for purchasers of a new car. The dealer then sells the used cars for a small profit. To ensure a reasonable profit, the dealer needs to be able to predict the price that the dealership will get for the used cars. For that reason, data were collected on all previous sales of used Toyota Corollas at the dealership. The data include the sales price and other information on the car, such as its age, mileage, fuel type, and engine size. A description of each of these variables is given in Table 6.1. A sample of this dataset is shown in Table 6.2.

The total number of records in the dataset is 1000 cars (we use the first 1000 cars from the dataset ToyotoCorolla.xls). After partitioning the data into training and validation sets (at a 60% : 40% ratio), we fit a multiple linear

TABLE 6.1   **VARIABLES IN THE TOYOTA COROLLA EXAMPLE**

Variable	Description
Price	Offer price in euros
Age	Age in months as of August 2004
Kilometers	Accumulated kilometers on odometer
Fuel Type	Fuel type (*Petrol, Diesel, CNG*)
HP	Horsepower
Metallic	Metallic color? (Yes = 1, No = 0)
Automatic	Automatic (Yes = 1, No = 0)
CC	Cylinder volume in cubic centimeters
Doors	Number of doors
QuartTax	Quarterly road tax in euros
Weight	Weight in kilograms

TABLE 6.2   PRICES AND ATTRIBUTES FOR A SAMPLE OF 30 USED TOYOTA COROLLA CARS

Price	Age	Kilometers	Fuel Type	HP	Metallic	Auto-matic	CC	Doors	Quart Tax	Weight
13500	23	46986	Diesel	90	1	0	2000	3	210	1165
13750	23	72937	Diesel	90	1	0	2000	3	210	1165
13950	24	41711	Diesel	90	1	0	2000	3	210	1165
14950	26	48000	Diesel	90	0	0	2000	3	210	1165
13750	30	38500	Diesel	90	0	0	2000	3	210	1170
12950	32	61000	Diesel	90	0	0	2000	3	210	1170
16900	27	94612	Diesel	90	1	0	2000	3	210	1245
18600	30	75889	Diesel	90	1	0	2000	3	210	1245
21500	27	19700	Petrol	192	0	0	1800	3	100	1185
12950	23	71138	Diesel	69	0	0	1900	3	185	1105
20950	25	31461	Petrol	192	0	0	1800	3	100	1185
19950	22	43610	Petrol	192	0	0	1800	3	100	1185
19600	25	32189	Petrol	192	0	0	1800	3	100	1185
21500	31	23000	Petrol	192	1	0	1800	3	100	1185
22500	32	34131	Petrol	192	1	0	1800	3	100	1185
22000	28	18739	Petrol	192	0	0	1800	3	100	1185
22750	30	34000	Petrol	192	1	0	1800	3	100	1185
17950	24	21716	Petrol	110	1	0	1600	3	85	1105
16750	24	25563	Petrol	110	0	0	1600	3	19	1065
16950	30	64359	Petrol	110	1	0	1600	3	85	1105
15950	30	67660	Petrol	110	1	0	1600	3	85	1105
16950	29	43905	Petrol	110	0	1	1600	3	100	1170
15950	28	56349	Petrol	110	1	0	1600	3	85	1120
16950	28	32220	Petrol	110	1	0	1600	3	85	1120
16250	29	25813	Petrol	110	1	0	1600	3	85	1120
15950	25	28450	Petrol	110	1	0	1600	3	85	1120
17495	27	34545	Petrol	110	1	0	1600	3	85	1120
15750	29	41415	Petrol	110	1	0	1600	3	85	1120
11950	39	98823	CNG	110	1	0	1600	5	197	1119

regression model between price (the dependent variable) and the other variables (as predictors) using the training set only. Figure 6.1 shows the estimated coefficients (as computed by XLMiner). Notice that the Fuel Type predictor has three categories (Petrol, Diesel, and CNG), and we therefore have two dummy

Input variables	Coefficient	Std. Error	p-value	SS
Constant term	-2327.281494	1622.562866	0.15210986	81481950000
Age	-134.137619	4.77474403	0	5888770000
Mileage	-0.0199055	0.00236949	0	172544200
Fuel_Type_Diesel	129.2410126	536.7660523	0.80982733	2427870
Fuel_Type_Petrol	2670.873291	520.0211792	0.0000004	670008.4375
Horse_Power	33.95512009	5.37533283	0	339071900
Metalic_Color	-38.04909897	120.321022	0.75196105	716922.5
Automatic	224.9384003	269.0696716	0.40356547	10970180
CC	0.0209207	0.0959821	0.8275463	1553226
Doors	-3.00326943	61.79518509	0.96125734	17263280
Quarterly_Tax	22.90351105	2.48583364	0	221851400
Weight	12.9385519	1.51249933	0	136067800

Residual df	588
Multiple R-squared	0.861344575
Std. Dev. estimate	1363.600464
Residual SS	1093331000

FIGURE 6.1   ESTIMATED COEFFICIENTS FOR REGRESSION MODEL OF PRICE VS. CAR ATTRIBUTES

Predicted Value	Actual Value	Residual
16199	13750	-2449
16686	13950	-2736
16266	16900	634
16236	18600	2364
20534	20950	416
20520	19600	-920
19860	21500	1640
19504	22500	2996
20385	22000	1615
16993	16950	-43
16106	16950	844
16099	16250	151
15789	15750	-39
15590	15950	360
15660	14950	-710
15668	14750	-918
15300	16750	1450
17919	19000	1081
17242	17950	708
19148	21950	2802

(a)

Total sum of squared errors	RMS Error	Average Error
795600925.2	1410.319933	110.9145714

(b)

FIGURE 6.2 **(A) PREDICTED PRICES (AND ERRORS) FOR 20 CARS IN VALIDATION SET AND (B) SUMMARY PREDICTIVE MEASURES FOR ENTIRE VALIDATION SET**

variables in the model [e.g., Petrol (0/1) and Diesel (0/1); the third, CNG (0/1), is redundant given the information on the first two dummies]. These coefficients are then used to predict prices of used Toyota Corolla cars based on their age, mileage, and so on. Figure 6.2 shows a sample of 20 of the predicted prices for cars in the validation set, using the estimated model. It gives the predictions and their errors (relative to the actual prices) for these 20 cars. On the right we get overall measures of predictive accuracy. Note that the average error is $111. A boxplot of the residuals (Figure 6.3) shows that 50% of the errors are approximately between ±$850. This might be small relative to the car price but should be taken into account when considering the profit. Such measures are used to assess the predictive performance of a model and to compare models. We discuss such measures in the next section. This example also illustrates the point about the relaxation of the normality assumption. A histogram or probability plot of prices shows a right-skewed distribution. In a classical modeling case where the goal is to obtain a good fit to the data, the dependent variable would be transformed (e.g., by taking a natural log) to achieve a more "normal" variable. Although the fit of such a model to the training data is expected to be better, it will not necessarily yield a significant predictive improvement. In this

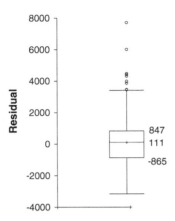

FIGURE 6.3   **BOXPLOT OF MODEL RESIDUALS (BASED ON VALIDATION SET)**

example the average error in a model of log(price) is −$160, compared to $111 in the original model for price.

6.4   Variable Selection in Linear Regression

Reducing the Number of Predictors

A frequent problem in data mining is that of using a regression equation to predict the value of a dependent variable when we have many variables available to choose as predictors in our model. Given the high speed of modern algorithms for multiple linear regression calculations, it is tempting in such a situation to take a kitchen-sink approach: Why bother to select a subset? Just use all the variables in the model. There are several reasons why this could be undesirable.

- It may be expensive or not feasible to collect a full complement of predictors for future predictions.
- We may be able to measure fewer predictors more accurately (e.g., in surveys).
- The more predictors there are, the higher the chance of missing values in the data. If we delete or impute cases with missing values, multiple predictors will lead to a higher rate of case deletion or imputation.
- *Parsimony* is an important property of good models. We obtain more insight into the influence of predictors in models with few parameters.
- Estimates of regression coefficients are likely to be unstable, due to *multicollinearity* in models with many variables. (Multicollinearity is the presence of two or more predictors sharing the same linear relationship with the outcome variable.) Regression coefficients are more stable for

parsimonious models. One very rough rule of thumb is to have a number of cases n larger than $5(p+2)$, where p is the number of predictors.

- It can be shown that using predictors that are uncorrelated with the dependent variable increases the variance of predictions.
- It can be shown that dropping predictors that are actually correlated with the dependent variable can increase the average error (bias) of predictions.

The last two points mean that there is a trade-off between too few and too many predictors. In general, accepting some bias can reduce the variance in predictions. This *bias–variance trade-off* is particularly important for large numbers of predictors because in that case it is very likely that there are variables in the model that have small coefficients relative to the standard deviation of the noise and also exhibit at least moderate correlation with other variables. Dropping such variables will improve the predictions, as it will reduce the prediction variance. This type of bias–variance trade-off is a basic aspect of most data mining procedures for prediction and classification. In light of this, methods for reducing the number of predictors p to a smaller set are often used.

How to Reduce the Number of Predictors

The first step in trying to reduce the number of predictors should always be to use domain knowledge. It is important to understand what the various predictors are measuring and why they are relevant for predicting the response. With this knowledge the set of predictors should be reduced to a sensible set that reflects the problem at hand. Some practical reasons for predictor elimination are expense of collecting this information in the future, inaccuracy, high correlation with another predictor, many missing values, or simply irrelevance. Also helpful in examining potential predictors are summary statistics and graphs, such as frequency and correlation tables, predictor-specific summary statistics and plots, and missing value counts.

The next step makes use of computational power and statistical significance. In general, there are two types of methods for reducing the number of predictors in a model. The first is an exhaustive search for the "best" subset of predictors by fitting regression models with all the possible combinations of predictors. The second is to search through a partial set of models. We describe these two approaches next.

Exhaustive Search   The idea here is to evaluate all subsets. Since the number of subsets for even moderate values of p is very large, we need some way to examine the most promising subsets and to select from them. Criteria for evaluating and comparing models are based on the fit to the training data. One popular criterion is the *adjusted* R^2, which is defined as

$$R^2_{adj} = 1 - \frac{n-1}{n-p-1}(1 - R^2),$$

where R^2 is the proportion of explained variability in the model (in a model with a single predictor, this is the squared correlation). Like R^2, higher values of adjusted R^2 indicate better fit. Unlike R^2, which does not account for the number of predictors used, adjusted R^2 uses a penalty on the number of predictors. This avoids the artificial increase in R^2 that can result from simply increasing the number of predictors but not the amount of information. It can be shown that using R^2_{adj} to choose a subset is equivalent to picking the subset that minimizes $\hat{\sigma}^2$.

Another criterion that is often used for subset selection is known as *Mallow's* C_p. This criterion assumes that the full model (with all predictors) is unbiased, although it may have predictors that if dropped would reduce prediction variability. With this assumption we can show that if a subset model is unbiased, the average C_p value equals the number of parameters $p + 1$ (= number of predictors $+1$), the size of the subset. So a reasonable approach to identifying subset models with small bias is to examine those with values of C_p that are near $p + 1$. C_p is also an estimate of the error[1] for predictions at the x values observed in the training set. Thus, good models are those that have values of C_p near $p + 1$ and that have small p (i.e., are of small size). C_p is computed from the formula

$$C_p = \frac{SSR}{\hat{\sigma}^2_{full}} + 2(p+1) - n$$

where $\hat{\sigma}^2_{full}$ is the estimated value of σ^2 in the full model that includes all predictors, and SSR is the sum of squares of Regression given in the ANOVA table. It is important to remember that the usefulness of this approach depends heavily on the reliability of the estimate of σ^2 for the full model. This requires that the training set contain a large number of observations relative to the number of predictors. Finally, a useful point to note is that for a fixed size of subset, R^2, R^2_{adj}, and C_p all select the same subset. In fact, there is no difference between them in the order of merit they ascribe to subsets of a fixed size.

Figure 6.4 gives the results of applying an exhaustive search on the Toyota Corolla price data (with the 11 predictors). It reports the best model with a single predictor, 2 predictors, and so on. It can be seen that the R^2_{adj} increases until 6 predictors are used (number of coefficients = 7) and then stabilizes. The C_p indicates that a model with 9–11 predictors is good. The dominant predictor in all models is the age of the car, with horsepower and mileage playing important roles as well.

---

[1] In particular, it is the sum of the MSE (mean squared error) standardized by dividing by σ^2.

#Coeffs	RSS	Cp	R-Sq	Adj. R-Sq	Model (Constant present in all models)											
					1	2	3	4	5	6	7	8	9	10	11	12
2	1996467712	477.71	0.75	0.75	Constant	Age										
3	1672546432	305.51	0.79	0.79	Constant	Age	HP									
4	1438242432	181.50	0.82	0.82	Constant	Age	HP	Weight								
5	1258062976	86.59	0.84	0.84	Constant	Age	Mileage	HP	Weight							
6	1181816320	47.59	0.85	0.85	Constant	Age	Mileage	Petrol	QuartTax	Weight						
7	1095153024	2.98	0.86	0.86	Constant	Age	Mileage	Petrol		HP	QuartTax	Weight				
8	1093753344	4.23	0.86	0.86	Constant	Age	Mileage	Petrol		HP	Automatic	QuartTax	Weight			
9	1093557120	6.12	0.86	0.86	Constant	Age	Mileage	Petrol		HP	Metallic	Automatic	QuartTax	Weight		
10	1093422592	8.05	0.86	0.86	Constant	Age	Mileage	Diesel	Petrol	HP	Metallic	Automatic	QuartTax	Weight		
11	1093335424	10.00	0.86	0.86	Constant	Age	Mileage	Diesel	Petrol	HP	Metallic	Automatic	CC	QuartTax	Weight	
12	1093331072	12.00	0.86	0.86	Constant	Age	Mileage	Diesel	Petrol	HP	Metallic	Automatic	CC	Doors	QuartTax	Weight

FIGURE 6.4   EXHAUSTIVE SEARCH RESULT FOR REDUCING PREDICTORS IN TOYOTA COROLLA EXAMPLE

Popular Subset Selection Algorithms   The second method of finding the best subset of predictors relies on a partial, iterative search through the space of all possible regression models. The end product is one best subset of predictors (although there do exist variations of these methods that identify several close-to-best choices for different sizes of predictor subsets). This approach is computationally cheaper, but it has the potential of missing "good" combinations of predictors. None of the methods guarantee that they yield the best subset for any criterion, such as adjusted R^2. They are reasonable methods for situations with large numbers of predictors, but for moderate numbers of predictors the exhaustive search is preferable.

Three popular iterative search algorithms are forward selection, backward elimination, and stepwise regression. In *forward selection* we start with no predictors and then add predictors one by one. Each predictor added is the one (among all predictors) that has the largest contribution to R^2 on top of the predictors that are already in it. The algorithm stops when the contribution of additional predictors is not statistically significant. The main disadvantage of this method is that the algorithm will miss pairs or groups of predictors that perform very well together but perform poorly as single predictors. This is similar to interviewing job candidates for a team project one by one, thereby missing groups of candidates who perform superiorly together, but poorly on their own.

In *backward elimination* we start with all predictors and then at each step eliminate the least useful predictor (according to statistical significance). The algorithm stops when all the remaining predictors have significant contributions. The weakness of this algorithm is that computing the initial model with all predictors can be time consuming and unstable. *Stepwise regression* is like forward selection except that at each step we consider dropping predictors that are not statistically significant, as in backward elimination.

Note: In XLMiner, unlike other popular software packages (SAS, Minitab, etc.), these three algorithms yield a table similar to the one that the exhaustive search yields rather than a single model. This allows the user to decide on the subset size after reviewing all possible sizes based on criteria such as R^2_{adj} and C_p.

#Coeffs	RSS	Cp	R-Sq	Adj. R-Sq	Probability	1	2	3	4	5	6	7	8	9	10	11
2	1996467712	477.712341	0.74681	0.7463861	0	Constant	Age									
3	1780184064	363.393707	0.77424	0.7734821	0	Constant	Age	Weight								
4	1482806272	205.462128	0.81195	0.8110051	0	Constant	Age	Petrol	Weight							
5	1310214400	114.64119	0.83384	0.8327225	0	Constant	Age	Petrol	QuartTax	Weight						
6	1181816320	47.5879288	0.85012	0.8488613	8E-08	Constant	Age	Mileage	Petrol	QuartTax	Weight					
7	1095153024	2.97988558	0.86111	0.8597082	0.962122	Constant	Age	Mileage	Petrol	HP	QuartTax	Weight				
8	1093753344	4.22712946	0.86129	0.8596509	0.993999	Constant	Age	Mileage	Petrol	HP	Automatic	QuartTax	Weight			
9	1093557120	6.12159872	0.86132	0.8594386	0.989111	Constant	Age	Mileage	Petrol	HP	Metallic	Automatic	QuartTax	Weight		
10	1093422592	8.0492487	0.86133	0.8592177	0.975677	Constant	Age	Mileage	Diesel	Petrol	HP	Metallic	Automatic	QuartTax	Weight	
11	1093335424	10.0023689	0.86134	0.8589899	0.961197	Constant	Age	Mileage	Diesel	Petrol	HP	Metallic	Automatic	CC	QuartTax	Weight
12	1093331072	12.0000286	0.86134	0.8587507	1	Constant	Age	Mileage	Diesel	Petrol	HP	Metallic	Automatic	CC	Doors	QuartTax

FIGURE 6.5   BACKWARD ELIMINATION RESULT FOR REDUCING PREDICTORS IN TOYOTA COROLLA EXAMPLE

For the Toyota Corolla price example, forward selection yields exactly the same results as those found in an exhaustive search: For each number of predictors the same subset is chosen (it therefore gives a table identical to the one in Figure 6.4). Notice that this will not always be the case. In comparison, backward elimination starts with the full model and then drops predictors one by one in this order: Doors, CC, Diesel, Metallic, Automatic, QuartTax, Petrol, Weight, and Age (see Figure 6.5). The R^2_{adj} and C_p measures indicate exactly the same subsets as those suggested by the exhaustive search. In other words, it correctly identifies Doors, CC, Diesel, Metallic, and Automatic as the least useful predictors. Backward elimination would yield a different model than that of the exhaustive search only if we decided to use fewer than six predictors. For instance, if we were limited to two predictors, backward elimination would choose Age and Weight, whereas an exhaustive search shows that the best pair of predictors is actually Age and HP.

The results for stepwise regression can be seen in Figure 6.6. It chooses the same subsets as forward selection for subset sizes of 1–7 predictors. However, for 8–10 predictors, it chooses a different subset than that chosen using the other methods: It decides to drop Doors, Quart Tax, and Weight. This means that it fails to detect the best subsets for 8–10 predictors. R^2_{adj} is largest at 6 predictors (the same 6 as were selected by the other models), but C_p indicates that the full model with 11 predictors is the best fit.

#Coeffs	RSS	Cp	R-Sq	Adj. R-Sq	Probability	1	2	3	4	5	6	7	8	9	10	11
2	1996467712	477.712341	0.74681	0.7463861	0	Constant	Age									
3	1672546432	305.505524	0.78789	0.7871783	0	Constant	Age	HP								
4	1438242560	181.495499	0.8176	0.816685	0	Constant	Age	HP	Weight							
5	1258062976	86.5938416	0.84045	0.8393808	0	Constant	Age	Mileage	HP	Weight						
6	1188944640	51.4215813	0.84922	0.8479497	2E-08	Constant	Age	Mileage	HP	QuartTax	Weight					
7	1095153024	2.97988558	0.86111	0.8597082	0.962122	Constant	Age	Mileage	Petrol	HP	QuartTax	Weight				
8	1093753344	4.22712946	0.86129	0.8596509	0.993999	Constant	Age	Mileage	Petrol	HP	Automatic	QuartTax	Weight			
9	1468513408	207.775345	0.81376	0.8112433	0	Constant	Age	Mileage	Diesel	Petrol	HP	Metallic	Automatic	CC		
10	1451250000	200.491074	0.81595	0.8131461	0	Constant	Age	Mileage	Diesel	Petrol	HP	Metallic	Automatic	CC	Doors	
11	1229398784	83.1780624	0.84409	0.8414415	0	Constant	Age	Mileage	Diesel	Petrol	HP	Metallic	Automatic	CC	Doors	QuartTax
12	1093331072	12.0000286	0.86134	0.8587507	1	Constant	Age	Mileage	Diesel	Petrol	HP	Metallic	Automatic	CC	Doors	QuartTax

FIGURE 6.6   STEPWISE SELECTION RESULT FOR REDUCING PREDICTORS IN TOYOTA COROLLA EXAMPLE

This example shows that the search algorithms yield fairly good solutions, but we need to carefully determine the number of predictors to retain. It also shows the merits of running a few searches and using the combined results to determine the subset to choose. There is a popular (but false) notion that stepwise regression is superior to backward elimination and forward selection because of its ability to add and to drop predictors. This example shows clearly that it is not always so.

Finally, additional ways to reduce the dimension of the data are by using principal components (Chapter 4) and regression trees (Chapter 9).

PROBLEMS

6.1 **Predicting Boston Housing Prices.** The file BostonHousing.xls contains information collected by the U.S. Bureau of the Census concerning housing in the area of Boston, Massachusetts. The dataset includes information on 506 census housing tracts in the Boston area. The goal is to predict the median house price in new tracts based on information such as crime rate, pollution, and number of rooms. The dataset contains 14 predictors, and the response is the median house price (MEDV). Table 6.3 describes each of the predictors and the response.

TABLE 6.3	DESCRIPTION OF VARIABLES FOR BOSTON HOUSING EXAMPLE
CRIM	Per capita crime rate by town
ZN	Proportion of residential land zoned for lots over 25,000 ft^2
INDUS	Proportion of nonretail business acres per town
CHAS	Charles River dummy variable (= 1 if tract bounds river; = 0 otherwise)
NOX	Nitric oxide concentration (parts per 10 million)
RM	Average number of rooms per dwelling
AGE	Proportion of owner-occupied units built prior to 1940
DIS	Weighted distances to five Boston employment centers
RAD	Index of accessibility to radial highways
TAX	Full-value property tax rate per $10,000
PTRATIO	Pupil/teacher ratio by town
B	1000(Bk − 0.63)^2 where Bk is the proportion of blacks by town
LSTAT	% Lower status of the population
MEDV	Median value of owner-occupied homes in $1000s

a. Why should the data be partitioned into training and validation sets? For what will the training set be used? For what will the validation set be used?

Fit a multiple linear regression model to the median house price (MEDV) as a function of CRIM, CHAS, and RM.

b. Write the equation for predicting the median house price from the predictors in the model.

c. What median house price is predicted for a tract in the Boston area that does not bound the Charles River, has a crime rate of 0.1, and where the average number of rooms per house is 6? What is the prediction error?

d. Reduce the number of predictors:

  i. Which predictors are likely to be measuring the same thing among the 14 predictors? Discuss the relationships among INDUS, NOX, and TAX.

  ii. Compute the correlation table for the 13 numerical predictors and search for highly correlated pairs. These have potential redundancy and can cause multicollinearity. Choose which ones to remove based on this table.

  iii. Use an exhaustive search to reduce the remaining predictors as follows: First, choose the top three models. Then run each of these models separately on the training set, and compare their predictive accuracy for the validation set. Compare RMSE and average error, as well as lift charts. Finally, describe the best model.

6.2 Predicting Software Reselling Profits. Tayko Software is a software catalog firm that sells games and educational software. It started out as a software manufacturer and then added third-party titles to its offerings. It recently revised its collection of items in a new catalog, which it mailed out to its customers. This mailing yielded 1000 purchases. Based on these data, Tayko wants to devise a model for predicting the spending amount that a purchasing customer will yield. The file Tayko.xls contains information on 1000 purchases. Table 6.4 describes the variables to be used in the problem (the Excel file contains additional variables).

TABLE 6.4   **DESCRIPTION OF VARIABLES FOR TAYKO SOFTWARE EXAMPLE**

FREQ	Number of transactions in the preceding year
LAST_UPDATE	Number of days since last update to customer record
WEB	Whether customer purchased by Web order at least once
GENDER	Male or female
ADDRESS_RES	Whether it is a residential address
ADDRESS_US	Whether it is a U.S. address
SPENDING (response)	Amount spent by customer in test mailing (in dollars)

a. Explore the spending amount by creating a pivot table for the categorical variables and computing the average and standard deviation of spending in each category.

b. Explore the relationship between spending and each of the two continuous predictors by creating two scatterplots (SPENDING vs. FREQ, and SPENDING vs. LAST_UPDATE). Does there seem to be a linear relationship?

c. To fit a predictive model for SPENDING:

  i. Partition the 1000 records into training and validation sets.

  ii. Run a multiple linear regression model for SPENDING versus all six predictors. Give the estimated predictive equation.

  iii. Based on this model, what type of purchaser is most likely to spend a large amount of money?

  iv. If we used backward elimination to reduce the number of predictors, which predictor would be dropped first from the model?

  v. Show how the prediction and the prediction error are computed for the first purchase in the validation set.

  vi. Evaluate the predictive accuracy of the model by examining its performance on the validation set.

  vii. Create a histogram of the model residuals. Do they appear to follow a normal distribution? How does this affect the predictive performance of the model?

6.3 Predicting Airfares on New Routes. Several new airports have opened in major cities, opening the market for new routes (a route refers to a pair of airports), and Southwest has not announced whether it will cover routes to/from these cities. In order to price flights on these routes, a major airline collected information on 638 air routes in the United States. Some factors are known about these new routes: the distance traveled, demographics of the city where the new airport is located, and whether this city is a vacation destination. Other factors are yet unknown (e.g., the number of passengers who will travel this route). A major unknown factor is whether Southwest or another discount airline will travel on these new routes. Southwest's strategy (point-to-point routes covering only major cities, use of secondary airports, standardized fleet, low fares) has been very different from the model followed by the older and bigger airlines (hub-and-spoke model extending to even smaller cities, presence in primary airports, variety in fleet, pursuit of high-end business travelers).The presence of discount airlines is therefore believed to reduce the fares greatly.

The file Airfares.xls contains real data that were collected for the third quarter of 1996. They consist of the predictors and response listed in Table 6.5. Note that some cities are served by more than one airport, and in those cases the airports are distinguished by their three-letter code.

TABLE 6.5   DESCRIPTION OF VARIABLES FOR AIRFARE EXAMPLE

S_CODE	Starting airport's code
S_CITY	Starting city
E_CODE	Ending airport's code
E_CITY	Ending city
COUPON	Average number of coupons (a one-coupon flight is a nonstop flight, a two-coupon flight is a one-stop flight, etc.) for that route
NEW	Number of new carriers entering that route between Q3-96 and Q2-97
VACATION	Whether (Yes) or not (No) a vacation route
SW	Whether (Yes) or not (No) Southwest Airlines serves that route
HI	Herfindahl index: measure of market concentration
S_INCOME	Starting city's average personal income
E_INCOME	Ending city's average personal income
S_POP	Starting city's population
E_POP	Ending city's population
SLOT	Whether or not either endpoint airport is slot controlled (this is a measure of airport congestion)
GATE	Whether or not either endpoint airport has gate constraints (this is another measure of airport congestion)
DISTANCE	Distance between two endpoint airports in miles
PAX	Number of passengers on that route during period of data collection
FARE	Average fare on that route

a. Explore the numerical predictors and response (FARE) by creating a correlation table and examining some scatterplots between FARE and those predictors. What seems to be the best single predictor of FARE?

b. Explore the categorical predictors (excluding the first four) by computing the percentage of flights in each category. Create a pivot table with the average fare in each category. Which categorical predictor seems best for predicting FARE?

c. Find a model for predicting the average fare on a new route:

  i. Convert categorical variables (e.g., SW) into dummy variables. Then, partition the data into training and validation sets. The model will be fit to the training data and evaluated on the validation set.

  ii. Use stepwise regression to reduce the number of predictors. You can ignore the first four predictors (S_CODE, S_CITY, E_CODE, E_CITY). Report the estimated model selected.

  iii. Repeat (ii) using exhaustive search instead of stepwise regression. Compare the resulting best model to the one you obtained in (ii) in terms of the predictors that are in the model.

    iv. Compare the predictive accuracy of both models (ii) and (iii) using measures such as RMSE and average error and lift charts.

    v. Using model (iii), predict the average fare on a route with the following characteristics: COUPON = 1.202, NEW = 3, VACATION = No, SW = No, HI = 4442.141, S_INCOME = $28,760, E_INCOME = $27,664, S_POP = 4,557,004, E_POP = 3,195,503, SLOT = Free, GATE = Free, PAX = 12782, DISTANCE = 1976 miles.

    vi. Predict the reduction in average fare on the route if in (b) Southwest decides to cover this route [using model (iii)].

    vii. In reality, which of the factors will not be available for predicting the average fare from a new airport (i.e., before flights start operating on those routes)? Which ones can be estimated? How?

    viii. Select a model that includes only factors that are available before flights begin to operate on the new route. Use an exhaustive search to find such a model.

    ix. Use the model in (viii) to predict the average fare on a route with characteristics COUPON = 1.202, NEW = 3, VACATION = No, SW = No, HI = 4442.141, S_INCOME = $28,760, E_INCOME = $27,664, S_ POP = 4,557,004, E_POP = 3,195,503, SLOT = Free, GATE = Free, PAX = 12782, DISTANCE = 1976 miles.

    x. Compare the predictive accuracy of this model with model (iii). Is this model good enough, or is it worthwhile reevaluating the model once flights begin on the new route?

d. In competitive industries, a new entrant with a novel business plan can have a disruptive effect on existing firms. If a new entrant's business model is sustainable, other players are forced to respond by changing their business practices. If the goal of the analysis was to evaluate the effect of Southwest Airlines' presence on the airline industry rather than predicting fares on new routes, how would the analysis be different? Describe technical and conceptual aspects.

6.4 **Predicting Prices of Used Cars.** The file ToyotaCorolla.xls contains data on used cars (Toyota Corolla) on sale during late summer of 2004 in The Netherlands. It has 1436 records containing details on 38 attributes, including Price, Age, Kilometers, HP, and other specifications. The goal is to predict the price of a used Toyota Corolla based on its specifications. (The example in Section 4.2.1 is a subset of this dataset.)

    Data Preprocessing. Create dummy variables for the categorical predictors (Fuel Type and Metallic). Split the data into training (50%), validation (30%), and test (20%) datasets.

    Run a multiple linear regression using the *Prediction* menu in XLMiner with the output variable Price and input variables Age_08_04, KM, Fuel_Type, HP, Automatic, Doors, Quarterly_Tax, Mfg_Guarantee, Guarantee_Period, Airco, Automatic_Airco, CD_Player, Powered_Windows, Sport_Model, and Tow_Bar.

a. What appear to be the three or four most important car specifications for predicting the car's price?

b. Using metrics you consider useful, assess the performance of the model in predicting prices.

k-Nearest Neighbors (*k*-NN)

In this chapter we describe the *k*-nearest neighbor algorithm that can be used for classification (of a categorical outcome) or prediction (of a numerical outcome). To classify or predict a new record, the method relies on finding "similar" records in the training data. These "neighbors" are then used to derive a classification or prediction for the new record by voting (for classification) or averaging (for prediction). We explain how similarity is determined, how the number of neighbors is chosen, and how a classification or prediction is computed. *k*-NN is a highly automated data-driven method. We discuss the advantages and weaknesses of the *k*-NN method in terms of performance and practical considerations such as computational time.

7.1 *k*-NN CLASSIFIER (CATEGORICAL OUTCOME)

The idea in *k*-nearest neighbor methods is to identify *k* records in the training dataset that are similar to a new record that we wish to classify. We then use these similar (neighboring) records to classify the new record into a class, assigning the new record to the predominant class among these neighbors. Denote by $(x_1, x_2, \ldots, x_p)$ the values of the predictors for this new record. We look for records in our training data that are similar or "near" the record to be classified in the predictor space (i.e., records that have values close to $x_1, x_2, \ldots, x_p$). Then, based on the classes to which those proximate records belong, we assign a class to the record that we want to classify.

Data Mining for Business Intelligence, By Galit Shmueli, Nitin R. Patel, and Peter C. Bruce
Copyright © 2010 John Wiley & Sons Inc.

Determining Neighbors

The *k*-nearest neighbor algorithm is a classification method that does not make assumptions about the form of the relationship between the class membership (Y) and the predictors $X_1, X_2, \ldots, X_p$. This is a nonparametric method because it does not involve estimation of parameters in an assumed function form, such as the linear form assumed in linear regression (Chapter 6). Instead, this method draws information from similarities between the predictor values of the records in the dataset.

The central issue here is how to measure the distance between records based on their predictor values. The most popular measure of distance is the Euclidean distance. The Euclidean distance between two records $(x_1, x_2, \ldots, x_p)$ and $(u_1, u_2, \ldots, u_p)$ is

$$\sqrt{(x_1 - u_1)^2 + (x_2 - u_2)^2 + \cdots + (x_p - u_p)^2}. \tag{7.1}$$

You will find a host of other distance metrics in Chapters 12 and 14 for both numerical and categorical variables. However, the *k*-NN algorithm relies on many distance computations (between each record to be predicted and every record in the training set), and, therefore, the Euclidean distance, which is computationally cheap, is the most popular in *k*-NN.

To equalize the scales that the various predictors may have, note that in most cases predictors should first be standardized before computing a Euclidean distance.

Classification Rule

After computing the distances between the record to be classified and existing records, we need a rule to assign a class to the record to be classified, based on the classes of its neighbors. The simplest case is $k = 1$, where we look for the record that is closest (the nearest neighbor) and classify the new record as belonging to the same class as its closest neighbor. It is a remarkable fact that this simple, intuitive idea of using a single nearest neighbor to classify records can be very powerful when we have a large number of records in our training set. In turns out that the misclassification error of the one-nearest-neighbor scheme has a misclassification rate that is no more than twice the error when we know exactly the probability density functions for each class.

The idea of the **one-nearest neighbor** can be extended to $k > 1$ neighbors as follows:

1. Find the nearest k neighbors to the record to be classified.
2. Use a majority decision rule to classify the record, where the record is classified as a member of the majority class of the k neighbors.

Example: Riding Mowers

A riding-mower manufacturer would like to find a way of classifying families in a city into those likely to purchase a riding mower and those not likely to buy one. A pilot random sample is undertaken of 12 owners and 12 nonowners in the city. The data are shown and plotted in Table 7.1. We first partition the data into training data (18 households) and validation data (6 households). Obviously, this dataset is too small for partitioning, which can result in unstable results, but we continue with this for illustration purposes. The training set is shown in Figure!7.1.

Now consider a new household with $60,000 income and lot size 20,000 ft^2 (also shown in Figure 7.1). Among the households in the training set, the one closest to the new household (in Euclidean distance after normalizing income and lot size) is household 4, with $61,500 income and lot size 20,800 ft^2. If we use a 1NN classifier, we would classify the new household as an owner, like household 4. If we use $k = 3$, the three nearest households are 4, 9, and 14. The

TABLE 7.1	LOT SIZE, INCOME, AND OWNERSHIP OF A RIDING MOWER FOR 24 HOUSEHOLDS		
Household Number	Income ($ooos)	Lot Size (ooos ft²)	Ownership of Riding Mower
1	60.0	18.4	Owner
2	85.5	16.8	Owner
3	4.8	21.6	Owner
4	61.5	20.8	Owner
5	87.0	23.6	Owner
6	110.1	19.2	Owner
7	108.0	17.6	Owner
8	82.8	22.4	Owner
9	69.0	20.0	Owner
10	93.0	20.8	Owner
11	51.0	22.0	Owner
12	81.0	20.0	Owner
13	75.0	19.6	Nonowner
14	52.8	20.8	Nonowner
15	64.8	17.2	Nonowner
16	43.2	20.4	Nonowner
17	84.0	17.6	Nonowner
18	49.2	17.6	Nonowner
19	59.4	16.0	Nonowner
20	66.0	18.4	Nonowner
21	47.4	16.4	Nonowner
22	33.0	18.8	Nonowner
23	51.0	14.0	Nonowner
24	63.0	14.8	Nonowner

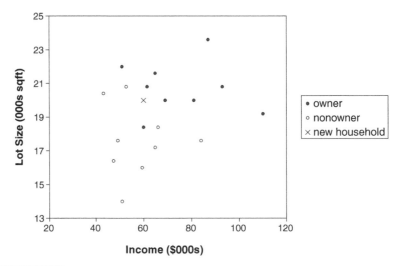

FIGURE 7.1 SCATTERPLOT OF LOT SIZE VS. INCOME FOR THE 18 HOUSEHOLDS IN THE TRAINING SET AND THE NEW HOUSEHOLD TO BE CLASSIFIED

first two are owners of riding mowers, and the last is a nonowner. The majority vote is therefore *owner*, and the new household would be classified as an owner.

Choosing *k*

The advantage of choosing $k > 1$ is that higher values of k provide smoothing that reduces the risk of overfitting due to noise in the training data. Generally speaking, if k is too low, we may be fitting to the noise in the data. However, if k is too high, we will miss out on the method's ability to capture the local structure in the data, one of its main advantages. In the extreme, $k = n = $ the number of records in the training dataset. In that case we simply assign all records to the majority class in the training data, irrespective of the values of $(x_1, x_2, \ldots, x_p)$, which coincides with the naive rule! This is clearly a case of oversmoothing in the absence of useful information in the predictors about the class membership. In other words, we want to balance between overfitting to the predictor information and ignoring this information completely. A balanced choice depends greatly on the nature of the data. The more complex and irregular the structure of the data, the lower the optimum value of k. Typically, values of k fall in the range of 1–20. Often, an odd number is chosen to avoid ties.

So how is k chosen? Answer: We choose the k that has the best classification performance. We use the training data to classify the records in the validation data, then compute error rates for various choices of k. For our example, if we choose $k = 1$, we will classify in a way that is very sensitive to the local characteristics of the training data. On the other hand, if we choose a large value of k, such as $k = 18$, we would simply predict the most frequent class in the dataset in all cases. This is a very stable prediction, but it completely ignores the

Value of *k*	% Error Training	% Error Validation	
1	0.00	33.33	
2	16.67	33.33	
3	11.11	33.33	
4	22.22	33.33	
5	11.11	33.33	
6	27.78	33.33	
7	22.22	33.33	
8	22.22	16.67	<--- Best k
9	22.22	16.67	
10	22.22	16.67	
11	16.67	33.33	
12	16.67	16.67	
13	11.11	33.33	
14	11.11	16.67	
15	5.56	33.33	
16	16.67	33.33	
17	11.11	33.33	
18	50.00	50.00	

FIGURE 7.2   **MISCLASSIFICATION RATE OF VALIDATION SET FOR VARIOUS CHOICES OF *K***

information in the predictors. To find a balance, we examine the misclassification rate (of the validation set) that results for different choices of *k* between 1 and 18. This is shown in Figure 7.2. We would choose *k* = 8, which minimizes the misclassification rate in the validation set.[1] Note, however, that now the validation set is used as an addition to the training set and does not reflect a holdout set as before. Ideally, we would want a third test set to evaluate the performance of the method on data that it did not see.

Once *k* is chosen, the algorithm uses it to generate classifications of new records. An example is shown in Figure 7.3, where eight neighbors are used to classify the new household.

Setting the Cutoff Value

k-NN uses a majority decision rule to classify a new record, where the record is classified as a member of the majority class of the *k* neighbors. The definition of "majority" is directly linked to the notion of a cutoff value applied to the class membership probabilities. Let us consider a binary outcome case. For a new record, the proportion of class 1 members among its neighbors is an estimate

---

[1] Partitioning such a small dataset is unwise in practice, as results will heavily rely on the particular partition. For instance, if you use a different partitioning, you might obtain a different "optimal" *k*. We use this example for illustration only.

Data range	['KNN Riding Mowers.xlsx']'Data_Partition1'!$G$20:$H$20	

Cut off Prob.Val. for Success (Updatable)	0.5	(Updating the value here will NOT update)

Row Id.	Predicted Class	Prob. for Owner	Actual #Nearest Neighbors	Income	Lot Size
1	owner	0.625	8	60	20

FIGURE 7.3   **CLASSIFYING A NEW HOUSEHOLD USING THE "BEST *K*" = 8**

of its probability of belonging to class 1. In the riding-mower example with $k = 3$, we found that the three nearest neighbors to the new household (with income = \$60,000 and lot size = 20,000 ft^2) are households 4, 9, and 14. Since 4 and 9 are owners and 14 is a nonowner, we can estimate for the new household a probability of 2/3 of being an owner (and 1/3 for being a nonowner). Using a simple majority rule is equivalent to setting the cutoff value to 0.5. Another example can be seen in Figure 7.3, where $k = 8$ was used to classify the new household. The "Prob for Owner" of 0.625 was obtained because 5 of the 8 neighbors were owners. Using a cutoff of 0.5 leads to a classification of "owner" for the new household.

As mentioned in Chapter 5, changing the cutoff value affects the classification matrix (i.e., the error rates). Hence, in some cases we might want to choose a cutoff other than the default 0.5 for the purpose of maximizing accuracy or for incorporating misclassification costs. In XLMiner this can be done by directly changing the cutoff value (as can be seen in Figure 7.3), which automatically changes the "Predicted Class" and the related classification matrices.

k-NN with More Than Two Classes

The *k*-NN classifier can easily be applied to an outcome with *m* classes, where $m > 2$. The "majority rule" means that a new record is classified as a member of the majority class of its *k* neighbors. An alternative, when there is a specific class that we are interested in identifying (and are willing to "overidentify" records as belonging to this class), is to calculate the proportion of the *k* neighbors that belong to this class of interest, use that as an estimate of the probability that the new record belongs to that class, and then refer to a user-specified cutoff value to decide whether to assign the new record to that class. For more on the use of a cutoff value in classification where there is a single class of interest, see Chapter 5.

7.2   *k*-NN FOR A NUMERICAL RESPONSE

The idea of *k*-NN can readily be extended to predicting a continuous value (as is our aim with multiple linear regression models). The first step of determining neighbors by computing distances remains unchanged. The second step, where a majority vote of the neighbors is used to determine class, is modified such that we take the average response value of the k-nearest neighbors to determine the prediction. Often, this average is a weighted average, with the weight decreasing with increasing distance from the point at which the prediction is required.

Another modification is in the error metric use for determining the "best k." Rather than the overall error rate used in classification, RMSE (or another prediction error metric) is used in prediction (see Chapter 5).

PANDORA

Pandora is an Internet music radio service that allows users to build customized "stations" that play music similar to a song or artist that they have specified. Pandora uses a k-NN style clustering/classification process called the Music Genome Project to locate new songs or artists that are close to the user-specified song or artist. In simplified terms, the process works roughly as follows for songs:

1. Pandora has established hundreds of variables on which a song can be measured on a scale from 0 to 5. Four such variables from the beginning of the list are
   o Acid Rock Qualities
   o Accordion Playing
   o Acousti-Lectric Sonority
   o Acousti-Synthetic Sonority

2. Pandora pays musicians to analyze tens of thousands of songs and rate each song on each of these attributes. Each song will then be represented by a row vector of values between 0 and 5, for example, for Led Zeppelin's Kashmir:

    Kashmir 4 0 3 3 ... (high on acid rock attributes, no accordion, etc.)

    This step represents a costly investment and lies at the heart of Pandora's value because these variables have been tested and selected because they accurately reflect the essence of a song and provide a basis for defining highly individualized preferences.

3. The online user specifies a song that he/she likes (the song must be in Pandora's database).

4. Pandora then calculates the statistical distance[2] between the user's song and the songs in its database. It selects a song that is close to the user-specified song and plays it.

5. The user then has the option of saying "I like this song," "I don't like this song," or saying nothing.

6. If "like" is chosen, the original song, plus the new song are merged into a two-song cluster[3] that is represented by a single vector, comprised means of the variables in the original two-song vectors.

7. If "don't like" is chosen, the vector of the song that is not liked is stored for future reference. (If the user does not express an opinion about the song, in our simplified example here the new song is not used for further comparisons.)

8. Pandora looks in its database for a new song, one whose statistical distance is close to the "like" song cluster[4] and not too close to the "don't like" song. Depending on the user's reaction, this new song might be added to the "like" cluster or "don't like" cluster.

Over time, Pandora develops the ability to deliver songs that match a particular taste of a particular user. A single user might build up multiple stations around different song clusters. Clearly, this is a less limiting approach than selecting music in terms of which "genre" it belongs to (which is a more limiting approach).

While the process described above is a bit more complex than the basic "classification of new data" process described in this chapter, the fundamental process—classifying a record according to its proximity to other records—is the same at its core. Note the role of domain knowledge in this machine learning process—the variables have been tested and selected by the project leaders, and the measurements have been made by human experts.

Further Reading: See www.pandora.com, Wikipedia's article on the Music Genome Project, and Joyce John's article "Pandora and the Music Genome Project," in the September 2006 *Scientific Computing* [23(10): 14, 40–41].

---

[2] See Chapter 12 for an explanation of statistical distance.
[3] See Chapter 14 for more on clusters.
[4] See Case 18.4 "Segmenting Consumers of Bath Soap" for an exercise involving the identification of clusters, which are then used for classification purposes.

7.3 ADVANTAGES AND SHORTCOMINGS OF *k*-NN ALGORITHMS

The main advantage of *k*-NN methods is their simplicity and lack of parametric assumptions. In the presence of a large enough training set, these methods perform surprisingly well, especially when each class is characterized by multiple combinations of predictor values. For instance, in real estate databases there are likely to be multiple combinations of {home type, number of rooms, neighborhood, asking price, etc.} that characterize homes that sell quickly versus those that remain for a long period on the market.

There are two difficulties with the practical exploitation of the power of the *k*-NN approach. First, although no time is required to estimate parameters from the training data (as would be the case for parametric models such as regression), the time to find the nearest neighbors in a large training set can be prohibitive. A number of ideas have been implemented to overcome this difficulty. The main ideas are:

- Reduce the time taken to compute distances by working in a reduced dimension using dimension reduction techniques such as principal components analysis (Chapter 4).
- Use sophisticated data structures such as search trees to speed up identification of the nearest neighbor. This approach often settles for an "almost nearest" neighbor to improve speed.
- Edit the training data to remove redundant or almost redundant points to speed up the search for the nearest neighbor. An example is to remove records in the training set that have no effect on the classification because they are surrounded by records that all belong to the same class.

Second, the number of records required in the training set to qualify as large increases exponentially with the number of predictors p. This is because the expected distance to the nearest neighbor goes up dramatically with p unless the size of the training set increases exponentially with p. This phenomenon is known as the *curse of dimensionality*, a fundamental issue pertinent to all classification, prediction, and clustering techniques. This is why we often seek to reduce the number of predictors through methods such as selecting subsets of the predictors for our model or by combining them using methods such as principal components analysis, singular value decomposition, and factor analysis (see Chapter 4).

PROBLEMS ▪

7.1 **Personal Loan Acceptance.** Universal Bank is a relatively young bank growing rapidly in terms of overall customer acquisition. The majority of these customers are liability customers (depositors) with varying sizes of relationship with the bank. The customer base of asset customers (borrowers) is quite small, and the bank is interested in expanding this base rapidly to bring in more loan business. In particular, it wants to explore ways of converting its liability customers to personal loan customers (while retaining them as depositors).

A campaign that the bank ran last year for liability customers showed a healthy conversion rate of over 9% success. This has encouraged the retail marketing department to devise smarter campaigns with better target marketing. The goal of our analysis is to model the previous campaign's customer behavior to analyze what combination of factors make a customer more likely to accept a personal loan. This will serve as the basis for the design of a new campaign.

The file UniversalBank.xls contains data on 5000 customers. The data include customer demographic information (age, income, etc.), the customer's relationship with the bank (mortgage, securities account, etc.), and the customer response to the last personal loan campaign (*Personal Loan*). Among these 5000 customers, only 480($= 9.6\%$) accepted the personal loan that was offered to them in the earlier campaign.

Partition the data into training (60%) and validation (40%) sets.

a. Consider the following customer:
Age$=40$, Experience$=10$, Income$=84$, Family$=2$, CCAvg$=2$, Education_2$=1$, Education_3$=0$, Mortgage$=0$, Securities Account$=0$, CD Account$=0$, Online$=1$ and Credit card $= 1$. Perform a *k*-NN classification with all predictors except ID and ZIP code using $k = 1$. Remember to transform categorical predictors with more than two categories into dummy variables first. Specify the *success* class as 1 (loan acceptance), and use the default cutoff value of 0.5. How would this customer be classified?

b. What is a choice of *k* that balances between overfitting and ignoring the predictor information?

c. Show the classification matrix for the validation data that results from using the best *k*.

d. Classify the customer using the best *k*.

e. Repartition the data, this time into training, validation, and test sets (50% : 30% : 20%). Apply the *k*-NN method with the *k* chosen above. Compare the classification matrix of the test set with that of the training and validation sets. Comment on the differences and their reason.

7.2 **Predicting Housing Median Prices.** The file BostonHousing.xls contains information on over 500 census tracts in Boston, where for each tract 14 variables are recorded. The last column (CAT.MEDV) was derived from MEDV, such that it obtains the value 1 if MEDV>30 and 0 otherwise. Consider the goal of predicting the median value (MEDV) of a tract, given the information in the first 13 columns.

Partition the data into training (60%) and validation (40%) sets.

a. Perform a *k*-NN prediction with all 13 predictors (ignore the CAT.MEDV column), trying values of *k* from 1 to 5. Make sure to normalize the data (click "normalize input data"). What is the best *k* chosen? What does it mean?

b. Predict the MEDV for a tract with the following information, using the best *k*:

CRIM	ZN	INDUS	CHAS	NOX	RM	AGE	DIS	RAD
0.2	0	7	0	0.538	6	62	4.7	4

TAX	PTRATIO	B	LSTAT
307	21	360	10

(Copy this table with the column names to a new worksheet and then in "Score new data" choose "from worksheet.")

c. Why is the error of the training data zero?

d. Why is the validation data error overly optimistic compared to the error rate when applying this k-NN predictor to new data?

e. If the purpose is to predict MEDV for several thousands of new tracts, what would be the disadvantage of using k-NN prediction? List the operations that the algorithm goes through in order to produce each prediction.

## CHAPTER 8

Naive Bayes

In this chapter we first present the complete, or exact, Bayesian classifier. We see that it can be used either to maximize overall classification accuracy or in the case where we are interested mainly in identifying records belonging to a particular class of interest. We next see how it is impractical in most cases and learn how to modify it (the "naive Bayesian classifier") so that it is generally applicable. The naive Bayesian classifier can be used only with categorical variables.

8.1 INTRODUCTION

The naive Bayes method (and, indeed, an entire branch of statistics) is named after the Reverend Thomas Bayes (1702–1761). To understand the naive Bayes classifier, we first look at the complete, or exact, Bayesian classifier. The basic principle is simple. For each record to be classified:

1. Find all the other records just like it (i.e., where the predictor values are the same).
2. Determine what classes they all belong to and which class is more prevalent.
3. Assign that class to the new record.

Alternatively (or in addition), it may be desirable to tweak the method so that it answers the question: What is an estimated probability of belonging to the class of interest? instead of Which class is the most probable? Obtaining class

Data Mining for Business Intelligence, By Galit Shmueli, Nitin R. Patel, and Peter C. Bruce
Copyright © 2010 John Wiley & Sons Inc.

probabilities allows using a sliding cutoff to classify a record as belonging to class i, even if i is not the most probable class for that record. This approach is useful when there is a specific class of interest that we are interested in identifying, and we are willing to "overidentify" records as belonging to this class. (See Chapter 5 for more details on the use of cutoffs for classification and on asymmetric misclassification costs)

Cutoff Probability Method

1. Establish a cutoff probability for the class of interest above which we consider that a record belongs to that class.
2. Find all the training records just like the new record (i.e., where the predictor values are the same).
3. Determine the probability that those records belong to the class of interest.
4. If that probability is above the cutoff probability, assign the new record to the class of interest.

Conditional Probability   Both procedures incorporate the concept of *conditional probability*, or the probability of event A given that event B has occurred [denoted $P(A|B)$]. In this case, we will be looking at the probability of the record belonging to class i given that its predictor values take on the values $x_1, x_2, \ldots, x_p$. In general, for a response with m classes $C_1, C_2, \ldots, C_m$, and the predictors $x_1, x_2, \ldots, x_p$, we want to compute

$$P(C_i|x_1, \ldots, x_p). \tag{8.1}$$

To classify a record, we compute its probability of belonging to each of the classes in this way, then classify the record to the class that has the highest probability or use the cutoff probability to decide whether it should be assigned to the class of interest.

From this definition, we see that the Bayesian classifier works only with categorical predictors. If we use a set of numerical predictors, then it is highly unlikely that multiple records will have identical values on these numerical predictors. Therefore, numerical predictors must be binned and converted to categorical predictors. *The Bayesian classifier is the only classification or prediction method presented in this book that is especially suited for (and limited to) categorical predictor variables.* Consider Example 1.

Example 1: Predicting Fraudulent Financial Reporting

An accounting firm has many large companies as customers. Each customer submits an annual financial report to the firm, which is then audited by the accounting firm. For simplicity, we will designate the outcome of the audit

TABLE 8.1	PIVOT TABLE FOR FINANCIAL REPORTING EXAMPLE		
	Prior Legal (X = 1)	No Prior Legal (X = 0)	Total
Fraudulent (C_1)	50	50	100
Truthful (C_2)	180	720	900
Total	230	770	1000

as "fraudulent" or "truthful," referring to the accounting firm's assessment of the customer's financial report. The accounting firm has a strong incentive to be accurate in identifying fraudulent reports—if it passes a fraudulent report as truthful, it would be in legal trouble.

The accounting firm notes that, in addition to all the financial records, it also has information on whether or not the customer has had prior legal trouble (criminal or civil charges of any nature filed against it). This information has not been used in previous audits, but the accounting firm is wondering whether it could be used in the future to identify reports that merit more intensive review. Specifically, it wants to know whether having had prior legal trouble is predictive of fraudulent reporting.

In this case, each customer is a record, and the response of interest, $Y = \{$fraudulent,truthful$\}$, has two classes into which a company can be classified: $C_1 =$ fraudulent and $C_2 =$ truthful. The predictor variable—"prior legal trouble"—has two values: 0 (no prior legal trouble) and 1 (prior legal trouble).

The accounting firm has data on 1500 companies that it has investigated in the past. For each company, it has information on whether the financial report was judged fraudulent or truthful and whether the company had prior legal trouble. After partitioning the data into a training set (1000 firms) and a validation set (500 firms), the counts in the training set are shown in Table 8.1.

8.2 APPLYING THE FULL (EXACT) BAYESIAN CLASSIFIER

Now consider the financial report from a new company, which we wish to classify as fraudulent or truthful by using these data. To do this, we compute the probabilities, as above, of belonging to each of the two classes.

If the new company had had prior legal trouble, the probability of belonging to the fraudulent class would be $P($fraudulent $\mid$ prior legal$) = 50/230$ (there were 230 companies with prior legal trouble in the training set, and 50 of them had fraudulent financial reports). The probability of belonging to the other class, truthful, is, of course, the remainder $= 180/230$.

Using the "Assign to the Most Probable Class" Method   If a company had prior legal trouble, we assign it to the "not-fraudulent" class. Similar calculations for the truthful case are left as an exercise to the reader. In this example, using the assign to the most probable class method, all records are assigned to the not-fraudulent class. This is the same result as the naive rule of "assign all records to the majority class."

Using the Cutoff Probability Method   In this example, we are more interested in identifying the fraudulent reports—those are the ones that can land the auditor in jail. We recognize that, in order to identify the fraudulent reports, some truthful reports will be misidentified as fraudulent, and the overall classification accuracy may decline. Our approach is, therefore, to establish a cutoff value for the probability of being fraudulent, and classify all records above that value as fraudulent. The technical formula for the calculation of this probability that a record belongs to class C_i is as follows:

$$P(C_i|x_1, \ldots, x_p) = \frac{P(x_1, \ldots, x_p|C_i)P(C_i)}{P(x_1, \ldots, x_p|C_1)P(C_1) + \cdots + P(x_1, \ldots, x_p|C_m)P(C_m)}.$$

(8.2)

In this example (where frauds are more rare), if the cutoff were established at 0.20, we would classify a prior legal trouble record as fraudulent because $P(\text{fraudulent} \mid \text{prior legal}) = 50/230 = 0.22$. The user can treat this cutoff as a "slider" to be adjusted to optimize performance, like other parameters in any classification model.

Practical Difficulty with the Complete (Exact) Bayes Procedure

The approach outlined above amounts to finding all the records in the sample that are exactly like the new record to be classified in the sense that the predictor values are all the same. This was easy in the small examples presented above, where there was just one predictor.

When the number of predictors gets larger (even to a modest number like 20), many of the records to be classified will be without exact matches. This can be understood in the context of a model to predict voting on the basis of demographic variables. Even a sizable sample may not contain even a single match for a new record who is a male Hispanic with high income from the U.S. Midwest who voted in the last election, did not vote in the prior election, has three daughters and one son, and is divorced. And this is just eight variables, a small number for most data mining exercises. The addition of just a single new variable with five equally frequent categories reduces the probability of a match by a factor of 5.

Solution: Naive Bayes

In the naive Bayes solution, we no longer restrict the probability calculation to those records that match the record to be classified. Instead we use the entire dataset.

Returning to our original basic classification procedure outlined at the beginning of the chapter, we recall that this procedure for classifying a new record was:

1. Find all the other records just like it (i.e., where the predictor values are the same).
2. Determine what classes they all belong to and which class is more prevalent.
3. Assign that class to the new record.

The naive Bayes modification (for the basic classification procedure) is as follows:

1. For class 1, find the individual probabilities that each predictor value in the record to be classified $(x_1, \ldots, x_p)$ occurs in class 1.
2. Multiply these probabilities times each other, then times the proportion of records belonging to class 1.
3. Repeat steps 1 and 2 for all the classes.
4. Estimate a probability for class i by taking the value calculated in step 2 for class i and dividing it by the sum of such values for all classes.
5. Assign the record to the class with the highest probability value for this set of predictor values.

The naive Bayes formula to calculate the probability that a record with a given set of predictor values $x_1, \ldots, x_p$ belongs to class 1 (C_1) among m classes is as follows:

$$P_{nb}(C_1|x_1, \ldots x_p)$$
$$= \frac{P(C_1) \left[P(x_1|C_1)P(x_2|C_1)\cdots P(x_p|C_1)\right]}{P(C_1) \left[P(x_1|C_1)P(x_2|C_1)\cdots P(x_p|C_1)\right] + \cdots + P(C_m) \left[P(x_1|C_m)P(x_2|C_m)\cdots P(x_p|C_m)\right]}.$$

$$(8.3)$$

This is a somewhat formidable formula; see Example 2 for a simpler numerical version. In probability terms, we have made a simplifying assumption that the exact *conditional probability* (the probability of belonging to a given class, given a set of predictor values) is well approximated by the product of the *unconditional probabilities* that those predictor values occur in the given class, overall, times the probability that a record belongs to that class, divided by the product of the *unconditional probabilities* that those predictor values occur across all classes. If predictor values are independent of one another, this approximation is the same as the exact value. In practice, the procedure works quite well—primarily because what is usually needed is not a probability value for each record that is accurate in absolute terms but just a reasonably accurate *rank ordering*.

Note that if all we are interested in is a rank ordering, and the denominator remains the same for all classes, it is sufficient to concentrate only on the numerator. The disadvantage of this approach is that the probability values it yields, while ordered correctly, are not on the same scale as the exact values that the user would anticipate.

The above procedure is for the basic case where we seek maximum classification accuracy for all classes. In the case of the relatively *rare class of special interest*, the procedure is:

1. Establish a cutoff probability for the class of interest above which we consider that a record belongs to that class.

2. For the class of interest, compute the probability that each predictor value in the record to be classified $(x_1, \ldots, x_p)$ occurs in the training data.

3. Multiply these probabilities times each other, then times the proportion of records belonging to the class of interest.

4. Estimate the probability for the class of interest by taking the value calculated in step 3 for the class of interest and dividing it by the sum of the similar values for all classes.

5. If this value falls above the cutoff, assign the new record to the class of interest, otherwise not.

6. Adjust the cutoff value as needed, as a parameter of the model.

Example 2: Predicting Fraudulent Financial Reports, Two Predictors

Let us expand the financial reports example to two predictors, and, using a small subset of data, compare the complete (exact) Bayes calculations to the naive Bayes calculations.

Consider the 10 customers of the accounting firm listed in Table 8.2. For each company, we have information on whether it had prior legal trouble, whether it is a small or large company, and whether the financial report was found to be fraudulent or truthful. Using this information, we will calculate the conditional probability of fraud, given each of the four possible combinations $\{y, \text{small}\}, \{y, \text{large}\}, \{n, \text{small}\}, \{n, \text{large}\}$.

TABLE 8.2	INFORMATION ON 10 COMPANIES	
Prior Legal Trouble	**Company Size**	**Status**
Yes	Small	Truthful
No	Small	Truthful
No	Large	Truthful
No	Large	Truthful
No	Small	Truthful
No	Small	Truthful
Yes	Small	Fraudulent
Yes	Large	Fraudulent
No	Large	Fraudulent
Yes	Large	Fraudulent

Complete (Exact) Bayes Calculations   The four probabilities are computed as:

$$P(\text{fraudulent}|\text{PriorLegal} = y, \text{Size} = \text{small}) = 1/2 = 0.5.$$

$$P(\text{fraudulent}|\text{PriorLegal} = y, \text{Size} = \text{large}) = 2/2 = 1.$$

$$P(\text{fraudulent}|\text{PriorLegal} = n, \text{Size} = \text{small}) = 0/3 = 0.$$

$$P(\text{fraudulent}|\text{PriorLegal} = n, \text{Size} = \text{large}) = 1/3 = 0.33.$$

Naive Bayes Calculations   Now we compute the naive Bayes probabilities. For the conditional probability of fraudulent behaviors given {Prior Legal = y, Size = small}, the numerator is a multiplication of the proportion of {Prior Legal = y} instances among the fraudulent companies, times the proportion of {Size = small} instances among the fraudulent companies, times the proportion of fraudulent companies: $(3/4)(1/4)(4/10) = 0.075$. However, to get the actual probabilities, we must also compute the numerator for the conditional probability of truth given {Prior Legal = y, Size = small}: $(1/6)(4/6)(6/10) = 0.067$. The denominator is then the sum of these two conditional probabilities ($0.075 + 0.067 = 0.14$). The conditional probability of fraudulent behaviors given {Prior Legal = y, Size = small} is therefore $0.075/0.14 = 0.53$. In a similar fashion, we compute all four conditional probabilities:

$$P_{nb}(\text{fraudulent}|\text{PriorLegal} = y, \text{Size} = \text{small}) = \frac{(3/4)(1/4)(4/10)}{(3/4)(1/4)(4/10) + (1/6)(4/6)(6/10)}$$
$$= 0.53.$$

$$P_{nb}(\text{fraudulent}|\text{PriorLegal} = y, \text{Size} = \text{large}) = 0.87.$$

$$P_{nb}(\text{fraudulent}|\text{PriorLegal} = n, \text{Size} = \text{small}) = 0.07.$$

$$P_{nb}(\text{fraudulent}|\text{PriorLegal} = n, \text{Size} = \text{large}) = 0.31.$$

Note how close these naive Bayes probabilities are to the exact Bayes probabilities. Although they are not equal, both would lead to exactly the same classification for a cutoff of 0.5 (and many other values). It is often the case that the rank ordering of probabilities is even closer to the exact Bayes method than are the probabilities themselves, and for classification purposes it is the rank orderings that matter.

We now consider a larger numerical example, where information on flights is used to predict flight delays.

Example 3: Predicting Delayed Flights

Predicting flight delays can be useful to a variety of organizations: airport authorities, airlines, and aviation authorities. At times, joint task forces have been formed to address the problem. If such an organization were to provide ongoing real-time assistance with flight delays, it would benefit from some advance notice about flights that are likely to be delayed.

In this simplified illustration, we look at six predictors (see Table 8.3). The outcome of interest is whether or not the flight is delayed (*delayed* means arrived more than 15 minutes late). Our data consist of all flights from the Washington, D.C., area into the New York City area during January 2004. The percentage of delayed flights among these 2346 flights is 18%. The data were obtained from the Bureau of Transportation Statistics (available on the Web at www.transtats.bts.gov). The goal is to accurately predict whether or not a new flight (not in this dataset), will be delayed.

A record is a particular flight. The response is whether the flight was delayed, and thus it has two classes ($1 = Delayed$ and $0 = On\ time$). In addition, information is collected on the predictors listed in Table 8.3.

The data were first partitioned into training and validation sets (with a 60% : 40% ratio), and then a naive Bayes classifier was applied to the training set.

The top table in Figure 8.1 shows the ratios of delayed flights and on-time flights in the training set (called Prior Class Probabilities). The bottom

TABLE 8.3	DESCRIPTION OF VARIABLES FOR FLIGHT DELAY EXAMPLE
Day of Week	Coded as: 1 = Monday, 2 = Tuesday,. . ., 7 = Sunday
Departure Time	Broken down into 18 intervals between 6:00 AM and 10:00 PM
Origin	Three airport codes: DCA (Reagan National), IAD (Dulles), BWI (Baltimore–Washington Int'l)
Destination	Three airport codes: JFK (Kennedy), LGA (LaGuardia), EWR (Newark)
Carrier	Eight airline codes: CO (Continental), DH (Atlantic Coast), DL (Delta), MQ (American Eagle), OH (Comair), RU (Continental Express), UA (United), and US (USAirways)
Weather	Coded as 1 if there was a weather-related delay

Prior Class Probabilities

According to relative occurrences in training data		

Class	Prob.	
1	0.193792581	<-- Success Class
0	0.806207419	

Conditional Probabilities

Classes-->				
Input Variables	**1**		**0**	
	Value	Prob	Value	Prob
CARRIER	CO	0.06640625	CO	0.038497653
	DH	0.33984375	DH	0.243192488
	DL	0.109375	DL	0.2
	MQ	0.1796875	MQ	0.112676056
	OH	0.01171875	OH	0.017840376
	RU	0.21484375	RU	0.170892019
	UA	0.0078125	UA	0.016901408
	US	0.0703125	US	0.2
DAY_OF_WEEK	1	0.203125	1	0.128638498
	2	0.16015625	2	0.139906103
	3	0.12890625	3	0.152112676
	4	0.12890625	4	0.159624413
	5	0.1640625	5	0.181220657
	6	0.0703125	6	0.131455399
	7	0.14453125	7	0.107042254
DEP_TIME_BLK	0600-0659	0.03515625	0600-0659	0.061971831
	0700-0759	0.05078125	0700-0759	0.060093897
	0800-0859	0.0546875	0800-0859	0.071361502
	0900-0959	0.0234375	0900-0959	0.053521127
	1000-1059	0.01953125	1000-1059	0.057276995
	1100-1159	0.01953125	1100-1159	0.038497653
	1200-1259	0.0546875	1200-1259	0.062910798
	1300-1359	0.05078125	1300-1359	0.068544601
	1400-1459	0.15234375	1400-1459	0.110798122
	1500-1559	0.08203125	1500-1559	0.064788732
	1600-1659	0.07421875	1600-1659	0.078873239
	1700-1759	0.15625	1700-1759	0.094835681
	1800-1859	0.03125	1800-1859	0.043192488
	1900-1959	0.08984375	1900-1959	0.040375587
	2000-2059	0.01953125	2000-2059	0.030985915
	2100-2159	0.0859375	2100-2159	0.061971831
DEST	EWR	0.38671875	EWR	0.283568075
	JFK	0.1875	JFK	0.176525822
	LGA	0.42578125	LGA	0.539906103
ORIGIN	BWI	0.09375	BWI	0.068544601
	DCA	0.484375	DCA	0.635680751
	IAD	0.421875	IAD	0.295774648
Weather	0	0.92578125	0	1
	1	0.07421875	1	0

FIGURE 8.1   OUTPUT FROM NAIVE BAYES CLASSIFIER APPLIED TO FLIGHT DELAYS (TRAINING) DATA

table shows the conditional probabilities for each class, as a function of the predictor values. Note that the conditional probabilities in the output can be computed simply by using pivot tables in Excel, looking at the percentage of records in a cell relative to the entire class. This is illustrated in

TABLE 8.4	PIVOT TABLE OF DELAYED AND ON-TIME FLIGHTS BY DESTINATION AIRPORT (ROWS)		
	Delayed %	On Time %	Total %
EWR	38.67	28.36	30.36
JFK	18.75	17.65	17.87
LGA	42.58	53.99	51.78
Total	100.00	100.00	100.00

Table 8.2, which displays the percent of delayed (or on-time) flights by destination airport as a percentage of the total delayed (or on-time) flights.

Note that in this example there are no predictor values that were not represented in the training data except for on-time flights (Class = 0) when the
weather was bad (Weather = 1). When the weather was bad, all flights in the
training set were delayed.

To classify a new flight, we compute the probability that it will be delayed and
the probability that it will be on time. Recall that since both will have the same
denominator, we can just compare the numerators. Each numerator is computed
by multiplying all the conditional probabilities of the relevant predictor values
and, finally, multiplying by the proportion of that class [in this case $\hat{P}$(delayed) =
0.19]. For example, to classify a Delta flight from DCA to LGA between 10 and
11 AM on a Sunday with good weather, we compute the numerators:

$$\hat{P}(\text{delayed} \mid \text{Carrier} = \text{DL}, \text{DayofWeek} = 7, \text{DepartureTime} = 1000 - 1059,$$

$$\text{Destination} = \text{LGA}, \text{Origin} = \text{DCA}, \text{Weather} = 0)$$

$$\propto (0.11)(0.14)(0.020)(0.43)(0.48)(0.93)(0.19) = 0.000011$$

$$\hat{P}(\text{ontime} \mid \text{Carrier} = \text{DL}, \text{DayofWeek} = 7, \text{DepartureTime} = 1000 - 1059,$$

$$\text{Destination} = \text{LGA}, \text{Origin} = \text{DCA}, \text{Weather} = 0)$$

$$\propto (0.2)(0.11)(0.06)(0.54)(0.64)(1)(0.81) = 0.00034$$

The symbol $\propto$ means "is proportional to," reflecting the fact that this calculation deals only with the numerator in the naive Bayes formula (8.3).

It is, therefore, more likely that the flight will be on time. Note that a record
with such a combination of predictors does not exist in the training set, and
therefore we use the naive Bayes rather than the exact Bayes.

To compute the actual probability, we divide each of the numerators by their
sum:

$$\hat{P}(\text{delayed} \mid \text{Carrier} = \text{DL}, \text{DayofWeek} = 7, \text{DepartureTime} = 1000 - 1059,$$

$$\text{Destination} = \text{LGA}, \text{Origin} = \text{DCA}, \text{Weather} = 0)$$

$$= \frac{0.000011}{0.000011 + 0.00034} = 0.03$$

$$\hat{P}(\text{ontime} \mid \text{Carrier} = \text{DL}, \text{DayofWeek} = 7, \text{DepartureTime} = 1000 - 1059,$$

$$\text{Destination} = \text{LGA}, \text{Origin} = \text{DCA}, \text{Weather} = 0)$$

$$= \frac{0.00034}{0.000011 + 0.00034} = 0.97$$

Of course, we rely on software to compute these probabilities for any records of interest (in the training set, the validation set, or for scoring new data). Figure 8.2 shows the estimated probabilities and classifications for a sample of flights in the validation set.

Finally, to evaluate the performance of the naive Bayes classifier for our data, we use the classification matrix, lift charts, and all the measures that were described in Chapter 5. For our example, the classification matrices for the training and validation sets are shown in Figure 8.3. We see that the overall error level is around 18% for both the training and validation data. In comparison, a naive rule that would classify all 880 flights in the validation set as on time would have missed the 172 delayed flights, resulting in a 20% error level. In other words, the naive Bayes is only slightly less accurate. However, examining the lift chart (Figure 8.4) shows the strength of the naive Bayes in capturing the delayed flights well.

XLMiner : Naive Bayes - Classification of Validation Data

Data range ['Flight Delays.xls']'Data_Partition1'!$C$1340: $H$2219   Back to Navigator

Cut off Prob. Val. for Success (Updatable)   0.5   (Updating the value here will NOT update value in summary report)

Rowld.	Predicted Class	Actual Class	Prob. for 1 (success)	CARRIERF	DAY_O WEEK	DEP_ TI ME _BLK	DEST	ORIGIN	Weather
2	0	0	0.160552079	DH	4	1600- 1659	JF K	DCA	0
3	0	0	0.197147877	DH	4	1200- 1259	LGA	IAD	0
7	0	0	0.248536067	DH	4	1200- 1259	JF K	IAD	0
8	0	0	0.263631618	DH	4	1600- 1659	JF K	IAD	0
11	0	0	0.281467602	DH	4	2100- 2159	LGA	IAD	0
13	0	0	0.025209812	DL	4	0900- 0959	LGA	DCA	0
14	0	0	0.048830719	DL	4	1200- 1259	LGA	DCA	0
15	0	0	0.07510312	DL	4	1400- 1459	LGA	DCA	0
16	0	0	0.088673655	DL	4	1700- 1759	LGA	DCA	0
22	0	0	0.113149152	MQ	4	1300- 1359	LGA	DCA	0
24	0	0	0.179013723	MQ	4	1500- 1559	LGA	DCA	0
25	0	0	0.277045255	MQ	4	1900- 1959	LGA	DCA	0
28	0	0	0.018897189	US	4	1100- 1159	LGA	DCA	0
33	0	0	0.366861013	RU	4	1400- 1459	EW R	BWI	0
34	0	0	0.409792076	RU	4	1700- 1759	EW R	BWI	0
40	0	0	0.44593539	DH	4	1700- 1759	EW R	IAD	0
42	0	0	0.403842858	DH	4	2100- 2159	EW R	IAD	0
46	0	0	0.343153176	RU	4	1900- 1959	EW R	DCA	0
47	0	0	0.244033602	RU	4	1400- 1459	EW R	DCA	0
50	0	0	0.18094682	RU	4	1600- 1659	EW R	DCA	0
57	0	1	0.097462126	DH	5	1000- 1059	LGA	IAD	0

FIGURE 8.2   ESTIMATED PROBABILITY OF DELAY FOR A SAMPLE OF THE VALIDATION SET

Training Data scoring - Summary Report

Cut off Prob.Val. for Success (Updatable)	0.5

Classification Confusion Matrix		
	Predicted Class	
Actual Class	1	0
1	43	213
0	35	1030

Error Report			
Class	**# Cases**	**# Errors**	**% Error**
1	256	213	83.20
0	1065	35	3.29
Overall	1321	248	18.77

Validation Data scoring - Summary Report

Cut off Prob.Val. for Success (Updatable)	0.5

Classification Confusion Matrix		
	Predicted Class	
Actual Class	1	0
1	30	142
0	15	693

Error Report			
Class	**# Cases**	**# Errors**	**% Error**
1	172	142	82.56
0	708	15	2.12
Overall	880	157	17.84

FIGURE 8.3   CLASSIFICATION MATRICES FOR FLIGHT DELAYS USING A NAIVE BAYES CLASSIFIER

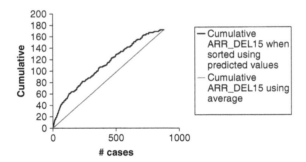

FIGURE 8.4   LIFT CHART OF NAIVE BAYES CLASSIFIER APPLIED TO FLIGHT DELAY DATA

8.3 ADVANTAGES AND SHORTCOMINGS OF THE NAIVE BAYES CLASSIFIER

The naive Bayes classifier's beauty is in its simplicity, computational efficiency, and good classification performance. In fact, it often outperforms more sophisticated classifiers even when the underlying assumption of (conditionally) independent predictors is far from true. This advantage is especially pronounced when the number of predictors is very large.

Three issues should be kept in mind, however. First, the naive Bayes classifier requires a very large number of records to obtain good results. Second, where a predictor category is not present in the training data, naive Bayes assumes that a new record with that category of the predictor has zero probability. This can be a problem if this rare predictor value is important. For example, consider the target variable *bought high-value life insurance* and the predictor category *own yacht*. If the training data have no records with *owns yacht* = 1, for any new records where *owns yacht* = 1, naive Bayes will assign a probability of 0 to the target variable *bought high-value life insurance*. With no training records with *owns yacht* = 1, of course, no data mining technique will be able to incorporate this potentially important variable into the classification model—it will be ignored. With naive Bayes, however, the absence of this predictor actively "out votes" any other information in the record to assign a 0 to the target value (when, in this case, it has a relatively good chance of being a 1). The presence of a large training set (and judicious binning of continuous variables, if required) helps mitigate this effect.

Finally, good performance is obtained when the goal is *classification* or *ranking* of records. However, when the goal is to estimate the *probability of class membership*, this method provides very biased results. For this reason the naive Bayes method is rarely used in credit scoring (Larsen, 2005).

SPAM FILTERING

Filtering spam is perhaps the most widely familiar application of data mining. Spam filtering, which is based in large part on natural language vocabulary, is a natural fit for a naive Bayesian classifier, which uses exclusively categorical variables. Most spam filters are based on this method, which works as follows:

1. Humans review a large number of e-mails, classify them as "spam" or "not spam," and from these select an equal (also large) number of spam e-mails and non-spam emails. This is the training data.

2. These e-mails will contain thousands of words; for each word compute the frequency with which it occurs in the spam class and the frequency with which it occurs in the non-spam class. Convert these frequencies into estimated probabilities (i.e., if the word "free" occurs in 500 out of 1000 spam e-mails, and only 100 out of 1000 non-spam e-mails, the probability that a spam email will contain the word "free" is 0.5, and the probability that a non-spam e-mail will contain the word "free" is 0.1).

3. If the only word in a new message that needs to be classified as spam or not spam is "free," we would classify the message as spam since the Bayesian posterior probability is 0.5/(0.5 + 01) or 5/6 that, given the appearance of "free," the message is spam.

4. Of course, we will have many more words to consider. For each such word, the probabilities described in step 2 are calculated and multiplied together, and formula (8.3) applied to determine the naive Bayes probability of belonging to the classes. In the simple version, class membership (spam or not spam) is determined by the higher probability.

5. In a more flexible interpretation, the "spam" probability is treated as a score for which the operator can establish (and change) a cutoff threshold—anything above that level is classified as spam.

6. Users have the option of building a personalized training database by classifying incoming messages as spam or non-spam, and adding them to the training database. One person's spam may be another person's substance.

It is clear that, even with the "naive" simplification, this is an enormous computational burden. Spam filters now typically operate at two levels—at servers (intercepting some spam that never makes it to your computer) and on individual computers (where you have the option of reviewing it). Spammers have also found ways to "poison" the vocabulary-based Bayesian approach, by including sequences of randomly selected irrelevant words. Since these words are randomly selected, they are unlikely to be systematically more prevalent in spam than in non-spam, and they dilute the effect of key spam terms such as "Viagra" and "free." For this reason, sophisticated spam classifiers also include variables based on elements other than vocabulary, such as the number of links in the message, the vocabulary in the subject line, determination of whether the "From:" e-mail address is the real originator (antispoofing), use of HTML and images, and origination at a dynamic or static IP address (the latter are more expensive and cannot be set up quickly).

PROBLEMS

8.1 **Personal Loan Acceptance.** The file UniversalBank.xls contains data on 5000 customers of Universal Bank. The data include customer demographic information (age, income, etc.), the customer's relationship with the bank (mortgage, securities account, etc.), and the customer response to the last personal loan campaign (Personal Loan). Among these 5000 customers, only 480 ($= 9.6\%$) accepted the personal loan that was offered to them in the earlier campaign. In this exercise we focus on two predictors: Online (whether or not the customer is an active user of online banking services) and Credit Card (abbreviated CC below) (does the customer hold a credit card issued by the bank), and the outcome Personal Loan (abbreviated Loan below).

Partition the data into training (60%) and validation (40%) sets.

a. Create a pivot table for the training data with Online as a column variable, CC as a row variable, and Loan as a secondary row variable. The values inside the cells should convey the count (how many records are in that cell).

b. Consider the task of classifying a customer that owns a bank credit card and is actively using online banking services. Looking at the pivot table, what is the probability that this customer will accept the loan offer? [This is the probability of loan acceptance (Loan $= 1$) conditional on having a bank credit card (CC $= 1$) and being an active user of online banking services (Online $= 1$).]

c. Create two separate pivot tables for the training data. One will have Loan (rows) as a function of Online (columns) and the other will have Loan (rows) as a function of CC.

d. Compute the following quantities [$P(A|B)$ means "the probability of A given B"]:

   i. $P(CC = 1|Loan = 1)$ (the proportion of credit card holders among the loan acceptors)

   ii. $P(Online = 1|Loan = 1)$

   iii. $P(Loan = 1)$ (the proportion of loan acceptors)

   iv. $P(CC = 1|Loan = 0)$

   v. $P(Online = 1|Loan = 0)$

   vi. $P(Loan = 0)$

e. Use the quantities computed above to compute the naive Bayes probability $P(Loan = 1|CC = 1, Online = 1)$.

f. Compare this value with the one obtained from the crossed pivot table in (b). Which is a more accurate estimate?

g. In XLMiner, run naive Bayes on the data. Examine the "Conditional probabilities" table, and find the entry that corresponds to $P(Loan = 1|CC = 1, Online = 1)$. Compare this to the number you obtained in (e).

8.2 **Automobile Accidents.** The file Accidents.xls contains information on 42,183 actual automobile accidents in 2001 in the United States that involved one of three levels of injury: NO INJURY, INJURY, or FATALITY. For each accident, additional information is recorded, such as day of week, weather conditions, and road type. A firm might be interested in developing a system for quickly classifying the severity of an accident based on initial reports and associated data in the system (some of which rely on GPS-assisted reporting).

Our goal here is to predict whether an accident just reported will involve an injury (MAX_SEV_IR = 1 or 2) or will not (MAX_SEV_IR = 0). For this purpose, create a dummy variable called INJURY that takes the value "yes" if MAX_SEV_IR = 1 or 2, and otherwise "no."

a. Using the information in this dataset, if an accident has just been reported and no further information is available, what should the prediction be? (INJURY = Yes or No?) Why?

b. Select the first 12 records in the dataset and look only at the response (INJURY) and the two predictors WEATHER_R and TRAF_CON_R.

    i. Create a pivot table that examines INJURY as a function of the 2 predictors for these 12 records. Use all 3 variables in the pivot table as rows/columns, and use counts for the cells.

    ii. Compute the exact Bayes conditional probabilities of an injury (INJURY = Yes) given the six possible combinations of the predictors.

    iii. Classify the 12 accidents using these probabilities and a cutoff of 0.5.

    iv. Compute manually the naive Bayes conditional probability of an injury given WEATHER_R = 1 and TRAF_CON_R = 1.

    v. Run a naive Bayes classifier on the 12 records and 2 predictors using XLMiner. Check *detailed report* to obtain probabilities and classifications for all 12 records. Compare this to the exact Bayes classification. Are the resulting classifications equivalent? Is the ranking (= ordering) of observations equivalent?

c. Let us now return to the entire dataset. Partition the data into training/validation sets (use XLMiner's "automatic" option for partitioning percentages).

    i. Assuming that no information or initial reports about the accident itself are available at the time of prediction (only location characteristics, weather conditions, etc.), which predictors can we include in the analysis? (Use the Data_Codes sheet.)

    ii. Run a naive Bayes classifier on the complete training set with the relevant predictors (and INJURY as the response). Note that all predictors are categorical. Show the classification matrix.

    iii. What is the overall error for the validation set?

    iv. What is the percent improvement relative to the naive rule (using the validation set)?

    v. Examine the conditional probabilities output. Why do we get a probability of zero for P(INJURY = No | SPD_LIM = 5)?

## CHAPTER 9

Classification and Regression Trees

This chapter describes a flexible data-driven method that can be used for both classification (called *classification tree*) and prediction (called *regression tree*). Among the data-driven methods, trees are the most transparent and easy to interpret. Trees are based on separating observations into subgroups by creating splits on predictors. These splits create logical rules that are transparent and easily understandable, for examples "If Age < 55 AND Education > 12 THEN class = 1." The resulting subgroups should be more homogeneous in terms of the outcome variable, thereby creating useful prediction or classification rules. We discuss the two key ideas underlying trees: recursive partitioning (for constructing the tree) and pruning (for cutting the tree back). In the context of tree construction, we also describe a few metrics of homogeneity that are popular in tree algorithms, for determining the homogeneity of the resulting subgroups of observations. We explain that pruning is a useful strategy for avoiding overfitting and show how it is done. We also describe alternative strategies for avoiding overfitting. As with other data-driven methods, trees require large amounts of data. However, once constructed, they are computationally cheap to deploy even on large samples. They also have other advantages such as being highly automated, robust to outliers, and can handle missing values. Finally, we describe how trees can be used for dimension reduction.

9.1 INTRODUCTION

If one had to choose a classification technique that performs well across a wide range of situations without requiring much effort from the analyst while being

Data Mining for Business Intelligence, By Galit Shmueli, Nitin R. Patel, and Peter C. Bruce

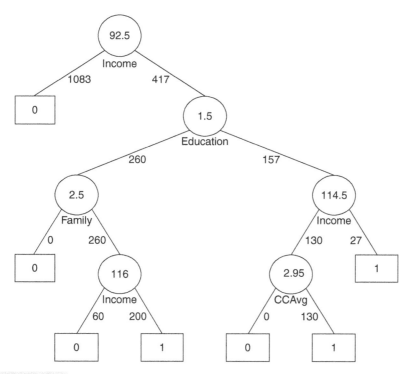

FIGURE 9.1 **BEST PRUNED TREE OBTAINED BY FITTING A FULL TREE TO THE TRAINING DATA AND PRUNING IT USING THE VALIDATION DATA**

readily understandable by the consumer of the analysis, a strong contender would be the tree methodology developed by Breiman et al. (1984). We discuss this classification procedure first; then in later sections we show how the procedure can be extended to prediction of a numerical outcome. The program that Breiman et al. created to implement these procedures was called CART (classification and regression trees). A related procedure is called C4.5.

What is a classification tree? Figure 9.1 describes a tree for classifying bank customers who receive a loan offer as either acceptors or nonacceptors, as a function of information such as their income, education level, and average credit card expenditure. One of the reasons that tree classifiers are very popular is that they provide easily understandable classification rules (at least if the trees are not too large). Consider the tree in the example. The square *terminal nodes* are marked with 0 or 1 corresponding to a nonacceptor (0) or acceptor (1). The values in the circle nodes give the splitting value on a predictor. This tree can easily be translated into a set of rules for classifying a bank customer. For example, the middle left square node in this tree gives us the following rule:

IF(Income > 92.5) AND (Education < 1.5) AND (Family ≤ 2.5)
THEN Class = 0 (nonacceptor).

In the following we show how trees are constructed and evaluated.

9.2 CLASSIFICATION TREES

There are two key ideas underlying classification trees. The first is the idea of recursive partitioning of the space of the predictor variables. The second is the idea of pruning using validation data. In the next few sections we describe recursive partitioning and in subsequent sections explain the pruning methodology.

Recursive Partitioning

Let us denote the dependent (response) variable by y and the predictor variables by $x_1, x_2, x_3, \ldots, x_p$. In classification, the outcome variable will be a categorical variable. Recursive partitioning divides up the p-dimensional space of the x variables into nonoverlapping multidimensional rectangles. The X variables here are considered to be continuous, binary, or ordinal. This division is accomplished recursively (i.e., operating on the results of prior divisions). First, one of the variables is selected, say x_i, and a value of x_i, say s_i, is chosen to split the p-dimensional space into two parts: one part that contains all the points with $x_i \le s_i$ and the other with all the points with $x_i > s_i$. Then one of these two parts is divided in a similar manner by choosing a variable again (it could be x_i or another variable) and a split value for the variable. This results in three (multidimensional) rectangular regions. This process is continued so that we get smaller and smaller rectangular regions. The idea is to divide the entire x space up into rectangles such that each rectangle is as homogeneous or "pure" as possible. By *pure* we mean containing points that belong to just one class. (Of course, this is not always possible, as there may be points that belong to different classes but have exactly the same values for every one of the predictor variables.)

Let us illustrate recursive partitioning with an example.

Example 1: Riding Mowers

We again use the riding-mower example presented in Chapter 7. A riding-mower manufacturer would like to find a way of classifying families in a city into those likely to purchase a riding mower and those not likely to buy one. A pilot random sample of 12 owners and 12 nonowners in the city is undertaken. The data are shown and plotted in Table 9.1 and Figure 9.2.

If we apply the classification tree procedure to these data, the procedure will choose Lot Size for the first split with a splitting value of 19. The (x_1, x_2) space is now divided into two rectangles, one with Lot Size$\le$ 19 and the other with Lot Size$>$ 19. This is illustrated in Figure 9.3.

Notice how the split has created two rectangles, each of which is much more homogeneous than the rectangle before the split. The upper rectangle contains points that are mostly owners (nine owners and three nonowners) and the lower rectangle contains mostly nonowners (nine nonowners and three owners).

| TABLE 9.1 | LOT SIZE, INCOME, AND OWNERSHIP OF A RIDING MOWER FOR 24 HOUSEHOLDS |

Household Number	Income ($000s)	Lot Size (000s ft^2)	Ownership of Riding Mower
1	60	18.4	Owner
2	85.5	16.8	Owner
3	64.8	21.6	Owner
4	61.5	20.8	Owner
5	87	23.6	Owner
6	110.1	19.2	Owner
7	108	17.6	Owner
8	82.8	22.4	Owner
9	69	20	Owner
10	93	20.8	Owner
11	51	22	Owner
12	81	20	Owner
13	75	19.6	Nonowner
14	52.8	20.8	Nonowner
15	64.8	17.2	Nonowner
16	43.2	20.4	Nonowner
17	84	17.6	Nonowner
18	49.2	17.6	Nonowner
19	59.4	16	Nonowner
20	66	18.4	Nonowner
21	47.4	16.4	Nonowner
22	33	18.8	Nonowner
23	51	14	Nonowner
24	63	14.8	Nonowner

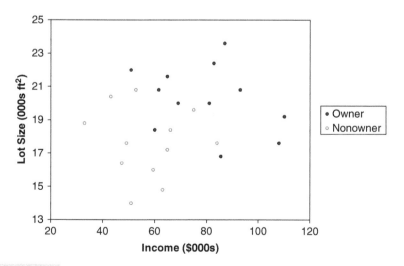

| FIGURE 9.2 | SCATTERPLOT OF LOT SIZE VS. INCOME FOR 24 OWNERS AND NONOWNERS OF RIDING MOWERS |

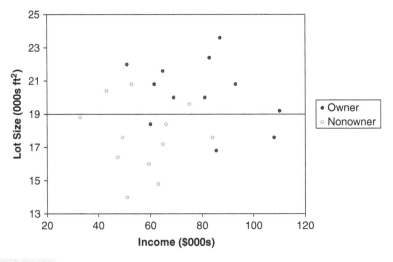

FIGURE 9.3   SPLITTING THE 24 OBSERVATIONS BY LOT SIZE VALUE OF 19

How was this particular split selected? The algorithm examined each variable (in this case, Income and Lot Size) and all possible split values for each variable to find the best split. What are the possible split values for a variable? They are simply the midpoints between pairs of consecutive values for the variable. The possible split points for Income are {38.1, 45.3, 50.1, ..., 109.5} and those for Lot Size are {14.4, 15.4, 16.2, ..., 23}. These split points are ranked according to how much they reduce impurity (heterogeneity) in the resulting rectangle. A pure rectangle is one that is composed of a single class (e.g., owners). The reduction in impurity is defined as overall impurity before the split minus the sum of the impurities for the two rectangles that result from a split.

Categorical Predictors   The above description used numerical predictors. However, categorical predictors can also be used in the recursive partitioning context. To handle categorical predictors, the split choices for a categorical predictor are all ways in which the set of categories can be divided into two subsets. For example, a categorical variable with four categories, say {a,b,c,d}, can be split in seven ways into two subsets: {a} and {b,c,d}; {b} and {a,c,d}; {c} and {a,b,d}; {d} and {a,b,c}; {a,b} and {c,d}; {a,c} and {b,d}; and {a,d} and {b,c}. When the number of categories is large, the number of splits becomes very large. XLMiner supports only binary categorical variables (coded as numbers). If you have a categorical predictor that takes more than two values, you will need to replace the variable with several dummy variables, each of which is binary in a manner that is identical to the use of dummy variables in regression.[1]

---

[1] This is a difference between CART and C4.5; the former performs only binary splits, leading to binary trees, whereas the latter performs splits that are as large as the number of categories, leading to "bushlike" structures.

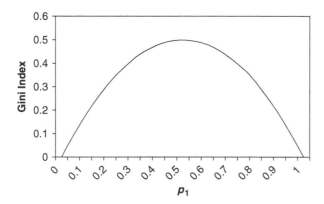

FIGURE 9.4   **VALUES OF THE GINI INDEX FOR A TWO-CLASS CASE AS A FUNCTION OF THE PROPORTION OF OBSERVATIONS IN CLASS 1 (P_1)**

9.3 MEASURES OF IMPURITY

There are a number of ways to measure impurity. The two most popular measures are the *Gini index* and an *entropy measure*. We describe both next. Denote the *m* classes of the response variable by $k = 1, 2, \ldots, m$.

The Gini impurity index for a rectangle *A* is defined by

$$I(A) = 1 - \sum_{k=1}^{m} p_k^2,$$

where p_k is the proportion of observations in rectangle *A* that belong to class *k*. This measure takes values between 0 (if all the observations belong to the same class) and $(m - 1)/m$ (when all *m* classes are equally represented). Figure 9.4 shows the values of the Gini index for a two-class case as a function of p_k. It can be seen that the impurity measure is at its peak when $p_k = 0.5$ (i.e., when the rectangle contains 50% of each of the two classes).[2]

A second impurity measure is the entropy measure. The entropy for a rectangle *A* is defined by

$$\text{Entropy}(A) = - \sum_{k=1}^{m} p_k \log_2(p_k)$$

[to compute $\log_2(x)$ in Excel, use the function $= \log(x, 2)$]. This measure ranges between 0 (most pure, all observations belong to the same class) and $\log_2(m)$ (when all *m* classes are represented equally). In the two-class case, the entropy measure is maximized (like the Gini index) at $p_k = 0.5$.

---

[2] XLMiner uses a variant of the Gini index called the *delta splitting rule*; for details, see XLMiner documentation.

Let us compute the impurity in the riding-mower example before and after the first split (using Lot Size with the value of 19). The unsplit dataset contains 12 owners and 12 nonowners. This is a two-class case with an equal number of observations from each class. Both impurity measures are therefore at their maximum value: Gini = 0.5 and entropy = $\log_2(2) = 1$. After the split, the upper rectangle contains nine owners and three nonowners. The impurity measures for this rectangle are Gini = $1 - 0.25^2 - 0.75^2 = 0.375$ and entropy = $-0.25 \log_2(0.25) - 0.75 \log_2(0.75) = 0.811$. The lower rectangle contains three owners and nine nonowners. Since both impurity measures are symmetric, they obtain the same values as for the upper rectangle.

The combined impurity of the two rectangles that were created by the split is a weighted average of the two impurity measures, weighted by the number of observations in each (in this case we ended up with 12 observations in each rectangle, but in general the number of observations need not be equal): Gini = $(12/24)(0.375) + (12/24)(0.375) = 0.375$ and entropy = $(12/24)(0.811) + (12/24)(0.811) = 0.811$. Thus, the Gini impurity index decreased from 0.5 before the split to 0.375 after the split. Similarly, the entropy impurity measure decreased from 1 before the split to 0.811 after the split.

By comparing the reduction in impurity across all possible splits in all possible predictors, the next split is chosen. If we continue splitting the mower data, the next split is on the Income variable at the value 84.75. Figure 9.5 shows that once again the tree procedure has astutely chosen to split a rectangle to increase the purity of the resulting rectangles. The left lower rectangle, which contains data points with Income ≤ 84.75 and Lot Size ≤ 19, has all points that are nonowners (with one exception); whereas the right lower rectangle,

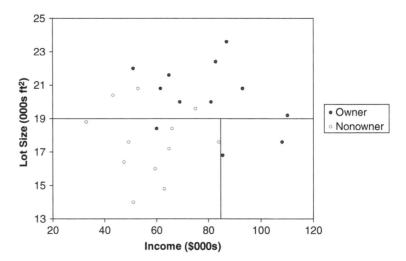

FIGURE 9.5   SPLITTING THE 24 OBSERVATIONS BY LOT SIZE VALUE OF $19K, AND THEN INCOME VALUE OF $84.75K

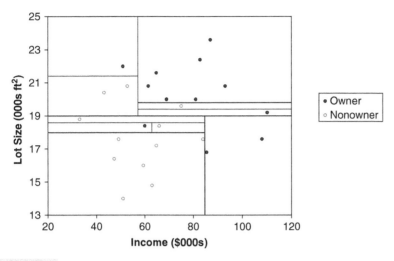

FIGURE 9.6   FINAL STAGE OF RECURSIVE PARTITIONING; EACH RECTANGLE CONSISTING OF A SINGLE CLASS (OWNERS OR NONOWNERS)

which contains data points with Income > 84.75 and Lot Size $\leq$ 19, consists exclusively of owners. We can see how the recursive partitioning is refining the set of constituent rectangles to become purer as the algorithm proceeds. The final stage of the recursive partitioning is shown in Figure 9.6.

Notice that each rectangle is now pure: It contains data points from just one of the two classes.

The reason the method is called a *classification tree algorithm* is that each split can be depicted as a split of a node into two successor nodes. The first split is shown as a branching of the root node of a tree in Figure 9.7. The tree representing the first three splits is shown in Figure 9.8. The full-grown tree is shown in Figure 9.9.

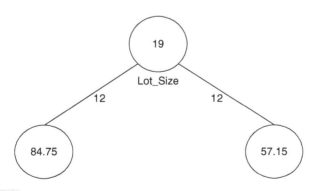

FIGURE 9.7   TREE REPRESENTATION OF FIRST SPLIT (CORRESPONDS TO FIGURE 9.3)

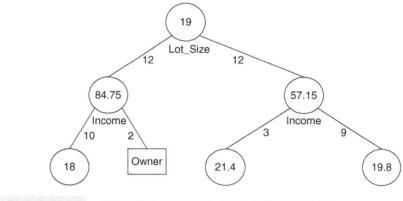

FIGURE 9.8   TREE REPRESENTATION OF FIRST THREE SPLITS

Tree Structure

We represent the nodes that have successors by circles. The numbers inside the circles are the splitting values and the name of the variable chosen for splitting at that node is shown below the node. The numbers on the left fork at a decision node are the number of records in the decision node that had values less than or equal to the splitting value, and the numbers on the right fork show the number

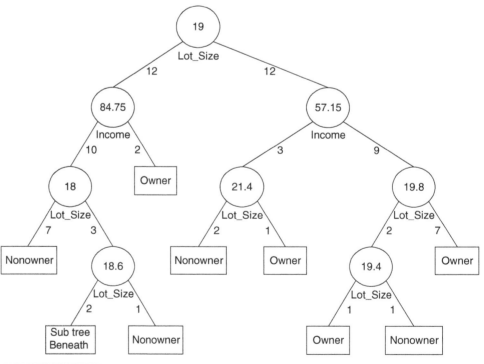

FIGURE 9.9   TREE REPRESENTATION AFTER ALL SPLITS (CORRESPONDS TO FIGURE 9.6)

that had a greater value. These are called *decision nodes* because if we were to use a tree to classify a new observation for which we knew only the values of the predictor variables, we would "drop" the observation down the tree in such a way that at each decision node the appropriate branch is taken until we get to a node that has no successors. Such terminal nodes are called the *leaves* of the tree. Each leaf node is depicted with a rectangle rather than a circle, and corresponds to one of the final rectangles into which the *x* space is partitioned.

Classifying a New Observation

To classify a new observation, it is "dropped" down the tree. When it has dropped all the way down to a terminal leaf, we can assign its class simply by taking a "vote" of all the training data that belonged to the leaf when the tree was grown. The class with the highest vote is assigned to the new observation. For instance, a new observation reaching the rightmost leaf node in Figure 9.9, which has a majority of observations that belong to the owner class, would be classified as "owner." Alternatively, if a single class is of interest, the algorithm counts the number of "votes" for this class, converts it to a proportion (estimated probability), then compares it to a user-specified cutoff value. See Chapter 5 for further discussion of the use of a cutoff value in classification, for cases where a single class is of interest.

In a binary classification situation (typically, with a *success* class that is relatively rare and of particular interest), we can also establish a lower cutoff to better capture those rare successes (at the cost of lumping in more nonsuccesses as successes). With a lower cutoff, the votes for the *success* class only need attain that lower cutoff level for the entire leaf to be classified as a *success*. The cutoff therefore determines the proportion of votes needed for determining the leaf class. It is useful to note that the type of trees grown by CART (called *binary trees*) have the property that the number of leaf nodes is exactly one more than the number of decision nodes.

9.4   EVALUATING THE PERFORMANCE OF A CLASSIFICATION TREE

To assess the accuracy of a tree in classifying new cases, we use the tools and criteria that were discussed in Chapter 5. We start by partitioning the data into training and validation sets. The training set is used to grow the tree, and the validation set is used to assess its performance. In the next section we discuss an important step in constructing trees that involves using the validation data. In that case, a third set of test data is preferable for assessing the accuracy of the final tree.

Each observation in the validation (or test) data is dropped down the tree and classified according to the leaf node it reaches. These predicted classes can then be compared to the actual memberships via a classification matrix. When a particular class is of interest, a lift chart is useful for assessing the model's ability to capture those members. We use the following example to illustrate this.

Example 2: Acceptance of Personal Loan

Universal Bank is a relatively young bank that is growing rapidly in terms of overall customer acquisition. The majority of these customers are liability customers with varying sizes of relationship with the bank. The customer base of asset customers is quite small, and the bank is interested in growing this base rapidly to bring in more loan business. In particular, it wants to explore ways of converting its liability customers to personal loan customers.

A campaign the bank ran for liability customers showed a healthy conversion rate of over 9% successes. This has encouraged the retail marketing department to devise smarter campaigns with better target marketing. The goal of our analysis is to model the previous campaign's customer behavior to analyze what combination of factors make a customer more likely to accept a personal loan. This will serve as the basis for the design of a new campaign.

The bank's dataset includes data on 5000 customers. The data include customer demographic information (age, income, etc.), customer response to the last personal loan campaign (Personal Loan), and the customer's relationship with the bank (mortgage, securities account, etc.). Among these 5000 customers, only 480 (= 9.6%) accepted the personal loan that was offered to them in the earlier campaign. Table 9.2 contains a sample of the bank's customer database for 20 customers, to illustrate the structure of the data.

After randomly partitioning the data into training (2500 observations), validation (1500 observations), and test (1000 observations) sets, we use the training data to construct a full-grown tree. The first four levels of the tree are shown in Figure 9.10, and the complete results are given in a form of a table in Figure 9.11.

Even with just four levels, it is difficult to see the complete picture. A look at the top tree node or the first row of the table reveals that the first predictor that is chosen to split the data is Income, with a value of 92.5 ($000s).

Since the full-grown tree leads to completely pure terminal leaves, it is 100% accurate in classifying the training data. This can be seen in Figure 9.12. In contrast, the classification matrix for the validation and test data (which were not used to construct the full-grown tree) show lower classification accuracy. The main reason is that the full-grown tree overfits the training data (to complete accuracy!). This motivates the next section, where we describe ways to avoid overfitting by either stopping the growth of the tree before it is fully grown or by pruning the full-grown tree.

TABLE 9.2 SAMPLE OF DATA FOR 20 CUSTOMERS OF UNIVERSAL BANK

ID	Age	Professional Experience	Income	Family Size	CC Avg	Education	Mortgage	Personal Loan	Securities Account	CD Account	Online Banking	Credit Card
1	25	1	49	4	1.60	UG	0	No	Yes	No	No	No
2	45	19	34	3	1.50	UG	0	No	Yes	No	No	No
3	39	15	11	1	1.00	UG	0	No	No	No	No	No
4	35	9	100	1	2.70	Grad	0	No	No	No	No	No
5	35	8	45	4	1.00	Grad	0	No	No	No	No	Yes
6	37	13	29	4	0.40	Grad	155	No	No	No	Yes	No
7	53	27	72	2	1.50	Grad	0	No	No	No	Yes	No
8	50	24	22	1	0.30	Prof	0	No	No	No	No	Yes
9	35	10	81	3	0.60	Grad	104	No	No	No	Yes	No
10	34	9	180	1	8.90	Prof	0	Yes	No	No	No	No
11	65	39	105	4	2.40	Prof	0	No	No	No	No	No
12	29	5	45	3	0.10	Grad	0	No	No	No	Yes	No
13	48	23	114	2	3.80	Prof	0	No	Yes	No	No	No
14	59	32	40	4	2.50	Grad	0	No	No	No	Yes	No
15	67	41	112	1	2.00	UG	0	No	Yes	No	Yes	No
16	60	30	22	1	1.50	Prof	0	No	No	No	Yes	Yes
17	38	14	130	4	4.70	Prof	134	Yes	No	No	No	No
18	42	18	81	4	2.40	UG	0	No	No	No	No	No
19	46	21	193	2	8.10	Prof	0	Yes	No	No	No	No
20	55	28	21	1	0.50	Grad	0	No	Yes	No	No	Yes

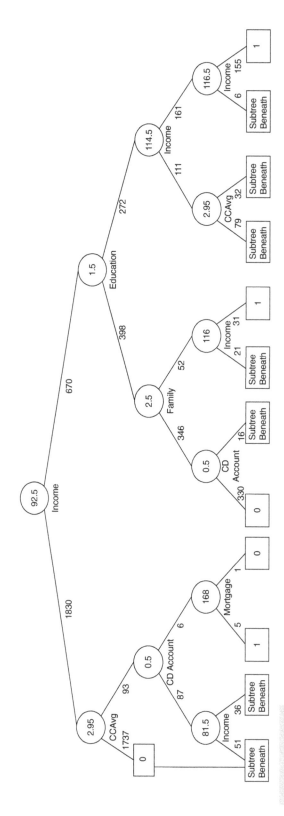

FIGURE 9.10   FIRST FOUR LEVELS OF THE FULL-GROWN TREE FOR THE LOAN ACCEPTANCE DATA USING THE TRAINING SET (2500 OBSERVATIONS)

#Decision Nodes	41		#Terminal Nodes	42					

Level	NodeID	ParentID	SplitVar	SplitValue	Cases	LeftChild	RightChild	Class	Node Type
0	0	N/A	Income	92.5	2500	1	2	0	Decision
1	1	0	CCAvg	2.95	1830	3	4	0	Decision
1	2	0	Education	1.5	670	5	6	0	Decision
2	3	1	N/A	N/A	1737	N/A	N/A	0	Terminal
2	4	1	CD Account	0.5	93	7	8	0	Decision
2	5	2	Family	2.5	398	9	10	0	Decision
2	6	2	Income	114.5	272	11	12	1	Decision
3	7	4	Income	81.5	87	13	14	0	Decision
3	8	4	Mortgage	168	6	15	16	1	Decision
3	9	5	CD Account	0.5	346	17	18	0	Decision
3	10	5	Income	116	52	19	20	1	Decision
3	11	6	CCAvg	2.95	111	21	22	0	Decision
3	12	6	Income	116.5	161	23	24	1	Decision
4	13	7	Age	28	51	25	26	0	Decision
4	14	7	CCAvg	3.75	36	27	28	0	Decision
4	15	8	N/A	N/A	5	N/A	N/A	1	Terminal
4	16	8	N/A	N/A	1	N/A	N/A	0	Terminal
4	17	9	N/A	N/A	330	N/A	N/A	0	Terminal
4	18	9	Mortgage	350.5	16	29	30	0	Decision
4	19	10	CCAvg	1.7	21	31	32	0	Decision
4	20	10	N/A	N/A	31	N/A	N/A	1	Terminal
4	21	11	Income	106.5	79	33	34	0	Decision
4	22	11	EducProf	0.5	32	35	36	1	Decision
4	23	12	CCAvg	1.1	6	37	38	1	Decision
4	24	12	N/A	N/A	155	N/A	N/A	1	Terminal
5	25	13	N/A	N/A	1	N/A	N/A	1	Terminal
5	26	13	N/A	N/A	50	N/A	N/A	0	Terminal
5	27	14	CCAvg	3.35	16	39	40	0	Decision
5	28	14	Mortgage	93.5	20	41	42	0	Decision
5	29	18	N/A	N/A	15	N/A	N/A	0	Terminal
5	30	18	N/A	N/A	1	N/A	N/A	1	Terminal
5	31	19	N/A	N/A	13	N/A	N/A	0	Terminal
5	32	19	Income	109.5	8	43	44	0	Decision
5	33	21	N/A	N/A	49	N/A	N/A	0	Terminal
5	34	21	CCAvg	1.75	30	45	46	0	Decision
5	35	22	Age	60	17	47	48	1	Decision
5	36	22	CCAvg	3.7	15	49	50	0	Decision
5	37	23	Online	0.5	2	51	52	1	Decision
5	38	23	N/A	N/A	4	N/A	N/A	1	Terminal
6	39	27	N/A	N/A	7	N/A	N/A	0	Terminal
6	40	27	CCAvg	3.65	9	53	54	1	Decision
6	41	28	N/A	N/A	16	N/A	N/A	0	Terminal
6	42	28	Mortgage	104.5	4	55	56	0	Decision
6	43	32	Experience	6.5	4	57	58	1	Decision
6	44	32	N/A	N/A	4	N/A	N/A	0	Terminal
6	45	34	Family	1.5	13	59	60	0	Decision
6	46	34	ZIP Code	94206.5	17	61	62	0	Decision
6	47	35	N/A	N/A	13	N/A	N/A	1	Terminal
6	48	35	Age	64	4	63	64	1	Decision
6	49	36	N/A	N/A	3	N/A	N/A	1	Terminal
6	50	36	Family	2.5	12	65	66	0	Decision
6	51	37	N/A	N/A	1	N/A	N/A	1	Terminal
6	52	37	N/A	N/A	1	N/A	N/A	0	Terminal
7	53	40	Age	61.5	5	67	68	1	Decision
7	54	40	EducProf	0.5	4	69	70	0	Decision
7	55	42	N/A	N/A	1	N/A	N/A	1	Terminal
7	56	42	N/A	N/A	3	N/A	N/A	0	Terminal
7	57	43	N/A	N/A	1	N/A	N/A	0	Terminal
7	58	43	N/A	N/A	3	N/A	N/A	1	Terminal
7	59	45	EducProf	0.5	7	71	72	0	Decision
7	60	45	ZIP Code	91409	6	73	74	1	Decision
7	61	46	Income	111	7	75	76	0	Decision
7	62	46	N/A	N/A	10	N/A	N/A	0	Terminal
7	63	48	N/A	N/A	2	N/A	N/A	0	Terminal
7	64	48	N/A	N/A	2	N/A	N/A	1	Terminal
7	65	50	N/A	N/A	9	N/A	N/A	0	Terminal
7	66	50	N/A	N/A	3	N/A	N/A	1	Terminal
8	67	53	N/A	N/A	4	N/A	N/A	1	Terminal
8	68	53	N/A	N/A	1	N/A	N/A	0	Terminal
8	69	54	N/A	N/A	1	N/A	N/A	1	Terminal
8	70	54	N/A	N/A	3	N/A	N/A	0	Terminal
8	71	59	Online	0.5	2	77	78	1	Decision
8	72	59	N/A	N/A	5	N/A	N/A	0	Terminal
8	73	60	Family	2.5	3	79	80	0	Decision
8	74	60	N/A	N/A	3	N/A	N/A	1	Terminal
8	75	61	Mortgage	54.5	3	81	82	1	Decision
8	76	61	N/A	N/A	4	N/A	N/A	0	Terminal
9	77	71	N/A	N/A	1	N/A	N/A	0	Terminal
9	78	71	N/A	N/A	1	N/A	N/A	1	Terminal
9	79	73	N/A	N/A	1	N/A	N/A	1	Terminal
9	80	73	N/A	N/A	2	N/A	N/A	0	Terminal
9	81	75	N/A	N/A	2	N/A	N/A	1	Terminal
9	82	75	N/A	N/A	1	N/A	N/A	0	Terminal

FIGURE 9.11 DESCRIPTION OF EACH SPLITTING STEP OF THE FULL-GROWN TREE FOR THE LOAN ACCEPTANCE DATA

Training Data Scoring - Summary Report (Using Full Tree)

Cut off Prob.Val. for Success (Updatable)	0.5

Classification Confusion Matrix

	Predicted Class	
Actual Class	1	0
1	235	0
0	0	2265

Error Report

Class	# Cases	# Errors	% Error
1	235	0	0.00
0	2265	0	0.00
Overall	2500	0	0.00

Validation Data Scoring - Summary Report (Using Full Tree)

Cut off Prob.Val. for Success (Updatable)	0.5

Classification Confusion Matrix

	Predicted Class	
Actual Class	1	0
1	128	15
0	17	1340

Error Report

Class	# Cases	# Errors	% Error
1	143	15	10.49
0	1357	17	1.25
Overall	1500	32	2.13

Test Data Scoring - Summary Report (Using Full Tree)

Cut off Prob.Val. for Success (Updatable)	0.5

Classification Confusion Matrix

	Predicted Class	
Actual Class	1	0
1	88	14
0	8	890

Error Report

Class	# Cases	# Errors	% Error
1	102	14	13.73
0	898	8	0.89
Overall	1000	22	2.20

FIGURE 9.12 CLASSIFICATION MATRIX AND ERROR RATES FOR THE TRAINING AND VALIDATION DATA

9.5   AVOIDING OVERFITTING

As the last example illustrated, using a full-grown tree (based on the training data) leads to complete overfitting of the data. As discussed in Chapter 5, overfitting will lead to poor performance on new data. If we look at the overall error at the various levels of the tree, it is expected to decrease as the number of levels grows until the point of overfitting. Of course, for the training data the overall error decreases more and more until it is zero at the maximum level of the tree. However, for new data, the overall error is expected to decrease until the point where the tree models the relationship between class and the predictors. After that, the tree starts to model the noise in the training set, and we expect the overall error for the validation set to start increasing. This is depicted in Figure 9.13. One intuitive reason for the overfitting at the high levels of the tree is that these splits are based on very small numbers of observations. In such cases, class difference is likely to be attributed to noise rather than predictor information.

Two ways to try and avoid exceeding this level, thereby limiting overfitting, are by setting rules to stop tree growth or, alternatively, by pruning the full-grown tree back to a level where it does not overfit. These solutions are discussed below.

Stopping Tree Growth: CHAID

One can think of different criteria for stopping the tree growth before it starts overfitting the data. Examples are tree depth (i.e., number of splits), minimum number of observations in a node, and minimum reduction in impurity. The problem is that it is not simple to determine what is a good stopping point using such rules.

Previous methods developed were based on the idea of recursive partitioning, using rules to prevent the tree from growing excessively and overfitting the training data. One popular method called *CHAID* (chi-squared automatic interaction detection) is a recursive partitioning method that predates classification

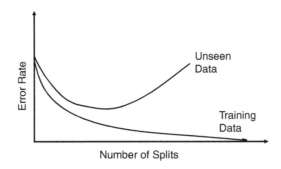

FIGURE 9.13   **ERROR RATE AS A FUNCTION OF THE NUMBER OF SPLITS FOR TRAINING VS. VALIDATION DATA: OVERFITTING**

and regression tree (CART) procedures by several years and is widely used in database marketing applications to this day. It uses a well-known statistical test (the chi-square test for independence) to assess whether splitting a node improves the purity by a statistically significant amount. In particular, at each node we split on the predictor that has the strongest association with the response variable. The strength of association is measured by the p-value of a chi-squared test of independence. If for the best predictor the test does not show a significant improvement, the split is not carried out, and the tree is terminated. This method is more suitable for categorical predictors, but it can be adapted to continuous predictors by binning the continuous values into categorical bins.

Pruning the Tree

An alternative solution that has proven to be more successful than stopping tree growth is pruning the full-grown tree. This is the basis of methods such as CART [developed by Breiman et al. (1984), implemented in multiple data mining software packages such as SAS Enterprise Miner, CART, MARS, and in XLMiner] and C4.5 implemented in packages such as IBM Modeler (previously Clementine by SPSS). In C4.5 the training data are used both for growing and pruning the tree. In CART the innovation is to use the validation data to prune back the tree that is grown from training data. CART and CART-like procedures use validation data to prune back the tree that has deliberately been overgrown using the training data. This approach is also used by XLMiner.

The idea behind pruning is to recognize that a very large tree is likely to be overfitting the training data and that the weakest branches, which hardly reduce the error rate, should be removed. In the mower example the last few splits resulted in rectangles with very few points (indeed, four rectangles in the full tree had just one point). We can see intuitively that these last splits are likely simply to be capturing noise in the training set rather than reflecting patterns that would occur in future data, such as the validation data. Pruning consists of successively selecting a decision node and redesignating it as a leaf node [lopping off the branches extending beyond that decision node (its *subtree*) and thereby reducing the size of the tree]. The pruning process trades off misclassification error in the validation dataset against the number of decision nodes in the pruned tree to arrive at a tree that captures the patterns but not the noise in the training data. Returning to Figure 9.13, we would like to find the point where the curve for the unseen data begins to increase.

To find this point, the CART algorithm uses a criterion called the *cost complexity* of a tree to generate a sequence of trees that are successively smaller to the point of having a tree with just the root node. (What is the classification rule for a tree with just one node?). This means that the first step is to find the best subtree of each size (1, 2, 3, . . .). Then, to chose among these, we want the tree that minimizes the error rate of the validation set. We then pick as our best tree the one tree in the sequence that gives the smallest misclassification error in the validation data.

Constructing the best tree of each size is based on the cost complexity (CC) criterion, which is equal to the misclassification error of a tree (based on the training data) plus a penalty factor for the size of the tree. For a tree T that has $L(T)$ leaf nodes, the cost complexity can be written as

$$CC(T) = \text{Err}(T) + \alpha L(T),$$

where err(T) is the fraction of training data observations that are misclassified by tree T and α is a penalty factor for tree size. When $\alpha = 0$, there is no penalty for having too many nodes in a tree, and the best tree using the cost complexity criterion is the full-grown unpruned tree. When we increase α to a very large value, the penalty cost component swamps the misclassification error component of the cost complexity criterion function, and the best tree is simply the tree with the fewest leaves: namely, the tree with simply one node. The idea is therefore to start with the full-grown tree and then increase the penalty factor α gradually until the cost complexity of the full tree exceeds that of a subtree. Then the same procedure is repeated using the subtree. Continuing in this manner, we generate a succession of trees with a diminishing number of nodes all the way to a trivial tree consisting of just one node.

From this sequence of trees it seems natural to pick the one that gave the minimum misclassification error on the validation dataset. We call this the *minimum error tree*. To illustrate this, Figure 9.14 shows the error rate for both the training and validation data as a function of the tree size. It can be seen that the training set error steadily decreases as the tree grows, with a noticeable drop in error rate between two and three nodes. The validation set error rate, however, reaches a minimum at 11 nodes and then starts to increase as the tree grows. At this point the tree is pruned and we obtain the minimum error tree.

A further enhancement is to incorporate the sampling error, which might cause this minimum to vary if we had a different sample. The enhancement uses the estimated standard error of the error rate to prune the tree even further (to the validation error rate, which is one standard error above the minimum.) In other words, the "best pruned tree" is the smallest tree in the pruning sequence that has an error within one standard error of the minimum error tree. The best pruned tree for the loan acceptance example is shown in Figure 9.15.

Returning to the loan acceptance example, we expect that the classification accuracy of the validation set using the pruned tree would be higher than using the full-grown tree (compare Figure 9.12 with Figure 9.16). However, the performance of the pruned tree on the validation data is not fully reflective of the performance on completely new data because the validation data were actually used for the pruning. This is a situation where it is particularly useful to evaluate

# Decision Nodes	% Error Training	% Error Validation			
41	0	2.133333			
40	0.04	2.2			
39	0.08	2.2			
38	0.12	2.2			
37	0.16	2.066667			
36	0.2	2.066667			
35	0.2	2.066667			
34	0.24	2.066667			
33	0.28	2.066667			
32	0.4	2.066667			
31	0.48	2.133333			
30	0.48	2.133333			
29	0.56	2.133333			
28	0.6	1.866667			
27	0.64	1.866667			
26	0.72	1.866667			
25	0.76	1.866667			
24	0.88	1.866667			
23	0.88	1.733333			
22	0.88	1.733333			
21	0.96	1.733333			
20	0.96	1.733333			
19	1	1.733333			
18	1	1.733333			
17	1.12	1.733333			
16	1.12	1.533333			
15	1.12	1.533333			
14	1.16	1.533333			
13	1.16	1.6			
12	1.2	1.6			
11	1.2	1.466667	<-- Min. Err. Tree	Std. Err.	0.003103929
10	1.6	1.666667			
9	2.2	1.666667			
8	2.2	1.866667			
7	2.24	1.866667			
6	2.24	1.6	<-- Best Pruned Tree		
5	4.44	1.8			
4	5.08	2.333333			
3	5.24	3.466667			
2	9.4	9.533333			
1	9.4	9.533333			
0	9.4	9.533333			

FIGURE 9.14 **ERROR RATE AS A FUNCTION OF THE NUMBER OF SPLITS FOR TRAINING VS. VALIDATION DATA FOR THE LOAN EXAMPLE**

the performance of the chosen model, whatever it may be, on a third set of data, the test set, which has not been used at all. In our example, the pruned tree applied to the test data yields an overall error rate of 1.7% (compared to 0% for the training data and 1.6% for the validation data). Although in this example the

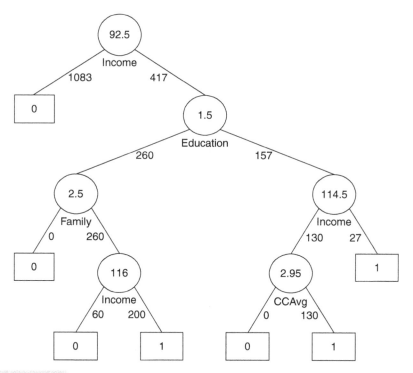

FIGURE 9.15   **BEST PRUNED TREE OBTAINED BY FITTING A FULL TREE TO THE TRAINING DATA AND PRUNING IT USING THE VALIDATION DATA**

performance on the validation and test sets is similar, the difference can be larger for other datasets.

9.6 CLASSIFICATION RULES FROM TREES

As described in Section 9.1, classification trees provide easily understandable *classification rules* (if the trees are not too large). Each leaf is equivalent to a classification rule. Returning to the example, the middle left leaf in the best pruned tree gives us the rule

IF(Income > 92.5) AND (Education < 1.5) AND (Family ≤ 2.5) THEN
Class = 0.

However, in many cases the number of rules can be reduced by removing redundancies. For example, the rule

IF(Income > 92.5) AND (Education > 1.5) AND (Income > 114.5) THEN
Class = 1

Training Data Scoring - Summary Report (Using Full Tree)

Cut off Prob.Val. for Success (Updatable)	0.5

Classification Confusion Matrix		
	Predicted Class	
Actual Class	1	0
1	235	0
0	0	2265

Error Report			
Class	# Cases	# Errors	% Error
1	235	0	0.00
0	2265	0	0.00
Overall	2500	0	0.00

Validation Data Scoring - Summary Report (Using Best Pruned Tree)

Cut off Prob.Val. for Success (Updatable)	0.5

Classification Confusion Matrix		
	Predicted Class	
Actual Class	1	0
1	127	16
0	8	1349

Error Report			
Class	# Cases	# Errors	% Error
1	143	16	11.19
0	1357	8	0.59
Overall	1500	24	1.60

Test Data Scoring - Summary Report (Using Best Pruned Tree)

Cut off Prob.Val. for Success (Updatable)	0.5

Classification Confusion Matrix		
	Predicted Class	
Actual Class	1	0
1	88	14
0	3	895

Error Report			
Class	# Cases	# Errors	% Error
1	102	14	13.73
0	898	3	0.33
Overall	1000	17	1.70

FIGURE 9.16   CLASSIFICATION MATRIX AND ERROR RATES FOR THE TRAINING, VALIDATION, AND TEST DATA BASED ON THE PRUNED TREE

can be simplified to

IF(Income > 114.5) AND (Education > 1.5) THEN Class = 1.

This transparency in the process and understandability of the algorithm that leads to classifying an observation as belonging to a certain class is very advantageous in settings where the final classification is not solely of interest. Berry and Linoff (2000) give the example of health insurance underwriting, where the insurer is required to show that coverage denial is not based on discrimination. By showing rules that led to denial (e.g., income < \$20K AND low credit history), the company can avoid lawsuits. Compared to the output of other classifiers, such as discriminant functions, tree-based classification rules are easily explained to managers and operating staff. Their logic is certainly far more transparent than that of weights in neural networks!

9.7 CLASSIFICATION TREES FOR MORE THAN TWO CLASSES

Classification trees can be used with an outcome that has more than two classes. In terms of measuring impurity, the two measures that were presented earlier (the Gini impurity index and the entropy measure) were defined for m classes and hence can be used for any number of classes. The tree itself would have the same structure, except that its terminal nodes would take one of the m-class labels.

9.8 REGRESSION TREES

The tree method can also be used for numerical response variables. Regression trees for prediction operate in much the same fashion as classification trees. The output variable, Y, is a numerical variable in this case, but both the principle and the procedure are the same: Many splits are attempted, and for each, we measure "impurity" in each branch of the resulting tree. The tree procedure then selects the split that minimizes the sum of such measures. To illustrate a regression tree, consider the example of predicting prices of Toyota Corolla automobiles (from Chapter 6). The dataset includes information on 1000 sold Toyota Corolla cars (we use the first 1000 cars from the dataset ToyotoCorolla.xls). The goal is to find a predictive model of price as a function of 10 predictors (including mileage, horsepower, number of doors, etc.). A regression tree for these data was built using a training set of 600. The best pruned tree is shown in Figure 9.17.

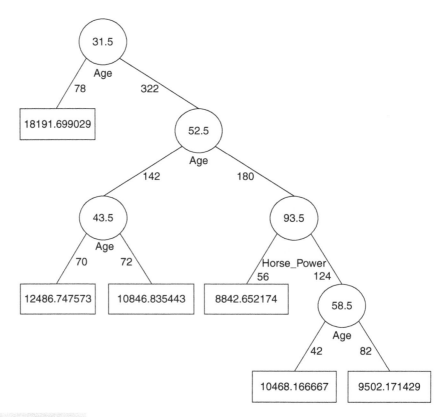

FIGURE 9.17   BEST PRUNED REGRESSION TREE FOR TOYOTA COROLLA PRICES

It can be seen that only two predictors show up as useful for predicting price: the age of the car and its horsepower. There are three details that are different in regression trees than in classification trees: prediction, impurity measures, and evaluating performance. We describe these next.

Prediction

Predicting the value of the response Y for an observation is performed in a fashion similar to the classification case: The predictor information is used for "dropping" down the tree until reaching a leaf node. For instance, to predict the price of a Toyota Corolla with Age = 55 and Horsepower = 86, we drop it down the tree and reach the node that has the value $8842.65. This is the price prediction for this car according to the tree. In classification trees the value of the leaf node (which is one of the categories) is determined by the "voting" of the training data that were in that leaf. In regression trees the value of the leaf node is determined by the average of the training data in that leaf. In the example above, the value $8842.6 is the average of the 56 cars in the training set that fall in the category of Age > 52.5 AND Horsepower < 93.5.

Measuring Impurity

We described two types of impurity measures for nodes in classification trees: the Gini index and the entropy-based measure. In both cases the index is a function of the ratio between the categories of the observations in that node. In regression trees a typical impurity measure is the sum of the squared deviations from the mean of the leaf. This is equivalent to the squared errors because the mean of the leaf is exactly the prediction. In the example above, the impurity of the node with the value $8842.6 is computed by subtracting $8842.6 from the price of each of the 56 cars in the training set that fell in that leaf, then squaring these deviations and summing them up. The lowest impurity possible is zero, when all values in the node are equal.

Evaluating Performance

As stated above, predictions are obtained by averaging the values of the responses in the nodes. We therefore have the usual definition of predictions and errors. The predictive performance of regression trees can be measured in the same way that other predictive methods are evaluated, using summary measures such as RMSE and charts such as lift charts.

9.9 ADVANTAGES, WEAKNESSES, AND EXTENSIONS

Tree methods are good off-the-shelf classifiers and predictors. They are also useful for variable selection, with the most important predictors usually showing up at the top of the tree. Trees require relatively little effort from users in the following senses: First, there is no need for transformation of variables (any monotone transformation of the variables will give the same trees). Second, variable subset selection is automatic since it is part of the split selection. In the loan example notice that the best pruned tree has automatically selected just four variables (Income, Education, Family, and CCAvg) out of the set 14 variables available.

Trees are also intrinsically robust to outliers since the choice of a split depends on the *ordering* of observation values and not on the absolute *magnitudes* of these values. However, they are sensitive to changes in the data, and even a slight change can cause very different splits!

Unlike models that assume a particular relationship between the response and predictors (e.g., a linear relationship such as in linear regression and linear discriminant analysis), classification and regression trees are nonlinear and non-parametric. This allows for a wide range of relationships between the predictors and the response. However, this can also be a weakness: Since the splits are done on single predictors rather than on combinations of predictors, the tree is likely to miss relationships between predictors, in particular linear structures such as those in linear or logistic regression models. Classification trees are useful classifiers in

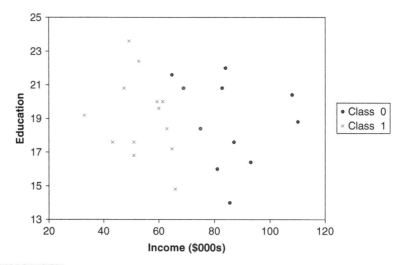

FIGURE 9.18   **SCATTERPLOT DESCRIBING A TWO-PREDICTOR CASE WITH TWO CLASSES**

cases where horizontal and vertical splitting of the predictor space adequately divides the classes. But consider, for instance, a dataset with two predictors and two classes, where separation between the two classes is most obviously achieved by using a diagonal line (as shown in Figure 9.18). A classification tree is therefore expected to have lower performance than methods such as discriminant analysis. One way to improve performance is to create new predictors that are derived from existing predictors, which can capture hypothesized relationships between predictors (similar to interactions in regression models).

Another performance issue with classification trees is that they require a large dataset in order to construct a good classifier. Recently, Breiman and Cutler introduced *random forests*,[3] an extension to classification trees that tackles these issues. The basic idea is to create multiple classification trees from the data (and thus obtain a "forest") and combine their output to obtain a better classifier.

An appealing feature of trees is that they handle missing data without having to impute values or delete observations with missing values. The method can be extended to incorporate an importance ranking for the variables in terms of their impact on the quality of the classification. From a computational aspect, trees can be relatively expensive to grow because of the multiple sorting involved in computing all possible splits on every variable. Pruning the data using the validation set adds further computation time. Finally, a very important practical advantage of trees is the transparent rules that they generate. Such transparency is often useful in managerial applications.

---

[3] For further details on random forests see www.stat.berkeley.edu/users/breiman/RandomForests/cc_home.htm.

PROBLEMS

9.1 Competitive Auctions on eBay.com. The file eBayAuctions.xls contains information on 1972 auctions transacted on eBay.com during May–June 2004. The goal is to use these data to build a model that will classify competitive auctions from noncompetitive ones. A *competitive auction* is defined as an auction with at least two bids placed on the item auctioned. The data include variables that describe the item (auction category), the seller (his/her eBay rating), and the auction terms that the seller selected (auction duration, opening price, currency, day-of-week of auction close). In addition, we have the price at which the auction closed. The goal is to predict whether or not the auction will be competitive.

Data Preprocessing. Create dummy variables for the categorical predictors. These include Category (18 categories), Currency (USD, GBP, Euro), EndDay (Monday–Sunday), and Duration (1, 3, 5, 7, or 10 days). Split the data into training and validation datasets using a 60% : 40% ratio.

a. Fit a classification tree using all predictors, using the best pruned tree. To avoid overfitting, set the minimum number of observations in a leaf node to 50. Also, set the maximum number of levels to be displayed at seven (the maximum allowed in XLminer). To remain within the limitation of 30 predictors, combine some of the categories of categorical predictors. Write down the results in terms of rules.

b. Is this model practical for predicting the outcome of a new auction?

c. Describe the interesting and uninteresting information that these rules provide.

d. Fit another classification tree (using the best-pruned tree, with a minimum number of observations per leaf node = 50 and maximum allowed number of displayed levels), this time only with predictors that can be used for predicting the outcome of a new auction. Describe the resulting tree in terms of rules. Make sure to report the smallest set of rules required for classification.

e. Plot the resulting tree on a scatterplot: Use the two axes for the two best (quantitative) predictors. Each auction will appear as a point, with coordinates corresponding to its values on those two predictors. Use different colors or symbols to separate competitive and noncompetitive auctions. Draw lines (you can sketch these by hand or use Excel) at the values that create splits. Does this splitting seem reasonable with respect to the meaning of the two predictors? Does it seem to do a good job of separating the two classes?

f. Examine the lift chart and the classification table for the tree. What can you say about the predictive performance of this model?

g. Based on this last tree, what can you conclude from these data about the chances of an auction obtaining at least two bids and its relationship to the auction settings set by the seller (duration, opening price, ending day, currency)? What would you recommend for a seller as the strategy that will most likely lead to a competitive auction?

9.2 Predicting Delayed Flights. The file FlightDelays.xls contains information on all commercial flights departing the Washington, D.C., area and arriving at New York during January 2004. For each flight there is information on the departure and arrival airports, the distance of the route, the scheduled time and date of the flight, and so on. The variable that we are trying to predict is whether or not a flight is delayed. A delay is defined as an arrival that is at least 15 minutes later than scheduled.

Data Preprocessing. Create dummies for day of week, carrier, departure airport, and arrival airport. This will give you 17 dummies. Bin the scheduled departure time into 2-hour bins (in XLMiner use *Data Utilities > Bin Continuous Data* and select 8 bins with equal width). After binning DEP_TIME into 8 bins, this new variable should be broken down into 7 dummies (because the effect will not be linear due to the morning and afternoon rush hours). This will avoid treating the departure time as a continuous predictor because it is reasonable that delays are related to rush-hour times. Partition the data into training and validation sets.

a. Fit a classification tree to the flight delay variable using all the relevant predictors. Do not include DEP_TIME (actual departure time) in the model because it is unknown at the time of prediction (unless we are doing our predicting of delays after the plane takes off, which is unlikely). In the third step of the classification tree menu, choose "Maximum # levels to be displayed = 6". Use the best pruned tree without a limitation on the minimum number of observations in the final nodes. Express the resulting tree as a set of rules.

b. If you needed to fly between DCA and EWR on a Monday at 7 AM, would you be able to use this tree? What other information would you need? Is it available in practice? What information is redundant?

c. Fit another tree, this time excluding the day-of-month predictor. (Why?) Select the option of seeing both the full tree and the best pruned tree. You will find that the best pruned tree contains a single terminal node.

   i. How is this tree used for classification? (What is the rule for classifying?)

   ii. To what is this rule equivalent?

   iii. Examine the full tree. What are the top three predictors according to this tree?

   iv. Why, technically, does the pruned tree result in a tree with a single node?

   v. What is the disadvantage of using the top levels of the full tree as opposed to the best pruned tree?

   vi. Compare this general result to that from logistic regression in the example in Chapter 10. What are possible reasons for the classification tree's failure to find a good predictive model?

9.3 **Predicting Prices of Used Cars (Regression Trees).** The file ToyotaCorolla.xls contains the data on used cars (Toyota Corolla) on sale during late summer of 2004 in The Netherlands. It has 1436 observations containing details on 38 attributes, including Price, Age, Kilometers, HP, and other specifications. The goal is to predict the price of a used Toyota Corolla based on its specifications. (The example in Section 9.8 is a subset of this dataset.)

Data Preprocessing. Create dummy variables for the categorical predictors (Fuel Type and Color). Split the data into training (50%), validation (30%), and test (20%) datasets.

a. Run a regression tree (RT) using the prediction menu in XLMiner with the output variable Price and input variables Age_08_04, KM, Fuel_Type, HP, Automatic, Doors, Quarterly_Tax, Mfg_Guarantee, Guarantee_Period, Airco, Automatic_Airco, CD_Player, Powered_Windows, Sport_Model, and Tow_Bar. Normalize the variables. Keep the minimum number of observations in a terminal node to 1 and the scoring option to Full Tree, to make the run least restrictive.

   i. Which appear to be the three or four most important car specifications for predicting the car's price?

ii. Compare the prediction errors of the training, validation, and test sets by examining their RMS error and by plotting the three boxplots. What is happening with the training set predictions? How does the predictive performance of the test set compare to the other two? Why does this occur?

iii. How can we achieve predictions for the training set that are not equal to the actual prices?

iv. If we used the best pruned tree instead of the full tree, how would this affect the predictive performance for the validation set? (*Hint:* Does the full tree use the validation data?)

b. Let us see the effect of turning the price variable into a categorical variable. First, create a new variable that categorizes price into 20 bins. Use *Data Utilities > Bin Continuous Data* to categorize Price into 20 bins of equal intervals (leave all other options at their default). Now repartition the data keeping Binned_ Price instead of Price. Run a classification tree (CT) using the Classification menu of XLMiner with the same set of input variables as in the RT, and with Binned_Price as the output variable. Keep the minimum number of observations in a terminal node to 1 and uncheck the Prune Tree option, to make the run least restrictive. Select "Normalize input data."

i. Compare the tree generated by the CT with the one generated by the RT. Are they different? (Look at structure, the top predictors, size of tree, etc.) Why?

ii. Predict the price, using the RT and the CT, of a used Toyota Corolla with the specifications listed in Table 9.3.

TABLE 9.3   SPECIFICATIONS FOR A PARTICULAR TOYOTA COROLLA

Variable	Value
Age_-08_-04	77
KM	117,000
Fuel_Type	Petrol
HP	110
Automatic	No
Doors	5
Quarterly_Tax	100
Mfg_Guarantee	No
Guarantee_Period	3
Airco	Yes
Automatic_Airco	No
CD_Player	No
Powered_Windows	No
Sport_Model	No
Tow_Bar	Yes

iii. Compare the predictions in terms of the variables that were used, the magnitude of the difference between the two predictions, and the advantages and disadvantages of the two methods.

Logistic Regression

In this chapter we describe the highly popular and powerful classification method called logistic regression. Like linear regression, it relies on a specific model relating the predictors with the outcome. The user must specify the predictors to include and their form (e.g., including any interaction terms). This means that even small datasets can be used for building logistic regression classifiers, and that once the model is estimated, it is computationally fast and cheap to classify even large samples of new observations. We describe the logistic regression model formulation and its estimation from data. We also explain the concepts of "logit," "odds," and "probability" of an event that arise in the logistic model context and the relations among the three. We discuss variable importance using coefficient and statistical significance and also mention variable selection algorithms for dimension reduction. All this is illustrated on an authentic dataset of flight information where the goal is to predict flight delays. Our presentation is strictly from a data mining perspective, where classification is the goal and performance is evaluated on a separate validation set. However, because logistic regression is heavily used also in statistical analyses for purposes of inference, we give a brief review of key concepts related to coefficient interpretation, goodness-of-fit evaluation, inference, and multiclass models in the Appendix at the end of this chapter.

10.1 INTRODUCTION

Logistic regression extends the ideas of linear regression to the situation where the dependent variable, Y, is categorical. We can think of a categorical variable

---

Data Mining for Business Intelligence, By Galit Shmueli, Nitin R. Patel, and Peter C. Bruce
Copyright © 2010 John Wiley & Sons Inc.

as dividing the observations into classes. For example, if Y denotes a recommendation on holding/selling/buying a stock, we have a categorical variable with three categories. We can think of each of the stocks in the dataset (the observations) as belonging to one of three classes: the *hold* class, the *sell* class, and the *buy* class. Logistic regression can be used for classifying a new observation, where the class is unknown, into one of the classes, based on the values of its predictor variables (called *classification*). It can also be used in data (where the class is known) to find similarities between observations within each class in terms of the predictor variables (called *profiling*). Logistic regression is used in applications such as:

1. Classifying customers as returning or nonreturning (classification)
2. Finding factors that differentiate between male and female top executives (profiling)
3. Predicting the approval or disapproval of a loan based on information such as credit scores (classification)

In this chapter we focus on the use of logistic regression for classification. We deal only with a binary dependent variable having two possible classes. At the end we show how the results can be extended to the case where Y assumes more than two possible outcomes. Popular examples of binary response outcomes are success/failure, yes/no, buy/don't buy, default/don't default, and survive/die. For convenience we often code the values of a binary response Y as 0 and 1.

Note that in some cases we may choose to convert continuous data or data with multiple outcomes into binary data for purposes of simplification, reflecting the fact that decision making may be binary (approve the loan/don't approve, make an offer/don't make an offer). As with multiple linear regression, the independent variables $X_1, X_2, \ldots, X_k$ may be categorical or continuous variables or a mixture of these two types. While in multiple linear regression the aim is to predict the value of the continuous Y for a new observation, in logistic regression the goal is to predict which class a new observation will belong to or simply to *classify* the observation into one of the classes. In the stock example, we would want to classify a new stock into one of the three recommendation classes: sell, hold, or buy.

In logistic regression we take two steps: the first step yields estimates of the *probabilities* of belonging to each class. In the binary case we get an estimate of $P(Y = 1)$, the probability of belonging to class 1 (which also tells us the probability of belonging to class 0). In the next step we use a cutoff value on these probabilities in order to classify each case in one of the classes. For example, in a binary case, a cutoff of 0.5 means that cases with an estimated probability of $P(Y = 1) > 0.5$ are classified as belonging to class 1, whereas cases with $P(Y = 1) < 0.5$ are classified as belonging to class 0. This cutoff need not be set at 0.5. When the event in question is a low-probability event, a higher-than-average cutoff value, although still below 0.5, may be sufficient to classify a case as belonging to class 1.

10.2 LOGISTIC REGRESSION MODEL

The logistic regression model is used in a variety of fields: whenever a structured model is needed to explain or predict categorical (in particular, binary) outcomes. One such application is in describing choice behavior in econometrics, which is useful in the context of the example above (see the accompanying box).

> **LOGISTIC REGRESSION AND CONSUMER CHOICE THEORY**
>
> In the context of choice behavior, the logistic model can be shown to follow from the *random utility theory* developed by Manski (1977) as an extension of the standard economic theory of consumer behavior. In essence, the consumer theory states that when faced with a set of choices, a consumer makes the choice that has the highest utility (a numerical measure of worth with arbitrary zero and scale). It assumes that the consumer has a preference order on the list of choices that satisfies reasonable criteria such as transitivity. The preference order can depend on the person (e.g., socioeconomic characteristics) as well as attributes of the choice. The random utility model considers the utility of a choice to incorporate a random element. When we model the random element as coming from a "reasonable" distribution, we can logically derive the logistic model for predicting choice behavior.

The idea behind logistic regression is straightforward: Instead of using Y as the dependent variable, we use a function of it, which is called the *logit*. The logit, it turns out, can be modeled as a linear function of the predictors. Once the logit has been predicted, it can be mapped back to a probability.

To understand the logit, we take several intermediate steps: First, we look at p, the probability of belonging to class 1 (as opposed to class 0). In contrast to Y, the class number, which only takes the values 0 and 1, p can take any value in the interval $[0, 1]$. However, if we express p as a linear function of the q predictors[1] in the form

$$p = \beta_0 + \beta_1 x_1 + \beta_2 x_2 + \cdots + \beta_q x_q, \tag{10.1}$$

---

[1] Unlike elsewhere in the book, where p denotes the number of predictors, in this chapter we indicate predictors by q, to avoid confusion with the probability p.

it is not guaranteed that the right-hand side will lead to values within the interval [0, 1]. The fix is to use a nonlinear function of the predictors in the form

$$p = \frac{1}{1 + e^{-(\beta_0 + \beta_1 x_1 + \beta_2 x_2 + \cdots + \beta_q x_q)}}. \tag{10.2}$$

This is called the *logistic response function*. For any values of $x_1, \ldots, x_q$, the right-hand side will always lead to values in the interval [0, 1]. Next, we look at a different measure of belonging to a certain class, known as *odds*. The odds of belonging to class 1 ($Y = 1$) is defined as the ratio of the probability of belonging to class 1 to the probability of belonging to class 0:

$$\text{Odds} = \frac{p}{1 - p}. \tag{10.3}$$

This metric is very popular in horse races, sports, gambling in general, epidemiology, and many other areas. Instead of talking about the *probability* of winning or contacting a disease, people talk about the *odds* of winning or contacting a disease. How are these two different? If, for example, the probability of winning is 0.5, the odds of winning are $0.5/0.5 = 1$. We can also perform the reverse calculation: Given the odds of an event, we can compute its probability by manipulating equation (10.3):

$$p = \frac{\text{odds}}{1 + \text{odds}}. \tag{10.4}$$

Substituting (10.2) into (10.4), we can write the relationship between the odds and the predictors as

$$\text{Odds} = e^{\beta_0 + \beta_1 x_1 + \beta_2 x_2 + \cdots + \beta_q x_q}. \tag{10.5}$$

This last equation describes a multiplicative (proportional) relationship between the predictors and the odds. Such a relationship is interpretable in terms of percentages, for example, a unit increase in predictor x_j is associated with an average increase of $\beta_j \times 100\%$ in the odds (holding all other predictors constant).

Now, if we take a log on both sides, ("log" refers to the natural logarithm (ln)) we get the standard formulation of a logistic model:

$$\log(\text{odds}) = \beta_0 + \beta_1 x_1 + \beta_2 x_2 + \cdots + \beta_q x_q. \tag{10.6}$$

The log(odds) is called the *logit*, and it takes values from $-\infty$ to ∞. Thus, our final formulation of the relation between the response and the predictors uses the logit as the dependent variable and models it as a *linear function* of the q predictors.

To see the relation between the probability, odds, and logit of belonging to class 1, look at Figure 10.1, which shows the odds (*a*) and logit (*b*) as a function of p. Notice that the odds can take any nonnegative value, and that the logit can take any real value. Let us examine some data to illustrate the use of logistic regression.

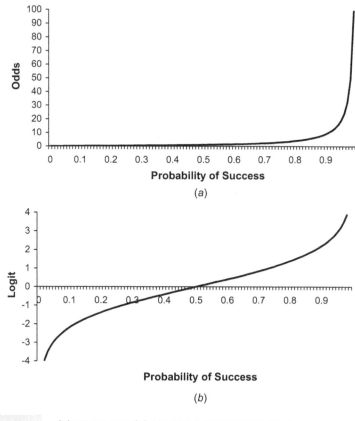

FIGURE 10.1   (*A*) ODDS AND (*B*) LOGIT AS A FUNCTION OF *P*

Example: Acceptance of Personal Loan

Recall the example described in Chapter 9, of acceptance of a personal loan by Universal Bank. The bank's dataset includes data on 5000 customers. The data include customer demographic information (Age, Income, etc.), customer response to the last personal loan campaign (Personal Loan), and the customer's relationship with the bank (mortgage, securities account, etc.). Among these 5000 customers, only 480 (=9.6%) accepted the personal loan that was offered to them in a previous campaign. The goal is to find characteristics of customers who are most likely to accept the loan offer in future mailings.

Data Preprocessing   We start by partitioning the data randomly using a standard 60% : 40% rate into training and validation sets. We use the training set to fit a model and the validation set to assess the model's performance.

Next, we create dummy variables for each of the categorical predictors. Except for education, which has three categories, the remaining four categorical variables have two categories. We therefore need $6 = 2 + 1 + 1 + 1 + 1$ dummy variables to describe these five categorical predictors. In XLMiner's classification functions, the response can remain in text form (Yes, No, etc.), but the predictor variables must be coded into dummy variables. We use the following coding:

$$\text{EducProf} = \begin{cases} 1 \text{ if education is Professional} \\ 0 \text{ otherwise} \end{cases}$$

$$\text{EducGrad} = \begin{cases} 1 \text{ if education is at Graduate level} \\ 0 \text{ otherwise} \end{cases}$$

$$\text{Securities} = \begin{cases} 1 \text{ if customer has securities account in bank} \\ 0 \text{ otherwise} \end{cases}$$

$$\text{CD} = \begin{cases} 1 \text{ if customer has CD account in bank} \\ 0 \text{ otherwise} \end{cases}$$

$$\text{Online} = \begin{cases} 1 \text{ if customer uses online banking} \\ 0 \text{ otherwise} \end{cases}$$

$$\text{CreditCard} = \begin{cases} 1 \text{ if customer holds Universal Bank credit card} \\ 0 \text{ otherwise} \end{cases}$$

Model with a Single Predictor

Consider first a simple logistic regression model with just one independent variable. This is analogous to the simple linear regression model in which we fit a straight line to relate the dependent variable, Y, to a single independent variable, X.

Let us construct a simple logistic regression model for classification of customers using the single predictor Income. The equation relating the dependent variable to the explanatory variable in terms of probabilities is

$$\text{Prob}(\text{Personal Loan} = \text{Yes} \mid \text{Income} = x) = \frac{1}{1 + e^{-(\beta_0 + \beta_1 x)}},$$

or equivalently, in terms of odds,

$$\text{Odds}(\text{Personal Loan} = \text{Yes}) = e^{\beta_0 + \beta_1 x}. \tag{10.7}$$

The estimated coefficients for the model are $b_0 = -6.3525$ and $b_1 = 0.0392$. So the fitted model is

$$P(\text{Personal Loan} = \text{Yes} \mid \text{Income} = x) = \frac{1}{1 + e^{6.3525 - 0.0392x}}. \tag{10.8}$$

Although logistic regression can be used for prediction in the sense that we predict the *probability* of a categorical outcome, it is most often used for classification. To see the difference between the two, consider predicting the probability of a customer accepting the loan offer as opposed to classifying the customer as an accepter/nonaccepter. From Figure 10.2 it can be seen that the loan acceptance can yield numbers between 0 and 1. To end up with classifications into either 0

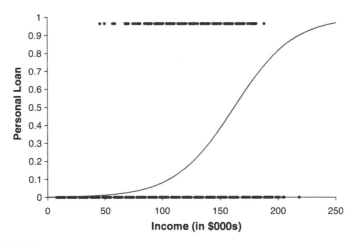

FIGURE 10.2   PLOT OF DATA POINTS (PERSONAL LOAN AS A FUNCTION OF INCOME) AND THE FITTED LOGISTIC CURVE

or 1 (e.g., a customer either accepts the loan offer or not), we need a threshold, or "cutoff value." This is true in the case of multiple predictor variables as well.

Cutoff Value   Given the values for a set of predictors, we can predict the probability that each observation belongs to class 1. The next step is to set a cutoff on these probabilities so that each observation is classified into one of the two classes. This is done by setting a cutoff value, c, such that observations with probabilities above c are classified as belonging to class 1, and observations with probabilities below c are classified as belonging to class 0.

In the Universal Bank example, in order to classify a new customer as an acceptor/nonacceptor of the loan offer, we use the information on his/her income by plugging it into the fitted equation (10.8). This yields an estimated probability of accepting the loan offer. We then compare it to the cutoff value. The customer is classified as an acceptor if the probability of his/her accepting the offer is above the cutoff. [2]

Different cutoff values lead to different classifications and, consequently, different classification matrices. There are several approaches to determining the "optimal" cutoff probability: A popular cutoff value for a two-class case is 0.5. The rationale is to assign an observation to the class in which its probability of membership is highest. A cutoff can also be chosen to maximize overall accuracy. This can be determined using a one-way data table in Excel (see Chapter 5).

---

[2] If we prefer to look at *odds* of accepting rather than the probability, an equivalent method is to use equation (10.7) and compare the odds to $c/(1 − c)$. If the odds are higher than this number, the customer is classified as an acceptor. If it is lower, we classify the customer as a nonacceptor.

The overall accuracy is computed for various values of the cutoff value, and the cutoff value that yields maximum accuracy is chosen. The danger is, of course, overfitting.

Alternatives to maximizing accuracy are to maximize sensitivity subject to some minimum level of specificity, or to minimize false positives subject to some maximum level of false negatives, and so on. Finally, a cost-based approach is to find a cutoff value that minimizes the expected cost of misclassification. In this case one must specify the misclassification costs and the prior probabilities of belonging to each class.

Estimating the Logistic Model from Data: Computing Parameter Estimates

In logistic regression, the relation between Y and the β parameters is nonlinear. For this reason the β parameters are not estimated using the method of least squares (as in multiple linear regression). Instead, a method called *maximum likelihood* is used. The idea, in brief, is to find the estimates that maximize the chance of obtaining the data that we have. This requires iterations using a computer program.[3]

Algorithms to compute the coefficient estimates are less robust than algorithms for linear regression. Computed estimates are generally reliable for well-behaved datasets where the number of observations with outcome variable values of both 0 and 1 are large, their ratio is "not too close" to either 0 or 1, and when the number of coefficients in the logistic regression model is small relative to the sample size (say, no more than 10%). As with linear regression, collinearity (strong correlation among the independent variables) can lead to computational difficulties. Computationally intensive algorithms have been developed recently that circumvent some of these difficulties. For technical details on the maximum-likelihood estimation in logistic regression, see Hosmer and Lemeshow (2000).

To illustrate a typical output from such a procedure, look at the output in Figure 10.3 for the logistic model fitted to the training set of 3000 Universal Bank customers. The dependent variable is Personal Loan, with Yes defined as the *success* (this is equivalent to setting the variable to 1 for an acceptor and 0 for a nonacceptor). Here we use all 12 predictors.

---

[3] The method of maximum likelihood ensures good asymptotic (large sample) properties for the estimates. Under very general conditions, maximum-likelihood estimators are: (1) *Consistent*—the probability of the estimator differing from the true value approaches zero with increasing sample size. (2) *Asymptotically efficient*—the variance is the smallest possible among consistent estimators. (3) *Asymptotically normally distributed*—this allows us to compute confidence intervals and perform statistical tests in a manner analogous to the analysis of linear multiple regression models, provided that the sample size is *large*.

Input variables	Coefficient	Std. Error	p-value	Odds
Constant term	-13.20165825	2.46772742	0.00000009	*
Age	-0.04453737	0.09096102	0.62439483	0.95643985
Experience	0.05657264	0.09005365	0.5298661	1.05820346
Income	0.0657607	0.00422134	0	1.06797111
Family	0.57155931	0.10119002	0.00000002	1.77102649
CCAvg	0.18724874	0.06153848	0.00234395	1.20592725
Mortgage	0.00175308	0.00080375	0.02917421	1.00175464
Securities Account	-0.85484785	0.41863668	0.04115349	0.42534789
CD Account	3.46900773	0.44893095	0	32.10486984
Online	-0.84355801	0.22832377	0.00022026	0.43017724
CreditCard	-0.96406376	0.28254223	0.00064463	0.38134006
EducGrad	4.58909273	0.38708162	0	98.40509796
EducProf	4.52272701	0.38425466	0	92.08635712

FIGURE 10.3   **LOGISTIC REGRESSION COEFFICIENT TABLE FOR PERSONAL LOAN ACCEPTANCE AS A FUNCTION OF 12 PREDICTORS**

Ignoring p-values for the coefficients, a model based on all 12 predictors would have the following estimated logistic equation:

$$\text{logit} = -13.201 - 0.045 \text{ Age} + 0.057 \text{ Experience} + 0.066 \text{ Income} + 0.572 \text{ Family}$$
$$+ 0.18724874 \text{ CCAvg} + 0.002 \text{ Mortgage} - 0.855 \text{ Securities} + 3.469 \text{ CD}$$
$$- 0.844 \text{ Online} - 0.964 \text{ Credit Card} + 4.589 \text{ EducGrad} + 4.523 \text{ EducProf}$$

$$(10.9)$$

The positive coefficients for the dummy variables CD, EducGrad, and EducProf mean that holding a CD account and having graduate or professional education (all marked by 1 in the dummy variables) are associated with higher probabilities of accepting the loan offer. On the other hand, having a securities account, using online banking, and owning a Universal Bank credit card are associated with lower acceptance rates. For the continuous predictors, positive coefficients indicate that a higher value on that predictor is associated with a higher probability of accepting the loan offer (e.g., income: higher income customers tend more to accept the offer). Similarly, negative coefficients indicate that a higher value on that predictor is associated with a lower probability of accepting the loan offer (e.g., Age: older customers are less likely to accept the offer).

If we want to talk about the *odds* of offer acceptance, we can use the last column (entitled "odds") to obtain the following equation:

$$\text{Odds(Personal Loan = Yes)} = e^{-13.201}(0.956)^{\text{Age}} (1.058)^{\textit{Experience}} (1.068)^{\text{Income}}$$
$$\times (1.771)^{\text{Family}} (1.206)^{\text{CCAvg}} (1.002)^{\text{Mortgage}}$$
$$\times (0.425)^{\text{Securities}} (32.105)^{\text{CD}} (0.430)^{\text{Online}}$$
$$\times (0.381)^{\text{CreditCard}}(98.405)^{\text{EducGrad}} (92.086)^{\text{EducProf}}.$$

$$(10.10)$$

Notice how positive coefficients in the logit model translate into coefficients larger than 1 in the odds model, and negative positive coefficients in the logit translate into coefficients smaller than 1 in the odds.

A third option is to look directly at an equation for the probability of acceptance, using equation (10.2). This is useful for estimating the probability of accepting the offer for a customer with given values of the 12 predictors.[4]

Interpreting Results in Terms of Odds

Recall that the odds are given by

$$\text{Odds} = e^{\beta_0+\beta_1 x_1+\beta_2 x_2+\cdots+\beta_k x_k}.$$

At first let us return to the single predictor example, where we model a customer's acceptance of a personal loan offer as a function of his/her income:

$$\text{Odds (Personal Loan = Yes)} = e^{\beta_0+\beta_1\cdot Income}.$$

We can think of the model as a multiplicative model of odds. The odds that a customer with income zero will accept the loan is estimated by $e^{-6.535+(0.039)(0)} = 0.0017$. These are the *base-case odds*. In this example it is obviously economically meaningless to talk about a zero income; the value zero and the corresponding base-case odds could be meaningful, however, in the context of other predictors. The odds of accepting the loan with an income of \$100K will increase by a multiplicative factor of $e^{(0.039)(100)} = 50.5$ over the base case, so the odds that such a customer will accept the offer are $e^{-6.535+(0.039)(100)} = 0.088$.

To generalize this to the multiple-predictor case, consider the 12 predictors in the personal loan offer example. The odds of a customer accepting the offer as a function of the 12 predictors are given in (10.10).

Suppose that the value of Income, or in general x_1, is increased by one unit from x_1 to $x_1 + 1$, while the other predictors (denoted $x_2, \ldots, x_{12}$) are held at their current value. We get the odds ratio

$$\frac{\text{odds}(x_1 + 1, x_2, \ldots, x_{12})}{\text{odds}(x_1, \ldots x_{12})} = \frac{e^{\beta_0+\beta_1(x_1+1)+\beta_2 x_2+\cdots+\beta_{12}x_{12}}}{e^{\beta_0+\beta_1 x_1+\beta_2 x_2+\cdots+\beta_{12}x_{12}}} = e^{\beta_1}.$$

This tells us that a single unit increase in x_1, holding $x_2, \ldots, x_{12}$ constant, is associated with an increase in the odds that a customer accepts the offer by a factor of e^{β_1}. In other words, β_1 is the multiplicative factor by which the odds (of belonging to class 1) increase when the value of x_1 is increased by 1 unit, *holding all other predictors constant*. If $\beta_1 < 0$, an increase in x_1 is associated with a decrease in the odds of belonging to class 1, whereas a positive value of β_1 is associated with an increase in the odds.

---

[4] If all q predictors are categorical, each having m_q categories, we need not compute probabilities/odds for each of the n observations. The number of different probabilities/odds is exactly $m_1 \times m_2 \times \cdots \times m_q$.

When a predictor is a dummy variable, the interpretation is technically the same but has a different practical meaning. For instance, the coefficient for CD was estimated from the data to be 3.469. Recall that the reference group is customers not holding a CD account. We interpret this coefficient as follows: $e^{3.469} = 32.105$ are the odds that a customer who has a CD account will accept the offer relative to a customer who does not have a CD account, holding all other factors constant. This means that customers who have CD accounts in Universal Bank are more likely to accept the offer than customers without a CD account (holding all other variables constant).

The advantage of reporting results in odds as opposed to probabilities is that statements such as those above are true for any value of x_1. Unless x_1 is a dummy variable, we cannot apply such statements about the effect of increasing x_1 by a single unit to probabilities. This is because the result depends on the actual value of x_1. So if we increase x_1 from, say, 3 to 4, the effect on p, the probability of belonging to class 1, will be different than if we increase x_1 from 30 to 31. In short, the change in the probability, p, for a unit increase in a particular predictor variable, while holding all other predictors constant, is not a constant—it depends on the specific values of the predictor variables. We therefore talk about probabilities only in the context of specific observations.

ODDS AND ODDS RATIOS

A common confusion is between odds and odds ratios. Since the odds are in fact a ratio (between the probability of belonging to class 1 and the probability of belonging to class 0), they are sometimes termed, erroneously, "odds ratios." However, odds ratios refer to the ratio of two odds! These are used to compare different classes of observations. For a categorical predictor, odds ratios are used to compare two categories. For example, we could compare loan offer acceptance for customers with professional education versus graduate education by looking at the ratio of odds of loan acceptance for customers with professional education divided by the odds of acceptance for customers with graduate education. This would yield an *odds ratio*. Ratios above 1 would indicate that the odds of acceptance for professionally educated customers are higher than for customers with graduate-level education.

10.3 EVALUATING CLASSIFICATION PERFORMANCE

The general measures of performance that were described in Chapter 5 are used to assess how well the logistic model does. Recall that there are several performance measures, the most popular being those based on the classification matrix (accuracy alone or combined with costs) and the lift chart. As in other classification methods, the goal is to find a model that accurately classifies observations to their class, using only the predictor information. A variant of this goal is to find a model that does a superior job of identifying the members of a particular class of interest (which might come at some cost to overall accuracy). Since the training data are used for selecting the model, we expect the model to perform quite well for those data, and therefore prefer to test its performance on the validation set. Recall that the data in the validation set were not involved in the model building, and thus we can use them to test the model's ability to classify data that it has not "seen" before.

To obtain the classification matrix from a logistic regression analysis, we use the estimated equation to predict the probability of class membership for each observation in the validation set, and use the cutoff value to decide on the class assignment of these observations. We then compare these classifications to the actual class memberships of these observations. In the Universal Bank case we use the estimated model in equation (10.9) to predict the probability of adoption in a validation set that contains 2000 customers (these data were not used in the modeling step). Technically, this is done by predicting the logit using the estimated model in equation (10.9) and then obtaining the probabilities p through the relation $p = e^{\text{logit}}/1 + e^{\text{logit}}$. We then compare these probabilities to our chosen cutoff value in order to classify each of the 2000 validation observations as acceptors or nonacceptors. XLMiner created the validation classification matrix automatically, and it is possible to obtain the detailed probabilities and classification for each observation. For example, Figure 10.4 shows a partial XLMiner output of scoring the validation set. It can be seen that the first four customers have a probability of accepting the offer that is lower than the cutoff of 0.5, and therefore they are classified as nonacceptors (0). The fifth customer's probability of acceptance is estimated by the model to exceed 0.5, and he or she is therefore classified as an acceptor (1), which in fact is a misclassification.

Another useful tool for assessing model classification performance is the lift (gains) chart (see Chapter 5). Figure 10.5(a) illustrates the lift chart obtained for the personal loan offer model using the validation set. The "lift" over the base curve indicates for a given number of cases (read on the x axis), the additional responders that you can identify by using the model. The same information is portrayed in in Figure 10.5(b): Taking the 10% of the records that are ranked by the model as "most probable 1's" yields 7.7 times as many 1's as would simply selecting 10% of the records at random.

Variable Selection

The next step includes searching for alternative models. As with multiple linear regression, we can build more complex models that reflect interactions between independent variables by including factors that are calculated from the interacting factors. For example, if we hypothesize that there is an interactive effect between income and family size, we should add an interaction

Back to Navigator

Data range: ['Universal Bank.xls']'Data_Partition1'!$C$3019:$Q$5018

Cut off Prob.Val. for Success (Updatable) **0.5** (Updating the value here will NOT update value in summary report)

Row Id.	Predicted Class	Actual Class	Prob. for 1 (success)	Log odds	Age	Experience	Income	Family
2	0	0	2.1351E-05	-10.75439275	45	19	34	3
3	0	0	3.34564E-06	-12.60785033	39	15	11	1
7	0	0	0.015822384	-4.13038073	53	27	72	2
8	0	0	0.000216511	-8.437650808	50	24	22	1
11	0	1	0.567824439	0.272980386	65	39	105	4

FIGURE 10.4 SCORING THE VALIDATION DATA: XLMINER'S OUTPUT FOR THE FIRST FIVE CUSTOMERS OF UNIVERSAL BANK (BASED ON 12 PREDICTORS).

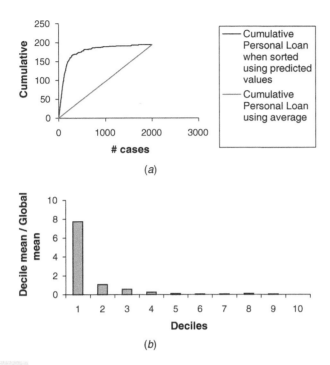

(a)

(b)

FIGURE 10.5   (A) LIFT AND (B) DECILE CHARTS OF VALIDATION DATA FOR UNIVERSAL BANK LOAN OFFER: COMPARING LOGISTIC MODEL CLASSIFICATION WITH CLASSIFICATION BY NAIVE MODEL

term of the form Income $\times$ Family. The choice among the set of alternative models is guided primarily by performance on the validation data. For models that perform roughly equally well, simpler models are generally preferred over more complex models. Note also that performance on validation data may be overly optimistic when it comes to predicting performance on data that have not been exposed to the model at all. This is because when the validation data are used to select a final model, we are selecting for how well the model performs with those data and therefore may be incorporating some of the random idiosyncracies of those data into the judgment about the best model. The model still may be the best among those considered, but it will probably not do as well with the unseen data. Therefore, one must consider practical issues such as costs of collecting variables, error proneness, and model complexity in the selection of the final model.

Impact of Single Predictors

As in multiple linear regression, for each predictor X_i we have an estimated coefficient b_i and an associated standard error σ_i. The associated p-value indicates the statistical significance of the predictor X_i, with very low p-values indicating a statistically significant relationship between the predictor and the outcome (given that the other predictors are accounted for), a relationship that is most likely not a result of chance. A statistically significant relationship, however, is not necessarily a *practically significant* one in which the predictor has great impact. If the sample is very large, the *p*-value will be very small simply because the chance uncertainty associated with a small sample is gone. We should also compare the odds of the different predictors, where we can see immediately which predictors have the most impact (given that the other predictors are accounted for) and which have the least impact.

10.4 EXAMPLE OF COMPLETE ANALYSIS: PREDICTING DELAYED FLIGHTS

Predicting flight delays would be useful to a variety of organizations: airport authorities, airlines, aviation authorities. At times, joint task forces have been formed to address the problem. Such an organization, if it were to provide on-going real-time assistance with flight delays, would benefit from some advance notice about flights that are likely to be delayed.

In this simplified illustration, we look at six predictors (see Table 10.1). The outcome of interest is whether the flight is delayed or not (*delayed* means more than 15 minutes late). Our data consist of all flights from the Washington, D.C., area into the New York City area during January 2004. The percent of delayed flights among these 2346 flights is 18%. The data were obtained from the Bureau of Transportation Statistics (available on the Web at www.transtats.bts.gov).

The goal is to predict accurately whether a new flight, not in this dataset, will be delayed or not. Our dependent variable is a binary variable called Delayed, coded as 1 for a delayed flight and 0 otherwise.

Other information that is available on the website, such as distance and arrival time, is irrelevant because we are looking at a certain route (distance, flight time, etc. should be approximately equal). A sample of the data for 20 flights is shown in Table 10.2.

TABLE 10.1   DESCRIPTION OF VARIABLES FOR FLIGHT DELAYS EXAMPLE

Day of Week	Coded as: 1 = Monday, 2 = Tuesday,..., 7 = Sunday
Departure Time	Broken down into 18 intervals between 6 AM and 10 PM
Origin	Three airport codes: DCA (Reagan National), IAD (Dulles), BWI (Baltimore–Washington Int'l)
Destination	Three airport codes: JFK (Kennedy), LGA (LaGuardia), EWR (Newark)
Carrier	Eight airline codes: CO (Continental), DH (Atlantic Coast), DL (Delta), MQ (American Eagle), OH (Comair), RU (Continental Express), UA (United), and US (USAirways)
Weather	Coded as 1 if there was a weather-related delay

TABLE 10.2   SAMPLE OF 20 FLIGHTS

Delayed	Carrier	Day of Week	Departure Time	Destination	Origin	Weather
0	DL	2	728	LGA	DCA	0
1	US	3	1600	LGA	DCA	0
0	DH	5	1242	EWR	IAD	0
0	US	2	2057	LGA	DCA	0
0	DH	3	1603	JFK	IAD	0
0	CO	6	1252	EWR	DCA	0
0	RU	6	1728	EWR	DCA	0
0	DL	5	1031	LGA	DCA	0
0	RU	6	1722	EWR	IAD	0
1	US	1	627	LGA	DCA	0
1	DH	2	1756	JFK	IAD	0
0	MQ	6	1529	JFK	DCA	0
0	US	6	1259	LGA	DCA	0
0	DL	2	1329	LGA	DCA	0
0	RU	2	1453	EWR	BWI	0
0	RU	5	1356	EWR	DCA	0
1	DH	7	2244	LGA	IAD	0
0	US	7	1053	LGA	DCA	0
0	US	2	1057	LGA	DCA	0
0	US	4	632	LGA	DCA	0

The number of flights in each cell for Thursday–Sunday flights is approximately double that of the number of Monday–Wednesday flights. The dataset includes four categorical variables: $X_1 =$ Origin, $X_2 =$ Carrier, $X_3 =$ Day Group (whether the flight was on Monday–Wednesday or Thursday–Sunday), and the response variable $Y =$ Flight Status (delayed or not delayed). In this example we have a binary response variable, or two classes. We start by looking at the pivot table for initial insight into the data (Table 10.3): It appears that more flights departing on Thursday–Sunday are delayed than those leaving on Monday–Wednesday. Also, the worst airport (in term of delays) seems to be IAD. The worst carrier, it

TABLE 10.3   NUMBER OF DELAYED FLIGHTS OUT OF WASHINGTON, D.C. AIRPORTS FOR FOUR CARRIERS BY DAY GROUP[a]

	Carrier				
	CO	DL	RU	US	Total
Airport					
BWI			**14** 19		**14** 19
DCA	**17** 9	**22** 25	**12** 16	**19** 16	**70** 66
IAD			**15** 18		**15** 18
Total	**17** 9	**22** 25	**41** 53	**19** 16	**99** 103

[a] Bold numbers are delayed flights on Monday–Wednesday, regular numbers are delayed flights on Thursday–Sunday.

appears, depends on the day group: on Monday–Wedneday Continental seems to have the most delays, whereas on Thursday–Sunday Delta has the most delays.

Our main goal is to find a model that can obtain accurate classifications of new flights based on their predictor information. In some cases we might be interested in finding a certain percentage of flights that are most/least likely to get delayed. In other cases we may be interested in finding out which factors are associated with a delay (not only in this sample but in the entire population of flights on this route), and for those factors we would like to quantify these effects. A logistic regression model can be used for all these goals.

Data Preprocessing

We first create dummy variables for each of the categorical predictors: 2 dummies for the departure airport (with IAD as the reference airport), 2 for the arrival airport (with JFK as the reference), 7 for the carrier (with USAirways as the reference carrier), 6 for the day (with Sunday as the reference group), 15 for the departure hour (hourly intervals between 6 AM and 10 PM), blocked into hours. This yields a total of 33 dummies. In addition, we have a single dummy for weather delays. This is a very large number of predictors. Some initial investigation and knowledge from airline experts led us to aggregate the day of the week in a more compact way: It is known that delays are much more prevalent on this route on Sundays and Mondays. We therefore use a single dummy that signifies whether or not it is a Sunday or a Monday (denoted by 1).

We then partition the data using a 60% : 40% ratio into training and validation sets. We use the training set to fit a model and the validation set to assess the model's performance.

Model Fitting and Estimation

The estimated model with 28 predictors is given in Figure 10.6. Notice how negative coefficients in the logit model (the Coefficient column) translate into odds coefficients lower than 1, and positive logit coefficients translate into odds coefficients larger than 1.

Model Interpretation

The coefficient for Arrival Airport JFK is estimated from the data to be -0.67. Recall that the reference group is LGA. We interpret this coefficient as follows: $e^{-0.67} = 0.51$ are the odds of a flight arriving at JFK being delayed relative to a flight to LGA being delayed (= the base-case odds), holding all other factors constant. This means that flights to LGA are more likely to be delayed than those to JFK (holding everything else constant). If we take into account statistical significance of the coefficients, we see that in general the departure airport is not

Input variables	Coefficient	Std. Error	p-value	Odds
Constant term	-2.76648855	0.60903645	0.00000556	*
Weather	16.94781685	472.3040772	0.97137541	22926812
ORIGIN_BWI	0.31663841	0.407509	0.43715307	1.37250626
ORIGIN_DCA	-0.52621925	0.37920129	0.1652271	0.59083456
DEP_TIME_BLK_0700-0759	0.17635399	0.52038968	0.73469388	1.19286025
DEP_TIME_BLK_0800-0859	0.37122276	0.4879483	0.44678667	1.44950593
DEP_TIME_BLK_0900-0959	-0.2891154	0.61024719	0.6356656	0.74892575
DEP_TIME_BLK_1000-1059	-0.84254718	0.65849793	0.20072155	0.4306123
DEP_TIME_BLK_1100-1159	0.26919952	0.62188113	0.66510242	1.30891633
DEP_TIME_BLK_1200-1259	0.39577994	0.47712085	0.40681183	1.48554242
DEP_TIME_BLK_1300-1359	0.23689635	0.49711299	0.63368666	1.26730978
DEP_TIME_BLK_1400-1459	0.94953001	0.4257178	0.02571949	2.58449459
DEP_TIME_BLK_1500-1559	0.81428736	0.47320139	0.08528619	2.25756645
DEP_TIME_BLK_1600-1659	0.73656398	0.46096623	0.11007198	2.08874631
DEP_TIME_BLK_1700-1759	0.80683631	0.42013136	0.05480258	2.24080753
DEP_TIME_BLK_1800-1859	0.65816337	0.56922781	0.2475834	1.93124211
DEP_TIME_BLK_1900-1959	1.40413988	0.47974923	0.00342446	4.07202291
DEP_TIME_BLK_2000-2059	0.94785261	0.63308424	0.1343417	2.580163
DEP_TIME_BLK_2100-2159	0.76115495	0.45146817	0.09180449	2.14074731
DEST_EWR	-0.33785093	0.31752595	0.28732395	0.7133016
DEST_JFK	-0.66931868	0.2657896	0.01179471	0.5120573
CARRIER_CO	1.81500936	0.53502011	0.0006928	6.14113379
CARRIER_DH	1.25616693	0.52265555	0.016242	3.51193428
CARRIER_DL	0.41380161	0.33544913	0.21736139	1.51255703
CARRIER_MQ	1.73093832	0.32989427	0.00000015	5.64594936
CARRIER_OH	0.15529965	0.85175836	0.8553251	1.16800785
CARRIER_RU	1.27398086	0.51098496	0.01266023	3.57505608
CARRIER_UA	-0.59911883	1.17384589	0.60977846	0.54929543
Sun-Mon	0.53890741	0.16421914	0.00103207	1.71413302

FIGURE 10.6   ESTIMATED LOGISTIC REGRESSION MODEL FOR DELAYED FLIGHTS (BASED ON THE TRAINING SET)

associated with the chance of delays. For carriers, it appears that four carriers are significantly different from the base carrier (USAirways), with odds of 3.5–6.6 of delays relative to the other airlines. Weather has an enormous coefficient, which is not statistically significant. This is due to the fact that weather delays occurred only on two days (January 26 and 27), and those affected only some of the flights. Flights leaving on Sunday or Monday have, on average, odds of 1.7 of delays relative to other days of the week. Also, odds of delays appear to change over the course of the day, with the most noticeable difference between 7 and 8 PM and the reference category, 6–7 AM.

Model Performance

How should we measure the performance of models? One possible measure is "percent of flights correctly classified." Accurate classification can be obtained from the classification matrix for the validation data. The classification matrix gives a sense of the classification accuracy and what type of misclassification is more frequent. From the classification matrix and error rates in Figure 10.7 it can be seen that the model does better in classifying nondelayed flights correctly and is less accurate in classifying flights that were delayed. (*Note*: The same pattern appears in the classification matrix for the training data, so it is not surprising to see it emerge for new data.) If there is a nonsymmetric cost structure such that one type of misclassification is more costly than the other, the cutoff value can be selected to minimize the cost. Of course, this tweaking should be carried out on the training data and assessed only using the validation data.

In most conceivable situations, it is likely that the purpose of the model will be to identify those flights most likely to be delayed so that resources can be directed toward either reducing the delay or mitigating its effects. Air traffic controllers might work to open up additional air routes, or allocate more controllers to a specific area for a short time. Airlines might bring on personnel to rebook passengers and to activate standby flight crews and aircraft. Hotels might allocate space for stranded travellers. In all cases, the resources available are going to be limited and might vary over time and from organization to organization. In this situation, the most useful model would provide an ordering of flights by their probability of delay, letting the model users decide how far down that list to go in taking action. Therefore, model lift is a useful measure of performance— as you move down that list of flights, ordered by their delay probability, how much better does the model do in predicting delay than would a naive model which is simply the average delay rate for all flights? From the lift curve for the validation data (Figure 10.7) we see that our model is superior to the baseline (simple random selection of flights).

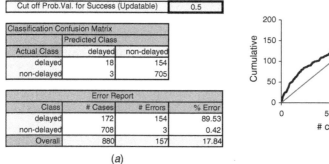

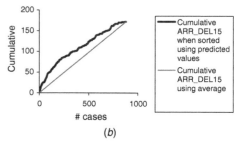

(a)

(b)

FIGURE 10.7   (A) CLASSIFICATION MATRIX AND ERROR RATES AND (B) LIFT CHART FOR THE FLIGHT DELAY VALIDATION DATA

Variable Selection

From the coefficient table for the flights delay model, it appears that several of the variables might be dropped or coded differently. We further explore alternative models by examining pivot tables and charts and using variable selection procedures. First, we find that most carriers depart from a single airport: For those that depart from all three airports, the delay rates are similar regardless of airport. This means that there are multiple combinations of carrier and departure airport that do not include any flights. We therefore drop the departure airport distinction and find that the model performance and fit is not harmed. We also drop the destination airport for a practical reason: Not all carriers fly to all airports. Our model would then be invalid for prediction in nonexistent combinations of carrier and destination airport. We also try regrouping the carriers and hour of day into fewer categories that are more distinguishable with respect to delays. Finally, we apply subset selection. Our final model includes only seven predictors and has the advantage of being more parsimonious. It does, however, include coefficients that are not statistically significant because our goal is prediction accuracy rather than model fit. Also, some of these variables have a practical importance (e.g., weather) and are therefore retained. Figure 10.8 displays the estimated model, with its goodness-of-fit measures, the training and validation classification matrices and error rates, and the lift charts. It can be seen that this model competes well with the larger model in terms of accurate classification and lift.

We therefore conclude with a seven-predictor model that required only the knowledge of the carrier, the day of week, the hour of the day, and whether it is likely that there will be a delay due to weather. The last piece of information is of course not known in advance, but we kept it in our model for purposes of interpretation. The impact of the other factors is estimated while holding weather constant [i.e., (approximately) comparing days with weather delays to days without weather delays]. If the aim is to predict in advance whether a particular flight will be delayed, a model without Weather should be used. To conclude, we can summarize that the highest chance of a nondelayed flight from DC to New York, based on the data from January 2004, would be a flight on Monday–Friday during the late morning hours, on Delta, United, USAirways, or Atlantic Coast Airlines. And clearly, good weather is advantageous!

10.5 APPENDIX: LOGISTIC REGRESSION FOR PROFILING

The presentation of logistic regression in this chapter has been primarily from a data mining perspective, where classification is the goal and performance is evaluated by reviewing results with a validation sample. For reference, some key concepts of a classical statistical perspective, are included below.

The Regression Model

Input variables	Coefficient	Std. Error	p-value	Odds
Constant term	-1.76942575	0.11373349	0	*
Weather	16.77862358	479.4146118	0.97208124	19358154
DEP_TIME_BLK_0600-0659	-0.62896502	0.36761174	0.08709048	0.53314334
DEP_TIME_BLK_0900-0959	-1.26741421	0.47863296	0.00809724	0.28155872
DEP_TIME_BLK_1000-1059	-1.37123489	0.52464402	0.00895813	0.25379336
DEP_TIME_BLK_1300-1359	-0.6303032	0.3188065	0.04803356	0.53243035
Sun-Mon	0.52237105	0.15871418	0.00099736	1.68602061
Carrier_CO_OH_MQ_RU	0.68775123	0.15049717	0.00000488	1.98923719

Residual df	1313
Std. Dev. Estimate	1186.54834
% Success in training data	19.37925814
# Iterations used	16
Multiple R-squared	0.08657298

Training Data Scoring - Summary Report

Cut off Prob.Val. for Success (Updatable)	0.5

Lift Chart (training dataset)

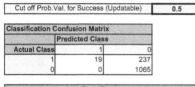

Classification Confusion Matrix		
	Predicted Class	
Actual Class	1	0
1	19	237
0	0	1065

Error Report			
Class	# Cases	# Errors	% Error
1	256	237	92.58
0	1065	0	0.00
Overall	1321	237	17.94

Validation Data Scoring - Summary Report

Cut off Prob.Val. for Success (Updatable)	0.5

Lift Chart (validation dataset)

Classification Confusion Matrix		
	Predicted Class	
Actual Class	1	0
1	13	159
0	0	708

Error Report			
Class	# Cases	# Errors	% Error
1	172	159	92.44
0	708	0	0.00
Overall	880	159	18.07

FIGURE 10.8 OUTPUT FOR LOGISTIC REGRESSION MODEL WITH ONLY SEVEN PREDICTORS.

Appendix A: Why Linear Regression Is Inappropriate for a Categorical Response

Now that you have seen how logistic regression works, we explain why linear regression is not suitable. Technically, one can apply a multiple linear regression model to this problem, treating the dependent variable Y as continuous. Of course, Y must be coded numerically (e.g., 1 for customers who did accept the loan offer and 0 for customers who did not accept it). Although software will yield an output that at first glance may seem usual (e.g., Figure 10.9), a closer look will reveal several anomalies:

1. Using the model to predict Y for each of the observations (or classify them) yields predictions that are not necessarily 0 or 1.

The Regression Model

Input variables	Coefficient	Std. Error	p-value	SS
Constant term	-0.23462872	0.01328709	0	27.26533127
Income	0.00318939	0.00009888	0	67.95861816
Family	0.03294198	0.00383914	0	4.53180361
CD Account	0.27016363	0.01788521	0	13.18045044

ANOVA

Source	df	SS	MS	F-statistic	p-value
Regression	3	85.67087221	28.5569574	494.364771	5.9883E-261
Error	2996	173.063797	0.057764952		
Total	2999	258.7346692			

FIGURE 10.9 **OUTPUT FOR MULTIPLE LINEAR REGRESSION MODEL OF PERSONAL LOAN ON THREE PREDICTORS.**

2. A look at the histogram or probability plot of the residuals reveals that the assumption that the dependent variable (or residuals) follows a normal distribution is violated. Clearly, if Y takes only the values 0 and 1, it cannot be normally distributed. In fact, a more appropriate distribution for the number of 1's in the dataset is the binomial distribution with $p = P(Y = 1)$.

3. The assumption that the variance of Y is constant across all classes is violated. Since Y follows a binomial distribution, its variance is $np(1 - p)$. This means that the variance will be higher for classes where the probability of adoption, p, is near 0.5 than where it is near 0 or 1.

Below you will find partial output from running a multiple linear regression of Personal Loan (PL, coded as PL = 1 for customers who accepted the loan offer and PL = 0 otherwise) on three of the predictors.
The estimated model is

$$\widehat{PL} = -0.2346 + 0.0032 \text{ Income} + 0.0329 \text{ Family} + 0.27016363 \text{ CD}.$$

To predict whether a new customer will accept the personal loan offer (PL = 1) or not (PL = 0), we input the information on its values for these three predictors. For example, we would predict the loan offer acceptance of a customer with an annual income of $50K with two family members who does not hold CD accounts in Universal Bank to be $-0.2346 + (0.0032)(50) + (0.0329)(2) = -0.009$. Clearly, this is not a valid "loan acceptance" value. Furthermore, the histogram of the residuals (Figure 10.10) reveals that the residuals are probably not normally distributed. Therefore, our estimated model is based on violated assumptions.

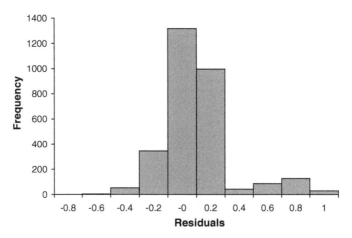

FIGURE 10.10 HISTOGRAM OF RESIDUALS FROM A MULTIPLE LINEAR REGRESSION MODEL OF LOAN ACCEPTANCE ON THE THREE PREDICTORS. THIS SHOWS THAT THE RESIDUALS DO NOT FOLLOW THE NORMAL DISTRIBUTION THAT THE MODEL ASSUMES.

Appendix B: Evaluating Goodness of Fit

When the purpose of the analysis is profiling (i.e., explaining the differences between the classes in terms of predictor values) we want to go beyond simply assessing how well the model classifies new data, and also assess how well the model fits the data it was trained on. For example, if we are interested in characterizing loan offer acceptors versus nonacceptors in terms of income, education, and so on, we want to find a model that fits the data best. However, since overfitting is a major danger in classification, a "too good" fit of the model to the training data should raise suspicions. In addition, questions regarding the usefulness of specific predictors can arise even in the context of classification models. We therefore mention some of the popular measures that are used to assess how well the model fits the data. Clearly, we look at the training set in order to evaluate goodness of fit.

Overall Fit   As in multiple linear regression, we first evaluate the overall fit of the model to the data before looking at single predictors. We ask: Is this group of predictors better than a simple naive model for explaining the different classes[5]?

The deviance D is a statistic that measures overall goodness of fit. It is similar to the concept of sum of squared errors (SSE) in the case of least-squares estimation (used in linear regression). We compare the deviance of our model, D (called Std Dev Estimate in XLMiner, e.g., in Figure 10.11), to the deviance

---

[5] In a naive model, no explanatory variables exist and each observation is classified as belonging to the majority class.

Residual df	2987
Std. Dev. Estimate	652.5175781
% Success in training data	9.533333333
# Iterations used	11
Multiple R-squared	0.65443069

FIGURE 10.11   **MEASURES OF GOODNESS OF FIT FOR UNIVERSAL BANK TRAINING DATA WITH A 12-PREDICTOR MODEL**

of the naive model, D_0. If the reduction in deviance is statistically significant (as indicated by a low p-value[6] or in XLMiner by a high *multiple R^2*), we consider our model to provide a good overall fit. XLMiner's *Multiple-R-Squared* measure is computed as $(D_0 - D)/D_0$. Given the model deviance and the multiple R^2, we can compute the null deviance by $D_0 = D/(1 - R^2)$.

Finally, the classification matrix and lift chart for the *training data* (Figure 10.12) give a sense of how accurately the model classifies the data. If the model fits the data well, we expect it to classify these data accurately into their actual classes.

Appendix C: Logistic Regression for More Than Two Classes

The logistic model for a binary response can be extended for more than two classes. Suppose that there are m classes. Using a logistic regression model, for each observation we would have m probabilities of belonging to each of the m classes. Since the m probabilities must add up to 1, we need estimate only $m - 1$ probabilities.

Ordinal Classes   Ordinal classes are classes that have a meaningful order. For example, in stock recommendations, the three classes *buy*, *hold*, and *sell* can be treated as ordered. As a simple rule, if classes can be numbered in a meaningful way, we consider them ordinal. When the number of classes is large (typically, more than 5), we can treat the dependent variable as continuous and perform multiple linear regression. When $m = 2$, the logistic model described above is used. We therefore need an extension of the logistic regression for a small number of ordinal classes ($3 \leq m \leq 5$). There are several ways to extend the binary class case. Here we describe the *proportional odds* or *cumulative logit method*. For other methods, see Hosmer and Lemeshow (2000).

---

[6] The difference between the deviance of a naive model and deviance of the model at hand approximately follows a chi-squared distribution with k degrees of freedom, where k is the number of predictors in the model at hand. Therefore, to get the p-value, compute the difference between the deviances (d) and then look up the probability that a chi-squared variable with k degrees of freedom is larger than d. This can be done using =CHIDIST(d, k) in Excel.

Cut off Prob.Val. for Success (Updatable)	0.5

Classification Confusion Matrix

Actual Class	Predicted Class	
	1	0
1	201	85
0	25	2689

Error Report			
Class	# Cases	# Errors	% Error
1	286	85	29.72
0	2714	25	0.92
Overall	3000	110	3.67

(a)

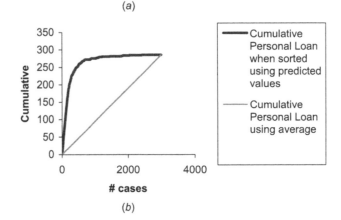

(b)

FIGURE 10.12   (A) CLASSIFICATION MATRIX AND (B) LIFT CHART FOR TRAINING DATA FOR UNIVERSAL BANK TRAINING DATA WITH 12 PREDICTORS

For simplicity of interpretation and computation, we look at *cumulative* probabilities of class membership. For example, in the stock recommendations we have $m = 3$ classes. Let us denote them by $1 = $ buy, $2 = $ hold, and $3 = $ sell. The probabilities that are estimated by the model are $P(Y \leq 1)$, (the probability of a buy recommendation) and $P(Y \leq 2)$ (the probability of a buy or hold recommendation). The three noncumulative probabilities of class membership can easily be recovered from the two cumulative probabilities:

$$P(Y = 1) = P(Y \leq 1),$$

$$P(Y = 2) = P(Y \leq 2) - P(Y \leq 1),$$

$$P(Y = 3) = 1 - P(Y \leq 2).$$

Next, we want to model each logit as a function of the predictors. Corresponding to each of the $m - 1$ cumulative probabilities is a logit. In our example we would have

$$\text{logit(buy)} = \log \frac{P(Y \le 1)}{1 - P(Y \le 1)},$$

$$\text{logit(buy or hold)} = \log \frac{P(Y \le 2)}{1 - P(Y \le 2)}.$$

Each of the logits is then modeled as a linear function of the predictors (as in the two-class case). If in the stock recommendations we have a single predictor x, we have two equations:

$$\text{logit(buy)} = \alpha_0 + \beta_1 x,$$

$$\text{logit(buy or hold)} = \beta_0 + \beta_1 x.$$

This means that both lines have the same slope (β_1) but different intercepts. Once the coefficients α_0, β_0, β_1 are estimated, we can compute the class membership probabilities by rewriting the logit equations in terms of probabilities. For the three-class case, for example, we would have

$$P(Y = 1) = P(Y \le 1) = \frac{1}{1 + e^{-(a_0 + b_1 x)}},$$

$$P(Y = 2) = P(Y \le 2) - P(Y \le 1) = \frac{1}{1 + e^{-(b_0 + b_1 x)}} - \frac{1}{1 + e^{-(a_0 + b_1 x)}},$$

$$P(Y = 3) = 1 - P(Y \le 2) = 1 - \frac{1}{1 + e^{-(b_0 + b_1 x)}},$$

where a_0, b_0, and b_1 are the estimates obtained from the training set.

For each observation we now have the estimated probabilities that it belongs to each of the classes. In our example, each stock would have three probabilities: for a buy recommendation, a hold recommendation, and a sell recommendation. The last step is to classify the observation into one of the classes. This is done by assigning it to the class with the highest membership probability. So if a stock had estimated probabilities $P(Y = 1) = 0.2$, $P(Y = 2) = 0.3$, and $P(Y = 3) = 0.5$, we would classify it as getting a sell recommendation.

This procedure is currently not implemented in XLMiner. Other non-Excel-based packages that do have such an implementation are Minitab and SAS.

Nominal Classes   When the classes cannot be ordered and are simply different from one another, we are in the case of nominal classes. An example is the choice between several brands of cereal. A simple way to verify that the classes are nominal is when it makes sense to tag them as A, B, C, ..., and the assignment of letters to classes does not matter. For simplicity, let us assume that there are $m = 3$ brands of cereal that consumers can choose from (assuming that each consumer chooses one). Then we estimate the probabilities $P(Y = A)$, $P(Y = B)$, and $P(Y = C)$. As before, if we know two of the probabilities, the third probability is determined. We therefore use one of the classes as the reference class. Let us use C as the reference brand.

The goal, once again, is to model the class membership as a function of predictors. So in the cereals example we might want to predict which cereal will be chosen if we know the cereal's price, x.

Next, we form $m - 1$ pseudologit equations that are linear in the predictors. In our example we would have

$$\text{logit}(A) = \log \frac{P(Y = A)}{P(Y = C)} = \alpha_0 + \alpha_1 x,$$

$$\text{logit}(B) = \log \frac{P(Y = B)}{P(Y = C)} = \beta_0 + \beta_1 x.$$

Once the four coefficients are estimated from the training set, we can estimate the class membership probabilities[7]:

$$P(Y = A) = \frac{e^{a_0 + a_1 x}}{1 + e^{a_0 + a_1 x} + e^{b_0 + b_1 x}},$$

$$P(Y = B) = \frac{e^{b_0 + b_1 x}}{1 + e^{a_0 + a_1 x} + e^{b_0 + b_1 x}},$$

$$P(Y = C) = 1 - P(Y = A) - P(Y = B),$$

where a_0, a_1, b_0, and b_1 are the coefficient estimates obtained from the training set. Finally, an observation is assigned to the class that has the highest probability.

---

[7] From the two logit equations we see that $\begin{aligned} P(Y = A) &= P(Y = C)e^{\alpha_0 + \alpha_1 x}, \\ P(Y = B) &= P(Y = C)e^{\beta_0 + \beta_1 x}, \end{aligned}$ and since $P(Y = A) +$ $P(Y = B) + P(Y = C) = 1$, we have $\begin{aligned} P(Y = C) &= 1 - P(Y = C)e^{\alpha_0 + \alpha_1 x} - P(Y = C)e^{\beta_0 + \beta_1 x} \\ &> P(Y = C) = \frac{1}{e^{\alpha_0 + \alpha_1 x} + e^{\beta_0 + \beta_1 x}}. \end{aligned}$ By plugging this form into the two equations above it, we also obtain the membership probabilities in classes A and B.

PROBLEMS ▪

10.1 Financial Condition of Banks. The file Banks.xls includes data on a sample of 20 banks. The Financial Condition column records the judgment of an expert on the financial condition of each bank. This dependent variable takes one of two possible values—*weak* or *strong*—according to the financial condition of the bank. The predictors are two ratios used in the financial analysis of banks: TotLns&Lses/Assets is the ratio of total loans and leases to total assets and TotExp/Assets is the ratio of total expenses to total assets. The target is to use the two ratios for classifying the financial condition of a new bank.

Run a logistic regression model (on the entire dataset) that models the status of a bank as a function of the two financial measures provided. Specify the *success* class as weak (this is similar to creating a dummy that is 1 for financially weak banks and 0 otherwise), and use the default cutoff value of 0.5.

a. Write the estimated equation that associates the financial condition of a bank with its two predictors in three formats:

  i. The logit as a function of the predictors
  ii. The odds as a function of the predictors
  iii. The probability as a function of the predictors

b. Consider a new bank whose total loans and leases/assets ratio = 0.6 and total expenses/assets ratio = 0.11. From your logistic regression model, estimate the following four quantities for this bank (use Excel to do all the intermediate calculations; show your final answers to four decimal places): the logit, the odds, the probability of being financially weak, and the classification of the bank.

c. The cutoff value of 0.5 is used in conjunction with the probability of being financially weak. Compute the threshold that should be used if we want to make a classification based on the odds of being financially weak, and the threshold for the corresponding logit.

d. Interpret the estimated coefficient for the total loans & leases to total assets ratio (TotLns&Lses/Assets) in terms of the odds of being financially weak.

e. When a bank that is in poor financial condition is misclassified as financially strong, the misclassification cost is much higher than when a financially strong bank is misclassified as weak. To minimize the expected cost of misclassification, should the cutoff value for classification (which is currently at 0.5) be increased or decreased?

10.2 Identifying Good System Administrators. A management consultant is studying the roles played by experience and training in a system administrator's ability to complete a set of tasks in a specified amount of time. In particular, she is interested in discriminating between administrators who are able to complete given tasks within a specified time and those who are not. Data are collected on the performance of 75 randomly selected administrators. They are stored in the file SystemAdministrators.xls.

The variable Experience measures months of full-time system administrator experience, while Training measures the number of relevant training credits. The dependent variable Completed is either Yes or No, according to whether or not the administrator completed the tasks.

    a. Create a scatterplot of Experience versus Training using color or symbol to differentiate programmers who complete the task from those who did not complete it. Which predictor(s) appear(s) potentially useful for classifying task completion?

    b. Run a logistic regression model with both predictors using the entire dataset as training data. Among those who complete the task, what is the percentage of programmers who are incorrectly classified as failing to complete the task?

    c. To decrease the percentage in part (c), should the cutoff probability be increased or decreased?

    d. How much experience must be accumulated by a programmer with 4 years of training before his or her estimated probability of completing the task exceeds 50%?

10.3 Sales of Riding Mowers. A company that manufactures riding mowers wants to identify the best sales prospects for an intensive sales campaign. In particular, the manufacturer is interested in classifying households as prospective owners or nonowners on the basis of Income (in $1000s) and Lot Size (in 1000 ft^2). The marketing expert looked at a random sample of 24 households, given in the file RidingMowers.xls. Use all the data to fit a logistic regression of ownership on the two predictors.

    a. What percentage of households in the study were owners of a riding mower?

    b. Create a scatterplot of Income versus Lot Size using color or symbol to differentiate owners from nonowners. From the scatterplot, which class seems to have the higher average income, owners or nonowners?

    c. Among nonowners, what is the percentage of households classified correctly?

    d. To increase the percentage of correctly classified nonowners, should the cutoff probability be increased or decreased?

    e. What are the odds that a household with a $60K income and a lot size of 20,000 ft^2 is an owner?

    f. What is the classification of a household with a $60K income and a lot size of 20,000 ft^2?

    g. What is the minimum income that a household with 16,000 ft^2 lot size should have before it is classified as an owner?

10.4 Competitive Auctions on eBay.com. The file eBayAuctions.xls contains information on 1972 auctions transacted on eBay.com during May–June 2004. The goal is to use these data to build a model that will distinguish competitive auctions from noncompetitive ones. A competitive auction is defined as an auction with at least two bids placed on the item being auctioned. The data include variables that describe the item (auction category), the seller (his or her eBay rating), and the auction terms that the seller selected (auction duration, opening price, currency, day of week of auction close). In addition, we have the price at which the auction closed. The goal is to predict whether or not the auction will be competitive.

    Data Preprocessing. Create dummy variables for the categorical predictors. These include Category (18 categories), Currency (USD, GBP, euro), EndDay (Monday–Sunday), and Duration (1, 3, 5, 7, or 10 days). Split the data into training and validation datasets using a 60% : 40% ratio.

    a. Create pivot tables for the average of the binary dependent variable (Competitive?) as a function of the various categorical variables (use the original variables, not the dummies). Use the information in the tables to reduce the number of dummies that will be used in the model. For example, categories that appear most similar with respect to the distribution of competitive auctions could be combined.

b. Run a logistic model with all predictors with a cutoff of 0.5. To remain within the limitation of 30 predictors, combine some of the categories of categorical predictors.

c. If we want to predict at the start of an auction whether it will be competitive, we cannot use the information on the closing price. Run a logistic model with all predictors as above, excluding price. How does this model compare to the full model with respect to accurate prediction?

d. Interpret the meaning of the coefficient for closing price. Does closing price have a practical significance? Is it statistically significant for predicting competitiveness of auctions? (Use a 10% significance level.)

e. Use stepwise selection and an exhaustive search to find the model with the best fit to the training data. Which predictors are used?

f. Use stepwise selection and an exhaustive search to find the model with the lowest predictive error rate (use the validation data). Which predictors are used?

g. What is the danger in the best predictive model that you found?

h. Explain why the best-fitting model and the best predictive models are the same or different.

i. If the major objective is accurate classification, what cutoff value should be used?

j. Based on these data, what auction settings set by the seller (duration, opening price, ending day, currency) would you recommend as being most likely to lead to a competitive auction?

## CHAPTER 11

Neural Nets

In this chapter we describe neural nets, a flexible data-driven method that can be used for classification or prediction. Although considered a "blackbox" in terms of interpretability, neural nets have been highly successful in terms of predictive accuracy. We discuss the concepts of "nodes" and "layers" (input layers, output layers, and hidden layers) and how they connect to form the structure of a network. We then explain how a neural network is fitted to data using a numerical example. Because overfitting is a major danger with neural nets, we present a strategy for avoiding it. We describe the different parameters that a user must specify and explain the effect of each on the process. Finally, we discuss the usefulness of neural nets and their limitations.

11.1 INTRODUCTION

Neural networks, also called *artificial neural networks*, are models for classification and prediction. The neural network is based on a model of biological activity in the brain, where neurons are interconnected and learn from experience. Neural networks mimic the way that human experts learn. The learning and memory properties of neural networks resemble the properties of human learning and memory, and they also have a capacity to generalize from particulars.

A number of successful applications have been reported in financial applications [see Trippi and Turban, (1996)] such as bankruptcy predictions, currency market trading, picking stocks and commodity trading, detecting fraud in credit card and monetary transactions, and customer relationship management (CRM). There have also been a number of very successful applications of neural nets in

Data Mining for Business Intelligence, By Galit Shmueli, Nitin R. Patel, and Peter C. Bruce
Copyright © 2010 John Wiley & Sons Inc.

engineering applications. One of the best known is ALVINN, an autonomous vehicle driving application for normal speeds on highways. Using as input a 30×32 grid of pixel intensities from a fixed camera on the vehicle, the classifier provides the direction of steering. The response variable is a categorical one with 30 classes, such as *sharp left*, *straight ahead*, and *bear right*.

The main strength of neural networks is their high predictive performance. Their structure supports capturing very complex relationships between predictors and a response, which is often not possible with other classifiers.

11.2 CONCEPT AND STRUCTURE OF A NEURAL NETWORK

The idea behind neural networks is to combine the input information in a very flexible way that captures complicated relationships among these variables and between them and the response variable. For instance, recall that in linear regression models the form of the relationship between the response and the predictors is assumed to be linear. In many cases the exact form of the relationship is much more complicated or is generally unknown. In linear regression modeling we might try different transformations of the predictors, interactions between predictors, and so on. In comparison, in neural networks the user is not required to specify the correct form. Instead, the network tries to learn about such relationships from the data. In fact, linear regression and logistic regression can be thought of as special cases of very simple neural networks that have only input and output layers and no hidden layers.

Although researchers have studied numerous different neural network architectures, the most successful applications in data mining of neural networks have been *multilayer feedforward networks*. These are networks in which there is an *input layer* consisting of nodes that simply accept the input values and successive layers of nodes that receive input from the previous layers. The outputs of nodes in a layer are inputs to nodes in the next layer. The last layer is called the *output layer*. Layers between the input and output layers are known as *hidden layers*. A feedforward network is a fully connected network with a one-way flow and no cycles. Figure 11.1 shows a diagram for this architecture (two hidden layers are shown in this example).

11.3 FITTING A NETWORK TO DATA

To illustrate how a neural network is fitted to data, we start with a very small illustrative example. Although the method is by no means operational in such a small example, it is useful for explaining the main steps and operations, for

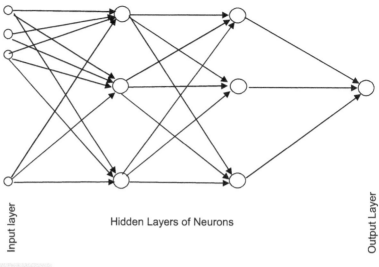

Input layer

Hidden Layers of Neurons

Output Layer

FIGURE 11.1   **MULTILAYER FEEDFORWARD NEURAL NETWORK**

showing how computations are done, and for integrating all the different aspects of neural network data fitting. We later discuss a more realistic setting.

Example 1: Tiny Dataset

Consider the following very small dataset. Table 11.1 includes information on a tasting score for a certain processed cheese. The two predictors are scores for fat and salt, indicating the relative presence of fat and salt in the particular cheese sample (where 0 is the minimum amount possible in the manufacturing process and 1 the maximum). The output variable is the cheese sample's consumer taste acceptance, where 1 indicates that a taste test panel likes the cheese and 0 that it does not like it.

Figure 11.2 describes an example of a typical neural net that could be used for predicting the acceptance for these data. We numbered the nodes in the example from 1 to 6. Nodes 1 and 2 belong to the input layer, nodes 3–5 belong to the hidden layer, and node 6 belongs to the output layer. The values on the connect-

TABLE 11.1   **TINY EXAMPLE ON TASTING SCORES FOR SIX CONSUMERS AND TWO PREDICTORS**

Obs.	Fat Score	Salt Score	Acceptance
1	0.2	0.9	1
2	0.1	0.1	0
3	0.2	0.4	0
4	0.2	0.5	0
5	0.4	0.5	1
6	0.3	0.8	1

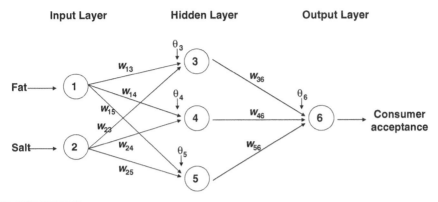

Input Layer   **Hidden Layer**   **Output Layer**

FIGURE 11.2 **NEURAL NETWORK FOR THE TINY EXAMPLE. CIRCLES REPRESENT NODES, $W_{i,j}$ ON ARROWS ARE WEIGHTS, AND θ_j ARE NODE BIAS VALUES**

ing arrows are called *weights*, and the weight on the arrow from node i to node j is denoted by $w_{i,j}$. The additional *bias* nodes, denoted by θ_j, serve as an intercept for the output from node j. These are all explained in further detail below.

Computing Output of Nodes

We discuss the input and output of the nodes separately for each of the three types of layers (input, hidden, and output). The main difference is the function used to map from the input to the output of the node.

Input nodes take as input the values of the predictors. Their output is the same as the input. If we have p predictors, the input layer will usually include p nodes. In our example there are two predictors, and therefore the input layer (shown in Figure 11.2) includes two nodes, each feeding into each node of the hidden layer. Consider the first observation: The input into the input layer is fat $= 0.2$ and salt $= 0.9$, and the output of this layer is also $x_1 = 0.2$ and $x_2 = 0.9$.

Hidden layer nodes take as input the output values from the input layer. The hidden layer in this example consists of three nodes, each receiving input from all the input nodes. To compute the output of a hidden layer node, we compute a weighted sum of the inputs and apply a certain function to it. More formally, for a set of input values $x_1, x_2, \ldots, x_p$, we compute the output of node j by taking the weighted sum[1] $\theta_j + \sum_{i=1}^{p} w_{ij}x_i$, where $\theta_j, w_{1,j}, \ldots, w_{p,j}$ are weights that are initially set randomly, then adjusted as the network "learns." Note that θ_j, also called the *bias* of node j, is a constant that controls the level of contribution of node j. In the next step we take a function g of this sum. The function g, also called a *transfer function*, is some monotone function, and examples

---

[1] Other options exist for combining inputs, such as taking the maximum or minimum of the weighted inputs rather than their sum, but they are much less popular.

are a linear function $[g(s) = bs]$, an exponential function $[g(s) = \exp(bs)]$, and a logistic/sigmoidal function $[g(s) = 1/1 + e^{-s}]$. This last function is by far the most popular one in neural networks. Its practical value arises from the fact that it has a squashing effect on very small or very large values but is almost linear in the range where the value of the function is between 0.1 and 0.9.

If we use a logistic function, we can write the output of node j in the hidden layer as

$$\text{Output}_j = g\left(\theta_j + \sum_{i=1}^{p} w_{ij} x_i\right) = \frac{1}{1 + e^{-(\theta_j + \sum_{i=1}^{p} w_{ij} x_i)}}. \tag{11.1}$$

Initializing the Weights   The values of θ_j and w_{ij} are typically initialized to small (generally, random) numbers in the range 0.00 ± 0.05. Such values represent a state of no knowledge by the network, similar to a model with no predictors. The initial weights are used in the first round of training.

Returning to our example, suppose that the initial weights for node 3 are $\theta_3 = -0.3$, $w_{1,3} = 0.05$, and $w_{2,3} = 0.01$ (as shown in Figure 11.3). Using the logistic function, we can compute the output of node 3 in the hidden layer (using the first observation) as

$$\text{Output}_3 = \frac{1}{1 + e^{-[-0.3+(0.05)(0.2)+(0.01)(0.9)]}} = 0.43.$$

Figure 11.3 shows the initial weights, inputs, and outputs for observation 1 in our tiny example. If there is more than one hidden layer, the same calculation applies, except that the input values for the second, third, and so on hidden layers would be the output of the preceding hidden layer. This means that the number of input values into a certain node is equal to the number of nodes in the preceding layer. (If there was an additional hidden layer in our

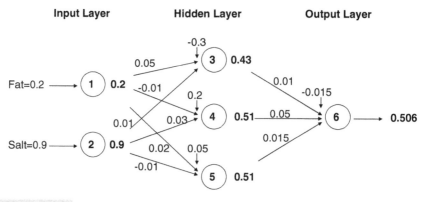

FIGURE 11.3   COMPUTING NODE OUTPUTS (IN BOLDFACE TYPE) USING THE FIRST OBSERVATION IN THE TINY EXAMPLE AND A LOGISTIC FUNCTION

example, its nodes would receive input from the three nodes in the first hidden layer.)

Finally, the output layer obtains input values from the (last) hidden layer. It applies the same function as above to create the output. In other words, it takes a weighted average of its input values and then applies the function g. This is the prediction of the model. For classification we use a cutoff value (for a binary response) or the output node with the largest value (for more than two classes).

Returning to our example, the single output node (node 6) receives input from the three hidden layer nodes. We can compute the prediction (the output of node 6) by

$$\text{Output}_6 = \frac{1}{1 + e^{-[-0.015+(0.01)(0.430)+(0.05)(0.507)+(0.015)(0.511)]}} = 0.506.$$

To classify this record, we use the cutoff of 0.5 and obtain the classification into class 1 (because $0.506 > 0.5$).

Relation to Linear and Logistic Regression   Consider a neural network with a single output node and no hidden layers. For a dataset with p predictors, the output node receives $x_1, x_2, \ldots, x_p$, takes a weighted sum of these, and applies the g function. The output of the neural network is therefore $g\left(\theta + \sum_{i=1}^{p} w_i x_i\right)$.

First, consider a numerical output variable y. If g is the identity function $[g(s) = s]$, the output is simply

$$\hat{y} = \theta + \sum_{i=1}^{p} w_i x_i.$$

This is exactly equivalent to the formulation of a multiple linear regression! This means that a neural network with no hidden layers, a single output node, and an identity function g searches only for linear relationships between the response and the predictors.

Now consider a binary output variable y. If g is the logistic function, the output is simply

$$\hat{P}(Y = 1) = \frac{1}{1 + e^{\theta + \sum_{i=1}^{p} w_i x_i}},$$

which is equivalent to the logistic regression formulation!

In both cases, although the formulation is equivalent to the linear and logistic regression models, the resulting estimates for the weights (*coefficients* in linear and logistic regression) can differ because the estimation method is different. The neural net estimation method is different from least squares, the method used to calculate coefficients in linear regression, or the maximum-likelihood method used in logistic regression. We explain below the method by which the neural network learns.

Preprocessing the Data

Neural networks perform best when the predictors and response variables are on a scale of [0,1]. For this reason, all variables should be scaled to a [0,1] interval before entering them into the network. For a numerical variable X that takes values in the range $[a, b]$ $(a < b)$, we normalize the measurements by subtracting a and dividing by $b - a$. The normalized measurement is then

$$X_{\text{norm}} = \frac{X - a}{b - a}.$$

Note that if $[a, b]$ is within the [0,1] interval, the original scale will be stretched.

If a and b are not known, we can estimate them from the minimal and maximal values of X in the data. Even if new data exceed this range by a small amount, yielding normalized values slightly lower than 0 or larger than 1, this will not affect the results much.

For binary variables, no adjustment needs to be made other than creating dummy variables. For categorical variables with m categories, if they are ordinal in nature, a choice of m fractions in [0,1] should reflect their perceived ordering. For example, if four ordinal categories are equally distant from each other, we can map them to [0, 0.25, 0.5, 1]. If the categories are nominal, transforming into $m - 1$ dummies is a good solution.

Another operation that improves the performance of the network is to transform highly skewed predictors. In business applications, there tend to be many highly right-skewed variables (such as income). Taking a log transform of a right-skewed variable will usually spread out the values more symmetrically.

Training the Model

Training the model means estimating the weights θ_j and w_{ij} that lead to the best predictive results. The process that we described earlier for computing the neural network output for an observation is repeated for all the observations in the training set. For each observation the model produces a prediction that is then compared with the actual response value. Their difference is the error for the output node. However, unlike least squares or maximum likelihood, where a global function of the errors (e.g., sum of squared errors) is used for estimating the coefficients, in neural networks the estimation process uses the errors iteratively to update the estimated weights.

In particular, the error for the output node is distributed across all the hidden nodes that led to it, so that each node is assigned "responsibility" for part of the error. Each of these node-specific errors is then used for updating the weights.

Back Propagation of Error   The most popular method for using model errors to update weights ("learning") is an algorithm called *back propagation*. As the name implies, errors are computed from the last layer (the output layer) back to the hidden layers.

Let us denote by $\hat{y}_k$ the output from output node k. The error associated with output node k is computed by

$$\text{err}_k = \hat{y}_k(1 - \hat{y}_k)(y_k - \hat{y}_k).$$

Notice that this is similar to the ordinary definition of an error $(y_k - \hat{y}_k)$ multiplied by a correction factor. The weights are then updated as follows:

$$\theta_j^{\text{new}} = \theta_j^{\text{old}} + l\text{err}_j, \tag{11.2}$$

$$w_{i,j}^{\text{new}} = w_{i,j}^{\text{old}} + l\text{err}_j,$$

where l is a *learning rate* or *weight decay* parameter, a constant ranging typically between 0 and 1, which controls the amount of change in weights from one iteration to the other.

In our example, the error associated with the output node for the first observation is $(0.506)(1 - 0.506)(1 - 0.506) = 0.123$. This error is then used to compute the errors associated with the hidden layer nodes, and those weights are updated accordingly using a formula similar to (11.2).

Two methods for updating the weights are case updating and batch updating. In *case updating*, the weights are updated after each observation is run through the network (called a *trial*). For example, if we used case updating in the tiny example, the weights would first be updated after running observation 1 as follows: Using a learning rate of 0.5, the weights θ_6 and $w_{3,6}$, $w_{4,6}$, and $w_{5,6}$ are updated to

$$\theta_6 = -0.015 + (0.5)(0.123) = 0.047,$$

$$w_{3,6} = 0.01 + (0.5)(0.123) = 0.072,$$

$$w_{4,6} = 0.05 + (0.5)(0.123) = 0.112,$$

$$w_{5,6} = 0.015 + (0.5)(0.123) = 0.077.$$

These new weights are next updated after the second observation is run through the network, the third, and so on, until all observations are used. This is called one *epoch*, *sweep*, or *iteration* through the data. Typically, there are many epochs.

In *batch updating*, the entire training set is run through the network before each updating of weights takes place. In that case, the errors err_k in the updating equation is the sum of the errors from all observations. In practice, case updating tends to yield more accurate results than batch updating but requires a longer runtime. This is a serious consideration since, even in batch updating, hundreds or even thousands of sweeps through the training data are executed.

When does the updating stop? The most common conditions are one of the following:

1. When the new weights are only incrementally different from those of the preceding iteration

2. When the misclassification rate reaches a required threshold

3. When the limit on the number of runs is reached

Let us examine the output from running a neural network on the tiny data. Following Figures 11.2 and 11.3, we used a single hidden layer with three nodes. The weights and classification matrix are shown in Figure 11.4. We can see that the network misclassifies all 1 observations and correctly classifies all 0 observations. This is not surprising since the number of observations is too small for estimating the 13 weights. However, for purposes of illustration we discuss the remainder of the output.

Training Data Scoring - Summary Report

Cut off Prob.Val. for Success (Updatable)	0.5

Classification Confusion Matrix

	Predicted Class	
Actual Class	1	0
1	0	3
0	0	3

Error Report

Class	# Cases	# Errors	% Error
1	3	3	100.00
0	3	0	0.00
Overall	6	3	50.00

Interlayer Connections Weights

Hidden Layer # 1	Input Layer		
	fat	salt	Bias Node
Node # 1	-0.110424	-0.0800683	0.011531
Node # 2	-0.10581	-0.10347	-0.0447816
Node # 3	-0.0400986	0.128012	-0.0534663

Output Layer	Hidden Layer # 1			
	Node # 1	Node # 2	Node # 3	Bias Node
1	-0.0964131	-0.1029	0.0763853	0.0470577
0	-0.00586047	0.100234	-0.0960382	0.0130296

Row Id.	Predicted Class	Actual Class	Prob. for 1 (success)	fat	salt
1	0	1	0.498658971	0.2	0.9
2	0	0	0.497477278	0.1	0.1
3	0	0	0.497954285	0.2	0.4
4	0	0	0.498202273	0.4	0.5
5	0	1	0.49800783	0.3	0.4
6	0	1	0.498571499	0.3	0.8

FIGURE 11.4 OUTPUT FOR NEURAL NETWORK WITH A SINGLE HIDDEN LAYER WITH THREE NODES FOR THE TINY DATA EXAMPLE

To run a neural net in XLMiner, choose the *Neural Network* option in either the *Prediction* or *Classification* menu, depending on whether your response variable is quantitative or categorical. There are several options that the user can choose, which are described in the software guide. Notice, however, two points:

1. *Normalize input data* applies standard normalization (subtracting the mean and dividing by the standard deviation). Scaling to a [0,1] interval should be done beforehand.

2. XLMiner employs case updating. The user can specify the number of epochs to run.

The final network weights are shown in two tables in the XLMiner output under *Inter-layer connection weights*. The upper table shows these weights in a format similar to that of our previous diagrams. The first table in Figure 11.4 shows the weights that connect the input layer and the hidden layer. The bias nodes in the last column are the weights θ_3, θ_4, and θ_5. The weights in this table are used to compute the output of the hidden layer nodes. Figure 11.5 shows these weights in a format similar to that of our previous diagrams. The weights were computed iteratively after choosing a random initial set of weights (like the set we chose in Figure 11.3). We use the weights in the way we described earlier to compute the hidden layer's output. For instance, for the first observation, the output of our previous node 3 (denoted *Node #1* in XLMiner) is

$$\text{Output}_3 = \frac{1}{1 + e^{-\{0.0115 + (-0.11)(0.2) + (-0.08)(0.9)\}}} = 0.48.$$

Similarly, we can compute the output from the two other hidden nodes for the same observation and get output$_4 = 0.46$ and output$_5 = 0.51$.

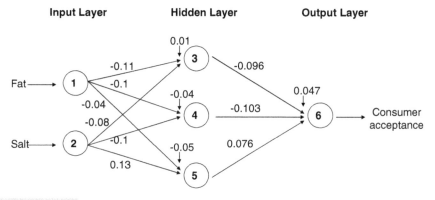

FIGURE 11.5   **NEURAL NETWORK FOR THE TINY EXAMPLE WITH WEIGHTS FROM XLMINER OUTPUT (FIGURE 11.4)**

The second table in Figure 11.4 gives the weights connecting the hidden and output layer nodes. Note that XLMiner uses two output nodes even though we have a binary response. This is equivalent to using a single node with a cutoff value. To compute the output from the (first) output node for the first observation, we use the outputs from the hidden layer that we computed above, and get

$$Output_6 = \frac{1}{1 + e^{-[0.047 + (-0.096)(0.48) + (-0.103)(0.46) + (0.076)(0.51)]}} = 0.498.$$

This is the probability shown in the bottom-most table in Figure 11.4 (for observation 1), which gives the neural network's predicted probabilities and the classifications based on these values. The probabilities for the other five observations are computed equivalently, replacing the input value in the computation of the hidden layer outputs and then plugging these outputs into the computation for the output layer.

Example 2: Classifying Accident Severity

Let us apply the network training process to some real data: U.S. automobile accidents that have been classified by their level of severity as *no injury*, *injury*, or *fatality*. A firm might be interested in developing a system for quickly classifying the severity of an accident, based on initial reports and associated data in the system (some of which rely on GPS-assisted reporting). Such a system could be used to assign emergency response team priorities. Figure 11.6 shows a small

Row Id.	Selected variables				
	ALCHL_I	PROFIL_I_R	SUR_COND	VEH_INVL	MAX_SEV_IR
1	2	0	1	1	0
4	2	0	2	2	1
5	2	1	1	2	1
6	2	0	1	1	0
9	2	1	1	1	1
10	2	0	1	1	0
12	1	1	1	2	1
17	2	1	1	2	1
18	2	1	4	1	1
19	2	0	1	1	0
20	2	1	1	2	2
21	2	1	1	2	1
23	2	1	3	2	1
26	1	0	1	1	2
27	1	1	1	2	2

FIGURE 11.6 SUBSET FROM THE ACCIDENT DATA, FOR A HIGH-FATALITY REGION

TABLE 11.2 **DESCRIPTION OF VARIABLES FOR AUTOMOBILE ACCIDENT EXAMPLE**

ALCHL_I	Presence (1) or absence (2) of alcohol
PROFIL_I_R	Profile of the roadway: level (1), grade (2), hillcrest (3), other (4), unknown (5)
SUR_COND	Surface condition of the road: dry (1), wet (2), snow/slush (3), ice (4), sand/dirt (5), other (6), unknown (7)
VEH_INVL	Number of vehicles involved
MAX_SEV_IR	Presence of injuries/fatalities: no injuries (0), injury (1), fatality (2)

extract (999 records, 4 predictor variables) from a U.S. government database. The explanation of the four predictor variables and response is given in Table 11.2.

With the exception of alcohol involvement and a few other variables in the larger database, most of the variables are ones that we might reasonably expect to be available at the time of the initial accident report, before accident details and severity have been determined by first responders. A data mining model that could predict accident severity on the basis of these initial reports would have value in allocating first responder resources.

To use a neural net architecture for this classification problem we use four nodes in the input layer, one for each of the four predictors, and three neurons (one for each class) in the output layer. We use a single hidden layer and experiment with the number of nodes. Increasing the number of nodes from four to eight and examining the resulting confusion matrices, we find that five nodes gave a good balance between improving the predictive performance on the training set without deteriorating the performance on the validation set. In fact, networks with more than five nodes in the hidden layer performed as well as the five-node network.

Note that there is a total of five connections from each node in the input layer to each node in the hidden layer, a total of $4 \times 5 = 20$ connections between the input layer and the hidden layer. In addition, there is a total of three connections from each node in the hidden layer to each node in the output layer, a total of $5 \times 3 = 15$ connections between the hidden layer and the output layer.

We train the network on the training partition of 600 records. Each iteration in the neural network process consists of presentation to the input layer of the predictors in a case, followed by successive computations of the outputs of the hidden layer nodes and the output layer nodes using the appropriate weights. The output values of the output layer nodes are used to compute the error. This error is used to adjust the weights of all the connections in the network using the backward propagation algorithm to complete the iteration. Since the training data has 600 cases, one sweep through the data, termed an *epoch*, consists of 600 iterations. We will train the network using 30 epochs, so there will be a total of 18,000 iterations. The resulting classification results, error rates, and weights following the last epoch of training the neural net on this data are shown in Figure 11.7.

Training Data Scoring - Summary Report

Classification Confusion Matrix			
	Predicted Class		
Actual Class	0	1	2
0	316	1	12
1	0	173	7
2	20	41	30

	Error Report		
Class	# Cases	# Errors	% Error
0	329	13	3.95
1	180	7	3.89
2	91	61	67.03
Overall	600	81	13.50

Validation Data Scoring - Summary Report

Classification Confusion Matrix			
	Predicted Class		
Actual Class	0	1	2
0	215	0	7
1	0	114	5
2	17	21	20

	Error Report		
Class	# Cases	# Errors	% Error
0	222	7	3.15
1	119	5	4.20
2	58	38	65.52
Overall	399	50	12.53

Interlayer Connections Weights

Hidden Layer # 1	Input Layer				
	ALCHL_I	PROFIL_I_R	SUR_COND	VEH_INVL	Bias Node
Node # 1	0.752536	-1.56418	-1.87962	-1.08218	-1.12008
Node # 2	1.00038	-1.59099	-0.545948	-0.818684	-0.835187
Node # 3	0.215221	-1.33309	-2.80806	-0.0157571	-0.646288
Node # 4	0.07074	-2.35661	-3.43757	-2.49196	-1.71293
Node # 5	1.00251	1.52574	4.30335	0.56337	-0.793611

Output Layer	Hidden Layer # 1					
	Node # 1	Node # 2	Node # 3	Node # 4	Node # 5	Bias Node
0	1.50246	1.27632	1.65596	3.12193	-2.7042	-2.87614
1	-1.26542	-0.211778	-2.13848	-2.71368	2.94968	-0.562839
2	-0.873338	-1.21267	0.601146	-2.78924	-2.79458	0.997683

FIGURE 11.7 XLMINER OUTPUT FOR NEURAL NETWORK FOR ACCIDENT DATA, WITH FIVE NODES IN THE HIDDEN LAYER, AFTER 30 EPOCHS

Note that had we stopped after only one pass of the data (600 iterations), the error would have been much worse, and none of the fatal accidents (coded as 2) would have been spotted, as can be seen in Figure 11.8. Our results can depend on how we set the different parameters, and there are a few pitfalls to avoid. We discuss these next.

Training Data Scoring - Summary Report

Classification Confusion Matrix			
	Predicted Class		
Actual Class	0	1	2
0	328	1	0
1	17	163	0
2	44	47	0

Error Report			
Class	# Cases	# Errors	% Error
0	329	1	0.30
1	180	17	9.44
2	91	91	100.00
Overall	600	109	18.17

Validation Data Scoring - Summary Report

Classification Confusion Matrix			
	Predicted Class		
Actual Class	0	1	2
0	221	1	0
1	16	103	0
2	29	29	0

Error Report			
Class	# Cases	# Errors	% Error
0	222	1	0.45
1	119	16	13.45
2	58	58	100.00
Overall	399	75	18.80

Interlayer Connections Weights

Hidden Layer # 1	Input Layer				
	ALCHL_I	PROFIL_I_R	SUR_COND	VEH_INVL	Bias Node
Node # 1	-0.0352883	-0.668327	-0.425221	-0.595552	-0.124272
Node # 2	0.0467488	-0.612836	-0.290081	-0.491289	0.00871788
Node # 3	-0.0386277	-0.660557	-0.621354	-0.683005	-0.0677495
Node # 4	0.0813384	-1.28064	-0.996499	-1.13877	-0.09437
Node # 5	0.206675	0.492427	0.477908	0.501264	0.0936535

Output Layer	Hidden Layer # 1					
	Node # 1	Node # 2	Node # 3	Node # 4	Node # 5	Bias Node
0	0.526532	0.347701	0.593953	1.57821	-0.97179	-0.871603
1	-0.57228	-0.463982	-0.75034	-1.32693	0.711036	0.283258
2	-0.33211	-0.456262	-0.303628	-0.509042	-0.417585	-0.646354

FIGURE 11.8 XLMINER OUTPUT FOR NEURAL NETWORK FOR ACCIDENT DATA, WITH FIVE NODES IN THE HIDDEN LAYER, AFTER ONLY ONE EPOCH

Avoiding Overfitting

A weakness of the neural network is that it can easily overfit the data, causing the error rate on validation data (and most important, on new data) to be too large. It is therefore important to limit the number of training epochs and not to overtrain the data. As in classification and regression trees, overfitting can be detected by examining the performance on the validation set and seeing when it starts deteriorating (while the training set performance is still improving). This approach is used in some algorithms (but not in XLMiner) to limit the number of training epochs: Periodically, the error rate on the validation dataset is computed while the network is being trained. The validation error decreases in the early epochs of the training, but after awhile it begins to increase. The point of minimum validation error is a good indicator of the best number of

Training Data Scoring - Summary Report

Classification Confusion Matrix			
	Predicted Class		
Actual Class	0	1	2
0	316	1	12
1	0	174	6
2	20	43	28

Error Report			
Class	# Cases	# Errors	% Error
0	329	13	3.95
1	180	6	3.33
2	91	63	69.23
Overall	600	82	13.67

Validation Data Scoring - Summary Report

Classification Confusion Matrix			
	Predicted Class		
Actual Class	0	1	2
0	215	0	7
1	0	114	5
2	17	23	18

Error Report			
Class	# Cases	# Errors	% Error
0	222	7	3.15
1	119	5	4.20
2	58	40	68.97
Overall	399	52	13.03

FIGURE 11.9   XLMINER OUTPUT FOR NEURAL NETWORK FOR ACCIDENT DATA WITH 25 NODES IN THE HIDDEN LAYER

epochs for training, and the weights at that stage are likely to provide the best error rate in new data.

To illustrate the effect of overfitting, compare the confusion matrices of the 5-node network (Figure 11.7) with those from a 25-node network (Figure 11.9). Both networks perform similarly on the training set, but the 25-node network does worse on the validation set.

Using the Output for Prediction and Classification

When the neural network is used for predicting a numerical response, the resulting output needs to be scaled back to the original units of that response. Recall that numerical variables (both predictor and response variables) are usually rescaled to a [0,1] interval before being used by the network. The output therefore will also be on a [0,1] scale. To transform the prediction back to the original y units, which were in the range [a, b], we multiply the network output by $b - a$ and add a.

When the neural net is used for classification and we have m classes, we will obtain an output from each of the m output nodes. How do we translate these m outputs into a classification rule? Usually, the output node with the largest value determines the net's classification.

In the case of a binary response ($m = 2$), we can use just one output node with a cutoff value to map a numerical output value to one of the two classes. Although we typically use a cutoff of 0.5 with other classifiers, in neural networks there is a tendency for values to cluster around 0.5 (from above and below). An alternative is to use the validation set to determine a cutoff that produces reasonable predictive performance.

11.4 REQUIRED USER INPUT

One of the time-consuming and complex aspects of training a model using back propagation is that we first need to decide on a network architecture. This means specifying the number of hidden layers and the number of nodes in each layer. The usual procedure is to make intelligent guesses using past experience and to do several trial-and-error runs on different architectures. Algorithms exist that grow the number of nodes selectively during training or trim them in a manner analogous to what is done in classification and regression trees (see Chapter 9). Research continues on such methods. As of now, no automatic method seems clearly superior to the trial-and-error approach. A few general guidelines for choosing an architecture follow.

> *Number of Hidden Layers*   The most popular choice for the number of hidden layers is one. A single hidden layer is usually sufficient to capture even very complex relationships between the predictors.

Size of Hidden Layer   The number of nodes in the hidden layer also determines the level of complexity of the relationship between the predictors that the network captures. The trade-off is between under- and overfitting. Using too few nodes might not be sufficient to capture complex relationships (recall the special cases of a linear relationship such as in linear and logistic regression, in the extreme case of zero nodes or no hidden layer). On the other hand, too many nodes might lead to overfitting. A rule of thumb is to start with p (number of predictors) nodes and gradually decrease or increase a bit while checking for overfitting.

Number of Output Nodes   For a binary response, a single node is sufficient, and a cutoff is used for classification. For a categorical response with $m > 2$ classes, the number of nodes should equal the number of classes. Finally, for a numerical response, typically a single output node is used unless we are interested in predicting more than one function.

In addition to the choice of architecture, the user should pay attention to the *choice of predictors*. Since neural networks are highly dependent on the quality of their input, the choice of predictors should be done carefully, using domain knowledge, variable selection, and dimension reduction techniques before using the network. We return to this point in the discussion of advantages and weaknesses below.

Other parameters that the user can control are the *learning rate* (a.k.a. weight decay), l, and the *momentum*. The first is used primarily to avoid overfitting, by down-weighting new information. This helps to tone down the effect of outliers on the weights and avoids getting stuck in local optima. This parameter typically takes a value in the range $[0, 1]$. Berry and Linoff (2000) suggest starting with a large value (moving away from the random initial weights, thereby "learning quickly" from the data) and then slowly decreasing it as the iterations progress and the weights are more reliable. Han and Kamber (2001) suggest the more concrete rule of thumb of setting $l = 1/$(current number of iterations). This means that at the start, $l = 1$, during the second iteration it is $l = 0.5$, and then it keeps decaying toward $l = 0$. Notice that in XLMiner the default is $l = 0$, which means that the weights do not decay at all.

The second parameter, called *momentum*, is used to "keep the ball rolling" (hence the term *momentum*) in the convergence of the weights to the optimum. The idea is to keep the weights changing in the same direction as they did in the preceding iteration. This helps to avoid getting stuck in a local optimum. High values of momentum mean that the network will be "reluctant" to learn from data that want to change the direction of the weights, especially when we consider case updating. In general, values in the range 0–2 are used.

11.5 EXPLORING THE RELATIONSHIP BETWEEN PREDICTORS AND RESPONSE

Neural networks are known to be "black boxes" in the sense that their output does not shed light on the patterns in the data that it models (like our brains). In fact, that is one of the biggest criticism of the method. However, in some cases it is possible to learn more about the relationships that the network captures, by conducting a sensitivity analysis on validation set. This is done by setting all predictor values to their mean and obtaining the network's prediction. Then the process is repeated by setting each predictor sequentially to its minimum (and then maximum) value. By comparing the predictions from different levels of the predictors, we can get a sense of which predictors affect predictions more and in what way.

11.6 ADVANTAGES AND WEAKNESSES OF NEURAL NETWORKS

As mentioned in Section 11.1, the most prominent advantage of neural networks is their good predictive performance. They are known to have high tolerance to noisy data and the ability to capture highly complicated relationships between the predictors and a response. Their weakest point is in providing insight into the structure of the relationship, hence their black-box reputation.

Several considerations and dangers should be kept in mind when using neural networks. First, although they are capable of generalizing from a set of examples, extrapolation is still a serious danger. If the network sees only cases in a certain range, its predictions outside this range can be completely invalid.

Second, neural networks do not have a built-in variable selection mechanism. This means that there is need for careful consideration of predictors. Combination with classification and regression trees (see Chapter 9) and other dimension reduction techniques (e.g., principal components analysis in Chapter 4) is often used to identify key predictors.

Third, the extreme flexibility of the neural network relies heavily on having sufficient data for training purposes. As our tiny example shows, a neural network performs poorly when the training set size is insufficient, even when the relationship between the response and predictors is very simple. A related issue is that in classification problems, the network requires sufficient records of the minority class in order to learn it. This is achieved by oversampling, as explained in Chapter 2.

Fourth, a technical problem is the risk of obtaining weights that lead to a local optimum rather than the global optimum, in the sense that the weights converge to values that do not provide the best fit to the training data. We described several parameters that are used to try to avoid this situation (such as controlling the learning rate and slowly reducing the momentum). However, there is no guarantee that the resulting weights are indeed the optimal ones.

Finally, a practical consideration that can determine the usefulness of a neural network is the timeliness of computation. Neural networks are relatively heavy on computation time, requiring a longer runtime than other classifiers. This runtime grows greatly when the number of predictors is increased (as there will be many more weights to compute). In applications where real-time or near-real-time prediction is required, runtime should be measured to make sure that it does not cause unacceptable delay in the decision making.

PROBLEMS

11.1 Credit Card Use. Consider the following hypothetical bank data on consumers' use of credit card credit facilities in Table 11.3. Create a small worksheet in Excel, like that used in Example 1, to illustrate one pass through a simple neural network.

11.2 Neural Net Evolution. A neural net typically starts out with random coefficients; hence, it produces essentially random predictions when presented with its first case. What is the key ingredient by which the net evolves to produce a more accurate prediction?

11.3 Car Sales. Consider again the data on used cars (ToyotaCorolla.xls) with 1436 records and details on 38 attributes, including Price, Age, KM, HP, and other specifications. The goal is to predict the price of a used Toyota Corolla based on its specifications.

    a. Use XLMiner's neural network routine to fit a model using the XLMiner default values for the neural net parameters, except normalizing the data. Record the RMS error for the training data and the validation data. Repeat the process, changing the number of epochs (and only this) to 300, 3000, and 10,000.

        i. What happens to the RMS error for the training data as the number of epochs increases?

        ii. What happens to the RMS error for the validation data?

        iii. Comment on the appropriate number of epochs for the model.

    b. Conduct a similar experiment to assess the effect of changing the number of layers in the network as well as the gradient descent step size.

11.4 Direct Mailing to Airline Customers. East-West Airlines has entered into a partnership with the wireless phone company Telcon to sell the latter's service via direct mail. The file EastWestAirlinesNN.xls contains a subset of a data sample of who has already received a test offer. About 13% accepted.

    You are asked to develop a model to classify East-West customers as to whether they purchased a wireless phone service contract (target variable Phone_Sale), a model that can be used to predict classifications for additional customers.

    a. Using XLMiner, run a neural net model on these data, using the option to normalize the data, setting the number of epochs at 3000, and requesting lift charts for both the training and validation data. Interpret the meaning (in business terms) of the leftmost bar of the validation lift chart (the bar chart).

TABLE 11.3   **DATA FOR CREDIT CARD EXAMPLE AND VARIABLE DESCRIPTIONS**[a]

Years	Salary	Used Credit
4	43	0
18	65	1
1	53	0
3	95	0
15	88	1
6	112	1

[a]Years = Number of years that a customer has been with the bank; Salary = customer's salary (in thousands of dollars); Used Credit 1 = customer has left an unpaid credit card balance at the end of at least one month in the prior year and 0 = balance was paid off at the end of each month.

b. Comment on the difference between the training and validation lift charts.

c. Run a second neural net model on the data, this time setting the number of epochs at 100. Comment now on the difference between this model and the model you ran earlier, and how overfitting might have affected results.

d. What sort of information, if any, is provided about the effects of the various variables?

## CHAPTER 12

Discriminant Analysis

In this chapter we describe the method of discriminant analysis, which is a model-based approach to classification. We discuss the main principle, where classification is based on the distance of an observation from each class average. We explain the underlying measure of "statistical distance," which takes into account the correlation between predictors. The output of a discriminant analysis procedure generates estimated "classification functions," which are then used to produce classification scores that can be translated into classifications or probabilities of class membership. One can also directly integrate misclassification costs into the discriminant analysis setup, and we explain how this is achieved. Finally, we discuss the underlying model assumptions, the practical robustness to some, and the advantages of discriminant analysis when the assumptions are reasonably met (e.g., the sufficiency of a small training sample).

12.1 INTRODUCTION

Discriminant analysis is another classification method. Like logistic regression, it is a classical statistical technique that can be used for classification and profiling. It uses continuous variable measurements on different classes of items to classify new items into one of those classes (*classification*). Common uses of the method have been in classifying organisms into species and subspecies; classifying applications for loans, credit cards, and insurance into low- and high-risk categories; classifying customers of new products into early adopters, early majority, late majority, and laggards; classifying bonds into bond rating categories;

Data Mining for Business Intelligence, By Galit Shmueli, Nitin R. Patel, and Peter C. Bruce
Copyright © 2010 John Wiley & Sons Inc.

classifying skulls of human fossils; as well as in research studies involving disputed authorship, decision on college admission, medical studies involving alcoholics and nonalcoholics, and methods to identify human fingerprints. Discriminant analysis can also be used to highlight aspects that distinguish the classes (*profiling*).

We return to two examples that were described in earlier chapters and use them to illustrate discriminant analysis.

Example 1: Riding Mowers

We return to the example from Chapter 7, where a riding-mower manufacturer would like to find a way of classifying families in a city into those likely to purchase a riding mower and those not likely to buy one. A pilot random sample of 12 owners and 12 nonowners in the city is undertaken. The data are given in Table 7.1, and a scatterplot is shown in Figure 12.1. We can think of a linear classification rule as a line that separates the two-dimensional region into two parts, with most of the owners in one half-plane and most nonowners in the complementary half-plane. A good classification rule would separate the data so that the fewest points are misclassified: The line shown in Figure 12.1 seems to do a good job in discriminating between the two classes as it makes 4 misclassifications out of 24 points. Can we do better?

Example 2: Personal Loan Acceptance

The riding-mower example is a classic example and is useful in describing the concept and goal of discriminant analysis. However, in today's business

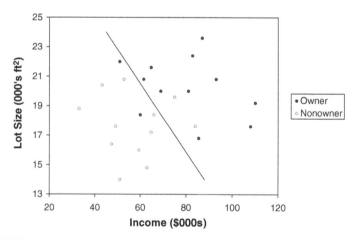

FIGURE 12.1   SCATTERPLOT OF LOT SIZE VS. INCOME FOR 24 OWNERS AND NONOWNERS OF RIDING MOWERS. THE (AD HOC) LINE TRIES TO SEPARATE OWNERS FROM NONOWNERS

applications, the number of records is much larger, and their separation into classes is much less distinct. To illustrate this, we return to the Universal Bank example described in Chapter 9, where the bank's goal is to find which factors make a customer more likely to accept a personal loan. For simplicity, we consider only two variables: the customer's annual income (Income, in $000s), and the average monthly credit card spending (CCAvg, in $000s). The first part of Figure 12.2 shows the acceptance of a personal loan by a subset of 200 customers from the bank's database as a function of Income and CCAvg. We use a log scale

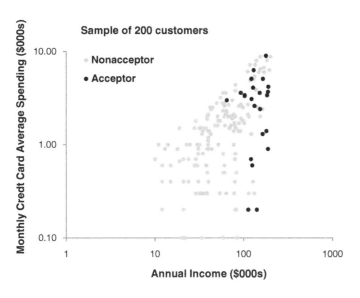

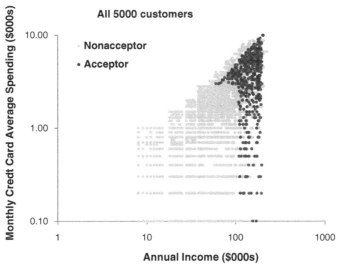

FIGURE 12.2 PERSONAL LOAN ACCEPTANCE AS A FUNCTION OF INCOME AND CREDIT CARD SPENDING FOR 5000 CUSTOMERS OF THE UNIVERSAL BANK (IN LOG SCALE)

on both axes to enhance visibility because there are many points condensed in the low-income, low-CC spending area. Even for this small subset, the separation is not clear. The second figure shows all 5000 customers and the added complexity of dealing with large numbers of observations.

12.2 DISTANCE OF AN OBSERVATION FROM A CLASS

Finding the best separation between items involves measuring their distance from their class. The general idea is to classify an item to the class to which it is closest. Suppose that we are required to classify a new customer of Universal Bank as being an acceptor or a nonacceptor of its personal loan offer, based on an income of x. From the bank's database we find that the average income for loan acceptors was \$144.75K and for nonacceptors \$66.24K. We can use Income as a predictor of loan acceptance via a simple *Euclidean distance rule*: If x is closer to the average income of the acceptor class than to the average income of the nonacceptor class, classify the customer as an acceptor; otherwise, classify the customer as a nonacceptor. In other words, if $|x - 144.75| < |x - 66.24|$, classification = acceptor; otherwise, nonacceptor. Moving from a single variable (income) to two or more variables, the equivalent of the mean of a class is the *centroid* of a class. This is simply the vector of means $\overline{\mathbf{x}} = [\overline{x}_1, \ldots, \overline{x}_p]$. The Euclidean distance between an item with p measurements $\mathbf{x} = [x_1, \ldots, x_p]$ and the centroid $\overline{\mathbf{x}}$ is defined as the root of the sum of the squared differences between the individual values and the means:

$$D_{\text{Euclidean}}(\mathbf{x}, \overline{\mathbf{x}}) = \sqrt{(x_1 - \overline{x}_1)^2 + \cdots + (x_p - \overline{x}_1)^p}. \tag{12.1}$$

Using the Euclidean distance has three drawbacks. First, the distance depends on the units we choose to measure the variables. We will get different answers if we decide to measure income in dollars, for instance, rather than in thousands of dollars.

Second, Euclidean distance does not take into account the variability of the variables. For example, if we compare the variability in income in the two classes, we find that for acceptors the standard deviation is lower than for nonacceptors (\$31.6K vs. \$40.6K). Therefore, the income of a new customer might be closer to the acceptors' average income in dollars, but because of the large variability in income for nonacceptors, this customer is just as likely to be a nonacceptor. We therefore want the distance measure to take into account the variance of the different variables and measure a distance in standard deviations rather than in the original units. This is equivalent to z scores.

Third, Euclidean distance ignores the correlation between the variables. This is often a very important consideration, especially when we are using many variables to separate classes. In this case there will often be variables, which, by themselves, are useful discriminators between classes, but in the presence of other variables are practically redundant, as they capture the same effects as the other variables.

A solution to these drawbacks is to use a measure called *statistical distance* (or *Mahalanobis distance*). Let us denote by S the covariance matrix between the p variables. The definition of a statistical distance is

$$D_{\text{Statistical}}(\mathbf{x}, \overline{\mathbf{x}}) = [\mathbf{x} - \overline{\mathbf{x}}]' S^{-1} [\mathbf{x} - \overline{\mathbf{x}}]$$

$$= [(x_1 - \overline{x}_1), (x_2 - \overline{x}_2), \ldots, (x_p - \overline{x}_p)] S^{-1} \begin{bmatrix} x_1 - \overline{x}_1 \\ x_2 - \overline{x}_2 \\ \vdots \\ x_p - \overline{x}_p \end{bmatrix}$$

$$(12.2)$$

(the notation' represents *transpose operation*, which simply turns the column vector into a row vector). S^{-1} is the inverse matrix of S, which is the p-dimension extension to division. When there is a single predictor ($p = 1$), this reduces to a z score since we subtract the mean and divide by the standard deviation. The statistical distance takes into account not only the predictor averages but also the spread of the predictor values and the correlations between the different predictors. To compute a statistical distance between an observation and a class, we must compute the predictor averages (the centroid) and the covariances between each pair of variables. These are used to construct the distances. The method of discriminant analysis uses statistical distance as the basis for finding a separating line (or, if there are more than two variables, a separating hyperplane) that is equally distant from the different class means.[1] It is based on measuring the statistical distances of an observation to each of the classes and allocating it to the closest class. This is done through *classification functions*, which are explained next.

12.3 FISHER'S LINEAR CLASSIFICATION FUNCTIONS

Linear classification functions were suggested in 1936 by the noted statistician R. A. Fisher as the basis for improved separation of observations into classes. The idea is to find linear functions of the measurements that maximize the ratio of between-class variability to within-class variability. In other words, we would obtain classes that are very homogeneous and differ the most from each other. For each observation, these functions are used to compute scores that measure the proximity of that observation to each of the classes. An observation is classified as

---

[1] An alternative approach finds a separating line or hyperplane that is "best" at separating the different clouds of points. In the case of two classes, the two methods coincide.

Variables	Classification Function	
	owner	nonowner
Constant	-73.16020203	-51.42144394
Income	0.42958561	0.32935533
Lot Size	5.46674967	4.68156528

FIGURE 12.3 DISCRIMINANT ANALYSIS OUTPUT FOR RIDING-MOWER DATA, DISPLAYING THE ESTIMATED CLASSIFICATION FUNCTIONS

belonging to the class for which it has the highest classification score (equivalent to the smallest statistical distance).

USING CLASSIFICATION FUNCTION SCORES TO CLASSIFY

For each record, we calculate the value of the classification function (one for each class); whichever class's function has the highest value (= score) is the class assigned to that record.

The classification functions are estimated using software (see Figure 12.3). Note that the number of classification functions is equal to the number of classes (in this case, 2).

To classify a family into the class of owners or nonowners, we use the functions above to compute the family's classification scores: A family is classified into the class of *owners* if the owner function is higher than the nonowner function and into *nonowners* if the reverse is the case. These functions are specified in a way that can be generalized easily to more than two classes. The values given for the functions are simply the weights to be associated with each variable in the linear function in a manner analogous to multiple linear regression. For instance, the first household has an income of \$60K and a lot size of 18.4K ft^2. Their owner score is therefore $-73.16 + (0.43)(60) + (5.47)(18.4) = 53.2$, and their nonowner score is $-51.42 + (0.33)(60) + (4.68)(18.4) = 54.48$. Since the second score is higher, the household is (mis)classified by the model as a nonowner. The scores for all 24 households are given in Figure 12.4.

An alternative way for classifying an observation into one of the classes is to compute the probability of belonging to each of the classes and assigning the observation to the most likely class. If we have two classes, we need only compute a single probability for each observation (e.g., of belonging to owners). Using a cutoff of 0.5 is equivalent to assigning the observation to the class with the highest classification score. The advantage of this approach is that we can sort the records in order of descending probabilities and generate lift curves. Let us assume that there are m classes. To compute the probability of belonging to a certain class k, for a certain observation i, we need to compute all the classification scores

Classification Scores

Row Id.	Predicted Class	Actual Class	Owners	Nonowners
1	nonowner	owner	53.2031285	54.48067701
2	owner	owner	55.4107621	55.38873348
3	owner	owner	72.7587384	71.04259149
4	owner	owner	66.9677061	66.21046668
5	owner	owner	93.2290383	87.71741038
6	owner	owner	79.0987673	74.72663127
7	owner	owner	69.449838	66.54448063
8	owner	owner	84.8646791	80.71623966
9	owner	owner	65.8161985	64.93537943
10	owner	owner	80.4996528	76.58515957
11	owner	owner	69.0171568	68.37011405
12	owner	owner	70.9712258	68.88764339
13	owner	nonowner	66.2070123	65.0388853
14	nonowner	nonowner	63.2303113	63.34507531
15	nonowner	nonowner	48.7050398	50.44370426
16	nonowner	nonowner	56.9195896	58.31063803
17	owner	nonowner	59.1397834	58.63995271
18	nonowner	nonowner	44.1902042	47.17838722
19	nonowner	nonowner	39.8251779	43.04730714
20	nonowner	nonowner	55.7806422	56.45680899
21	nonowner	nonowner	36.8568505	40.96766929
22	nonowner	nonowner	43.7910169	47.46070921
23	nonowner	nonowner	25.2831595	30.91759181
24	nonowner	nonowner	34.8115865	38.61510799

FIGURE 12.4 **CLASSIFICATION SCORES FOR RIDING-MOWER DATA**

$c_1(i), c_2(i), \ldots, c_m(i)$ and combine them using the following formula:

$$P[\text{observation } i(\text{with measurements } x_1, x_2, \ldots, x_p) \text{ belongs to class } k]$$

$$= \frac{e^{c_k(i)}}{e^{c_1(i)} + e^{c_2(i)} + \cdots + e^{c_m(i)}}.$$

In XLMiner these probabilities are computed automatically, as can be seen in Figure 12.5.

We now have three misclassifications, compared to four in our original (ad hoc) classifications. This can be seen in Figure 12.6, which includes the line resulting from the discriminant model.[2]

[2] The slope of the line is given by $-a_1/a_2$ and the intercept is $a_1/a_2\,\bar{x}_1 + \bar{x}_2$, where a_i is the difference between the ith classification function coefficients of owners and nonowners (e.g., here $a_{\text{income}} = 0.43 - 0.33$).

Cut off Prob.Val. for Success (Updatable)	0.5

Row Id.	Predicted Class	Actual Class	Prob. for - owner (success)	Income	Lot Size
1	nonowner	owner	0.217967810	60	18.4
2	owner	owner	0.505506928	85.5	16.8
3	owner	owner	0.847631864	64.8	21.6
4	owner	owner	0.680754087	61.5	20.8
5	owner	owner	0.995976726	87	23.6
6	owner	owner	0.987533139	110.1	19.2
7	owner	owner	0.948110638	108	17.6
8	owner	owner	0.984456382	82.8	22.4
9	owner	owner	0.706991915	69	20
10	owner	owner	0.980439587	93	20.8
11	owner	owner	0.656343749	51	22
12	owner	owner	0.889297203	81	20
13	owner	nonowner	0.762806287	75	19.6
14	nonowner	nonowner	0.471340450	52.8	20.8
15	nonowner	nonowner	0.149482655	64.8	17.2
16	nonowner	nonowner	0.199240432	43.2	20.4
17	owner	nonowner	0.622419543	84	17.6
18	nonowner	nonowner	0.047962588	49.2	17.6
19	nonowner	nonowner	0.038341401	59.4	16
20	nonowner	nonowner	0.337117362	66	18.4
21	nonowner	nonowner	0.016129906	47.4	16.4
22	nonowner	nonowner	0.024850999	33	18.8
23	nonowner	nonowner	0.003559986	51	14
24	nonowner	nonowner	0.021806029	63	14.8

FIGURE 12.5   DISCRIMINANT ANALYSIS OUTPUT FOR RIDING-MOWER DATA, DISPLAYING THE ESTIMATED PROBABILITY OF OWNERSHIP FOR EACH FAMILY

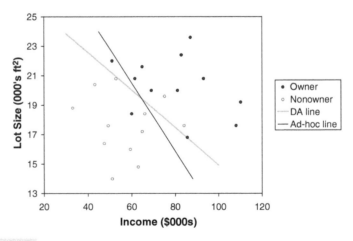

FIGURE 12.6   CLASS SEPARATION OBTAINED FROM THE DISCRIMINANT MODEL (COMPARED TO AD HOC LINE FROM FIGURE 12.1)

12.4 CLASSIFICATION PERFORMANCE OF DISCRIMINANT ANALYSIS

The discriminant analysis method relies on two main assumptions to arrive at classification scores: First, it assumes that the measurements in all classes come from a multivariate normal distribution. When this assumption is reasonably met, discriminant analysis is a more powerful tool than other classification methods, such as logistic regression. In fact, it is 30% more efficient than logistic regression if the data are multivariate normal, in the sense that we require 30% less data to arrive at the same results. In practice, it has been shown that this method is relatively robust to departures from normality in the sense that predictors can be nonnormal and even dummy variables. This is true as long as the smallest class is sufficiently large (approximately more than 20 cases). This method is also known to be sensitive to outliers in both the univariate space of single predictors and in the multivariate space. Exploratory analysis should therefore be used to locate extreme cases and determine whether they can be eliminated.

The second assumption behind discriminant analysis is that the correlation structure between the different measurements within a class is the same across classes. This can be roughly checked by estimating the correlation matrix for each class and comparing matrices. If the correlations differ substantially across classes, the classifier will tend to classify cases into the class with the largest variability. When the correlation structure differs significantly and the dataset is very large, an alternative is to use quadratic discriminant analysis.[3]

With respect to the evaluation of classification accuracy, we once again use the general measures of performance that were described in Chapter 5 (judging the performance of a classifier), with the principal ones based on the confusion matrix (accuracy alone or combined with costs) and the lift chart. The same argument for using the validation set for evaluating performance still holds. For example, in the riding-mower example, families 1, 13, and 17 are misclassified. This means that the model yields an error rate of 12.5% for these data. However, this rate is a biased estimate—it is overly optimistic because we have used the same data for fitting the classification parameters and for estimating the error. Therefore, as with all other models, we test performance on a validation set that includes data that were not involved in estimating the classification functions.

To obtain the confusion matrix from a discriminant analysis, we either use the classification scores directly or the probabilities of class membership that are computed from the classification scores. In both cases we decide on the

---

[3] In practice, quadratic discriminant analysis has not been found useful except when the difference in the correlation matrices is large and the number of observations available for training and testing is large. The reason is that the quadratic model requires estimating many more parameters that are all subject to error [for c classes and p variables, the total number of parameters to be estimated for all the different correlation matrices is $\varphi(p+1)/2$].

class assignment of each observation based on the highest score or probability. We then compare these classifications to the actual class memberships of these observations. This yields the confusion matrix. In the Universal Bank case we use the estimated classification functions in Figure 12.4 to predict the probability of loan acceptance in a validation set that contains 2000 customers (these data were not used in the modeling step).

12.5 PRIOR PROBABILITIES

So far we have assumed that our objective is to minimize the classification error. The method presented above assumes that the chances of encountering an item from either class requiring classification is the same. If the probability of encountering an item for classification in the future is not equal for the different classes, we should modify our functions to reduce our expected (long-run average) error rate. The modification is done as follows: Let us denote by p_j the prior or future probability of membership in class j (in the two-class case we have p_1 and $p_2 = 1 - p_1$). We modify the classification function for each class by adding $\log(p_j)$.[4] To illustrate this, suppose that the percentage of riding-mower owners in the population is 15% (compared to 50% in the sample). This means that the model should classify fewer households as owners. To account for this, we adjust the constants in the classification functions from Figure 12.3 and obtain the adjusted constants $-73.16 + \log(0.15) = -75.06$ for owners and $-51.42 + \log(0.85) = -50.58$ for nonowners. To see how this can affect classifications, consider family 13, which was misclassified as an owner in the case involving equal probability of class membership. When we account for the lower probability of owning a mower in the population, family 13 is classified properly as a nonowner (its owner classification score exceeds the nonowner score).

12.6 UNEQUAL MISCLASSIFICATION COSTS

A second practical modification is needed when misclassification costs are not symmetrical. If the cost of misclassifying a class 1 item is very different from the cost of misclassifying a class 2 item, we may want to minimize the expected cost of misclassification rather than the simple error rate (which does not take cognizance of unequal misclassification costs). In the two-class case, it is easy to manipulate the classification functions to account for differing misclassification costs (in

---

[4] XLMiner also has the option to set the prior probabilities as the ratios that are encountered in the dataset. This is based on the assumption that a random sample will yield a reasonable estimate of membership probabilities. However, for other prior probabilities the classification functions should be modified manually.

addition to prior probabilities). We denote by C_1 the cost of misclassifying a class 1 member (into class 2). Similarly, C_2 denotes the cost of misclassifying a class 2 member (into class 1). These costs are integrated into the constants of the classification functions by adding $\log(C_1)$ to the constant for class 1 and $\log(C_2)$ to the constant of class 2. To incorporate both prior probabilities and misclassification costs, add $\log(p_1 C_1)$ to the constant of class 1 and $\log(p_2 C_2)$ to that of class 2.

In practice, it is not always simple to come up with misclassification costs C_1 and C_2 for each class. It is usually much easier to estimate the ratio of costs C_2/C_1 (e.g., the cost of misclassifying a credit defaulter is 10 times more expensive than that of misclassifying a nondefaulter). Luckily, the relationship between the classification functions depends only on this ratio. Therefore, we can set $C_1 = 1$ and $C_2 = ratio$ and simply add $\log(C_2/C_1)$ to the constant for class 2.

12.7   CLASSIFYING MORE THAN TWO CLASSES

Example 3: Medical Dispatch to Accident Scenes

Ideally, every automobile accident call to 911 results in the immediate dispatch of an ambulance to the accident scene. However, in some cases the dispatch might be delayed (e.g., at peak accident hours or in some resource-strapped towns or shifts). In such cases, the 911 dispatchers must make decisions about which units to send based on sketchy information. It is useful to augment the limited information provided in the initial call with additional information in order to classify the accident as minor injury, serious injury, or death. For this purpose we can use data that were collected on automobile accidents in the United States in 2001 that involved some type of injury. For each accident, additional information is recorded, such as day of week, weather conditions, and road type. Figure 12.7 shows a small sample of records with 10 measurements of interest.

The goal is to see how well the predictors can be used to classify injury type correctly. To evaluate this, a sample of 1000 records was drawn and partitioned into training and validation sets, and a discriminant analysis was performed on the training data. The output structure is very similar to that for the two-class case. The only difference is that each observation now has three classification functions (one for each injury type), and the confusion and error matrices are 3×3 to account for all the combinations of correct and incorrect classifications (see Figure 12.8). The rule for classification is still to classify an observation to the class that has the highest corresponding classification score. The classification scores are computed, as before, using the classification function coefficients. This can be seen in Figure 12.9. For instance, the *no injury* classification score for the first accident in the training set is $-24.51 + 1.95(1) + 1.19(0) + \cdots + 16.36(1) = 30.93$. The *nonfatal* score is similarly computed as 31.42 and the *fatal* score as

Accident #	RushHour	WRK_ZONE	WKDY	INT_HWY	LGTCON	LEVEL	SPD_LIM	SUR_COND	TRAF_WAY	WEATHER	MAX_SEV
1	1	0	1	1	dark_light	1	70	ice	one_way	adverse	no-injury
2	1	0	1	0	dark_light	0	70	ice	divided	adverse	no-injury
3	1	0	1	0	dark_light	0	65	ice	divided	adverse	non-fatal
4	1	0	1	0	dark_light	0	55	ice	two_way	not_adverse	non-fatal
5	1	0	0	0	dark_light	0	35	snow	one_way	adverse	no-injury
6	1	0	1	0	dark_light	1	35	wet	divided	adverse	no-injury
7	0	0	1	1	dark_light	1	70	wet	divided	adverse	non-fatal
8	0	0	1	0	dark_light	1	35	wet	two_way	adverse	no-injury
9	1	0	1	0	dark_light	0	25	wet	one_way	adverse	non-fatal
10	1	0	1	0	dark_light	0	35	wet	divided	adverse	non-fatal
11	1	0	1	0	dark_light	0	30	wet	divided	adverse	non-fatal
12	1	0	1	0	dark_light	0	60	wet	divided	not_adverse	no-injury
13	1	0	1	0	dark_light	0	40	wet	two_way	not_adverse	no-injury
14	0	0	1	0	day	1	65	dry	two_way	not_adverse	fatal
15	1	0	0	0	day	0	55	dry	two_way	not_adverse	fatal
16	1	0	1	0	day	0	55	dry	two_way	not_adverse	non-fatal
17	1	0	0	0	day	0	55	dry	two_way	not_adverse	non-fatal
18	0	0	1	0	dark	0	55	ice	two_way	not_adverse	no-injury
19	0	0	0	0	dark	0	50	ice	two_way	adverse	no-injury
20	0	0	0	0	dark	1	55	snow	divided	adverse	no-injury

FIGURE 12.7   **SAMPLE OF 20 AUTOMOBILE ACCIDENTS FROM THE 2001 DEPARTMENT OF TRANSPORTATION DATABASE. EACH ACCIDENT IS CLASSIFIED AS ONE OF THREE INJURY TYPES (NO INJURY, NONFATAL, OR FATAL) AND HAS 10 MEASUREMENTS (EXTRACTED FROM A LARGER SET OF MEASUREMENTS)**

25.94. Since the *nonfatal* score is highest, this accident is (correctly) classified as having nonfatal injuries.

We can also compute for each accident the estimated probabilities of belonging to each of the three classes using the same relationship between classification scores and probabilities as in the two-class case. For instance, the probability of the above accident involving nonfatal injuries is estimated by the model as

$$\frac{e^{31.42}}{e^{31.42} + e^{30.93} + e^{25.94}} = 0.38. \tag{12.3}$$

The probabilities of an accident involving no injuries or fatal injuries are computed in a similar manner. For the first accident in the training set, the highest probability is that of involving no injuries, and therefore it is classified as a no injury accident. In general, membership probabilities can be obtained directly from XLMiner for the training set, the validation set, or for new observations.

12.8 ADVANTAGES AND WEAKNESSES

Discriminant analysis tends to be considered more of a statistical classification method than a data mining method. This is reflected in its absence or short mention in many data mining resources. However, it is very popular in social

(a) Classification Function

Variables	Classification Function		
	fatal	no-injury	non-fatal
Constant	-25.59584999	-24.51432228	-24.2336216
RushHour	0.92256236	1.95240343	1.9031992
WRK_ZONE	0.51786095	1.19506037	0.77056831
WKDY	4.78014898	6.41763353	6.11652184
INT_HWY	-1.84187829	-2.67303801	-2.53662229
LGTCON_day	3.70701218	3.66607523	3.7276206
LEVEL	2.62689376	1.56755066	1.71386576
SPD_LIM	0.50513172	0.46147966	0.45208475
SUR_COND_dry	9.99886131	15.8337965	16.25656509
TRAF_WAY_two_way	7.10797691	6.34214783	6.35494375
WEATHER_adverse	9.68802357	16.36388016	16.31727791

(b) Training Data scoring - Summary Report

Classification Confusion Matrix			
	Predicted Class		
Actual Class	fatal	no-injury	non-fatal
fatal	1	1	3
no-injury	6	114	172
non-fatal	6	95	202

Error Report			
Class	# Cases	# Errors	% Error
fatal	5	4	80.00
no-injury	292	178	60.96
non-fatal	303	101	33.33
Overall	600	283	47.17

FIGURE 12.8   XLMINER'S DISCRIMINANT ANALYSIS OUTPUT FOR THE THREE-CLASS INJURY EXAMPLE: (A) CLASSIFICATION FUNCTIONS AND (B) CONFUSION MATRIX FOR TRAINING SET

sciences and has shown good performance. The use and performance of discriminant analysis are similar to those of multiple linear regression. The two methods therefore share several advantages and weaknesses.

Like linear regression, discriminant analysis searches for the optimal weighting of predictors. In linear regression the weighting is with relation to the response, whereas in discriminant analysis it is with relation to separating the classes. Both use the same estimation method of least squares, and the resulting estimates are robust to local optima.

In both methods an underlying assumption is normality. In discriminant analysis we assume that the predictors are approximately from a multivariate normal distribution. Although this assumption is violated in many practical situations (such as with commonly used binary predictors), the method is surprisingly

Row Id.	Predicted Class	Actual Class	Score for fatal	Score for no-injury	Score for non-fatal	Prob. for class fatal	Prob. for class no-injury	Prob. for class non-fatal	RushHour	WRK_ZONE	WKDY	INT_HWY	LGTCON_day	LEVEL	SPD_LIM	SUR_COND_dry	TRAF_two_way	WEATHER_adverse
2	no-injury	no-injury	25.94	31.42	30.93	0.002583566	0.618769909	0.378646525	1	0	1	1	0	1	70	0	0	0
56	no-injury	non-fatal	15.00	15.58	15.01	0.263257586	0.471205318	0.265537095	1	0	1	0	0	0	55	0	1	1
79	no-injury	no-injury	2.69	9.95	9.81	0.000376892	0.535717942	0.463905165	1	0	0	0	0	0	35	0	0	0
141	no-injury	no-injury	10.10	17.94	17.64	0.000226522	0.574000564	0.425772914	1	0	1	0	0	1	35	0	0	1
203	no-injury	non-fatal	2.42	11.76	11.41	5.18896E-05	0.586851481	0.413096629	1	0	1	0	0	0	25	0	0	0
215	no-injury	non-fatal	7.47	16.37	15.93	8.33756E-05	0.609408333	0.390508291	1	0	1	0	0	0	35	0	0	1
281	no-injury	no-injury	10.41	11.54	10.91	0.174287408	0.539388146	0.286324446	1	0	1	0	0	1	60	0	1	0
660	non-fatal	non-fatal	22.57	28.37	28.46	0.001452278	0.476947897	0.521599825	1	0	0	1	1	0	45	0	1	0
838	non-fatal	no-injury	15.12	20.45	20.47	0.002408934	0.493295725	0.504295341	0	0	1	1	0	0	55	1	0	0
878	no-injury	non-fatal	23.62	29.32	29.15	0.00181542	0.540936208	0.457248372	1	0	1	1	0	0	70	1	0	0
882	no-injury	no-injury	20.17	25.06	24.99	0.003898532	0.515946916	0.480154552	0	0	1	1	0	0	65	1	0	0
907	no-injury	non-fatal	10.31	18.15	18.13	0.00019948	0.505559651	0.494240868	1	0	1	0	0	0	40	1	0	0
1072	non-fatal	non-fatal	12.84	20.46	20.39	0.000253881	0.517266017	0.482480102	1	0	1	0	0	0	45	1	0	0
1162	non-fatal	non-fatal	15.16	20.38	20.62	0.002378591	0.43818467	0.559436739	1	0	0	0	0	0	45	1	1	0
1245	no-injury	no-injury	19.94	26.80	26.74	0.000542513	0.513922638	0.485534849	1	0	1	1	0	0	45	1	1	0
1326	non-fatal	non-fatal	12.37	19.88	19.96	0.000262636	0.478861093	0.520876272	1	0	0	0	0	0	30	1	1	1
1468	no-injury	non-fatal	25.00	31.41	31.26	0.000877376	0.537149759	0.461972865	1	0	1	0	0	0	55	1	1	0

FIGURE 12.9 CLASSIFICATION SCORES, MEMBERSHIP PROBABILITIES, AND CLASSIFICATIONS FOR THE THREE-CLASS INJURY TRAINING DATASET.

robust. According to Hastie et al. (2001), the reason might be that data can usually support only simple separation boundaries, such as linear boundaries. However, for continuous variables that are found to be very skewed (e.g., through a histogram), transformations such as the log transform can improve performance. In addition, the method's sensitivity to outliers commands exploring the data for extreme values and removing those records from the analysis.

An advantage of discriminant analysis as a classifier (it is like logistic regression in this respect) is that it provides estimates of single-predictor contributions. This is useful for obtaining a ranking of the importance of predictors, and for variable selection.

Finally, the method is computationally simple, parsimonious, and especially useful for small datasets. With its parametric form, discriminant analysis makes the most out of the data and is therefore especially useful where the data are few (as explained in Section 12.4).

PROBLEMS

12.1 Personal Loan Acceptance. Universal Bank is a relatively young bank growing rapidly in terms of overall customer acquisition. The majority of these customers are liability customers with varying sizes of relationship with the bank. The customer base of asset customers is quite small, and the bank is interested in expanding this base rapidly to bring in more loan business. In particular, it wants to explore ways of converting its liability customers to personal loan customers.

A campaign the bank ran for liability customers last year showed a healthy conversion rate of over 9% successes. This has encouraged the retail marketing department to devise smarter campaigns with better target marketing. The goal of our analysis is to model the previous campaign's customer behavior to analyze what combination of factors make a customer more likely to accept a personal loan. This will serve as the basis for the design of a new campaign.

The file UniversalBank.xls contains data on 5000 customers. The data include customer demographic information (e.g., age, income), the customer's relationship with the bank (e.g., mortgage, securities account), and the customer response to the last personal loan campaign (Personal Loan). Among these 5000 customers, only 480 (= 9.6%) accepted the personal loan that was offered to them in the previous campaign.

Partition the data (60% training and 40% validation) and then perform a discriminant analysis that models Personal Loan as a function of the remaining predictors (excluding zip code). Remember to turn categorical predictors with more than two categories into dummy variables first. Specify the *success* class as 1 (loan acceptance), and use the default cutoff value of 0.5.

a. Compute summary statistics for the predictors separately for loan acceptors and nonacceptors. For continuous predictors, compute the mean and standard deviation. For categorical predictors, compute the percentages. Are there predictors where the two classes differ substantially?

b. Examine the model performance on the validation set.

    i. What is the misclassification rate?

    ii. Is one type of misclassification more likely than the other?

    iii. Select three customers who were misclassified as acceptors and three who were misclassified as nonacceptors. The goal is to determine why they are misclassified. First, examine their probability of being classified as acceptors: Is it close to the threshold of 0.5? If not, compare their predictor values to the summary statistics of the two classes to determine why they were misclassified.

c. As in many marketing campaigns, it is more important to identify customers who will accept the offer rather than customers who will not accept it. Therefore, a good model should be especially accurate at detecting acceptors. Examine the lift chart and decile chart for the validation set and interpret them in light of this goal.

d. Compare the results from the discriminant analysis with those from a logistic regression (both with cutoff 0.5 and the same predictors). Examine the confusion matrices, the lift charts, and the decile charts. Which method performs better on your validation set in detecting the acceptors?

e. The bank is planning to continue its campaign by sending its offer to 1000 additional customers. Suppose that the cost of sending the offer is $1 and the profit from an accepted offer is $50. What is the expected profitability of this campaign?

f. The cost of misclassifying a loan acceptor customer as a nonacceptor is much higher than the opposite misclassification cost. To minimize the expected cost of misclassification, should the cutoff value for classification (which is currently at 0.5) be increased or decreased?

12.2 **Identifying Good System Administrators.** A management consultant is studying the roles played by experience and training in a system administrator's ability to complete a set of tasks in a specified amount of time. In particular, she is interested in discriminating between administrators who are able to complete given tasks within a specified time and those who are not. Data are collected on the performance of 75 randomly selected administrators. They are stored in the file SystemAdministrators.xls.

Using these data, the consultant performs a discriminant analysis. The variable Experience measures months of full-time system administrator experience, while Training measures number of relevant training credits. The dependent variable Completed is either Yes or No, according to whether or not the administrator completed the tasks.

a. Create a scatterplot of Experience versus Training using color or symbol to differentiate administrators who completed the tasks from those who did not complete them. See if you can identify a line that separates the two classes with minimum misclassification.

b. Run a discriminant analysis with both predictors using the entire dataset as training data. Among those who completed the tasks, what is the percentage of administrators who are classified incorrectly as failing to complete the tasks?

c. Compute the two classification scores for an administrator with 4 years of higher education and 6 credits of training. Based on these, how would you classify this administrator?

d. How much experience must be accumulated by a administrator with 4 training credits before his or her estimated probability of completing the tasks exceeds 50%?

e. Compare the classification accuracy of this model to that resulting from a logistic regression with cutoff 0.5.

f. Compute the correlation between Experience and Training for administrators that completed the tasks and compare it to the correlation of administrators who did not complete the tasks. Does the equal correlation assumption seem reasonable?

12.3 **Detecting Spam E-mail (from the UCI Machine Learning Repository).** A team at Hewlett-Packard collected data on a large number of e-mail messages from their postmaster and personal e-mail for the purpose of finding a classifier that can separate e-mail messages that are spam versus nonspam (a.k.a. "ham"). The spam concept is diverse: It includes advertisements for products or websites, "make money fast" schemes, chain letters, pornography, and so on. The definition used here is "unsolicited commercial e-mail." The file Spambase.xls contains information on 4601 e-mail messages, among which 1813 are tagged "spam." The predictors include 57 attributes, most of them are the average number of times a certain word (e.g., mail, George) or symbol (e.g., #, !) appears in the e-mail. A few predictors are related to the number and length of capitalized words.

a. To reduce the number of predictors to a manageable size, examine how each predictor differs between the spam and nonspam e-mails by comparing the spam-class average and nonspam-class average. Which are the 11 predictors that appear to vary the most between spam and nonspam e-mails? From these 11, which words or signs occur more often in spam?

b. Partition the data into training and validation sets; then perform a discriminant analysis on the training data using only the 11 predictors.

c. If we are interested mainly in detecting spam messages, is this model useful? Use the confusion matrix, lift chart, and decile chart for the validation set for the evaluation.

d. In the sample, almost 40% of the e-mail messages were tagged as spam. However, suppose that the actual proportion of spam messages in these e-mail accounts is 10%. Compute the constants of the classification functions to account for this information.

e. A spam filter that is based on your model is used, so that only messages that are classified as nonspam are delivered, while messages that are classified as spam are quarantined. In this case, misclassifying a nonspam e-mail (as spam) has much heftier results. Suppose that the cost of quarantining a nonspam e-mail is 20 times that of not detecting a spam message. Compute the constants of the classification functions to account for these costs (assume that the proportion of spam is reflected correctly by the sample proportion).

## PART FIVE

Mining Relationships Among Records

## CHAPTER 13

Association Rules

In this chapter we describe the unsupervised learning methods of association rules (also called "affinity analysis"), where the goal is to identify item clusterings in transaction-type databases. Association rule discovery is popular in marketing, where it is called "market basket analysis" and is aimed at discovering which groups of products tend to be purchased together. We describe the two-stage process of rule generation and then assessment of rule strength to choose a subset. We describe the popular rule-generating Apriori algorithm and then criteria for judging the strength of rules. We also discuss issues related to the required data format and nonautomated methods for condensing the list of generated rules. The entire process is illustrated in a numerical example.

13.1 INTRODUCTION

Put simply, association rules, or *affinity analysis*, constitute a study of "what goes with what." For example, a medical researcher wants to learn what symptoms go with what confirmed diagnoses. This method is also called *market basket analysis* because it originated with the study of customer transactions databases to determine dependencies between purchases of different items.

13.2 DISCOVERING ASSOCIATION RULES IN TRANSACTION DATABASES

The availability of detailed information on customer transactions has led to the development of techniques that automatically look for associations between items that are stored in the database. An example is data collected using bar code scanners in supermarkets. Such *market basket databases* consist of a large number of transaction records. Each record lists all items bought by a customer on a single-purchase transaction. Managers would be interested to know if certain groups of items are consistently purchased together. They could use these data for store layouts to place items optimally with respect to each other, or they could use such information for cross selling, for promotions, for catalog design, and to identify customer segments based on buying patterns. Association rules provide information of this type in the form of "if–then" statements. These rules are computed from the data; unlike the if–then rules of logic, association rules are probabilistic in nature.

Such rules are commonly encountered in online *recommendation systems* (or *recommender systems*), where customers examining an item or items for possible purchase are shown other items that are often purchased in conjunction with the first item(s). The display from Amazon.com's online shopping system illustrates the application of rules such as this. In the example shown in Figure 13.1, a

See larger image
Share your own customer images

Bound Away
Last Train Home
★★★★★ (2 customer reviews)
More about this product

List Price: $16.98
Price: $16.98 & eligible for **FREE Super Saver Shipping** on orders over $25. Details

Availability: In Stock.
To ensure delivery by December 22, choose FREE Super Saver Shipping. See more on holiday shipping. Ships from and sold by **Amazon.com**. Gift-wrap available.

Want it delivered Tuesday, December 5? Order it in the next 9 hours and 5 minutes, and choose **One-Day Shipping** at checkout. See details

44 used & new available from $8.99

Better Together
Buy this album with Time and Water ~ Last Train Home today!

 + **Buy Together Today: $33.96**

FIGURE 13.1   RECOMMENDATIONS BASED ON ASSOCIATION RULES

TABLE 13.1 TRANSACTIONS FOR PURCHASES OF DIFFERENT-COLORED CELLULAR PHONE FACEPLATES

Transaction	Faceplate	Colors	Purchased
1	red	white	green	
2	white	orange		
3	white	blue		
4	red	white	orange	
5	red	blue		
6	white	blue		
7	white	orange		
8	red	white	blue	green
9	red	white	blue	
10	yellow			

purchaser of Last Train Home's *Bound Away* audio CD is shown the other CDs most frequently purchased by other Amazon purchasers of this CD.

We introduce a simple artificial example and use it throughout the chapter to demonstrate the concepts, computations, and steps of affinity analysis. We end by applying affinity analysis to a more realistic example of book purchases.

Example 1: Synthetic Data on Purchases of Phone Faceplates

A store that sells accessories for cellular phones runs a promotion on faceplates. Customers who purchase multiple faceplates from a choice of six different colors get a discount. The store managers, who would like to know what colors of faceplates customers are likely to purchase together, collected the transaction database as shown in Table 13.1.

13.3 GENERATING CANDIDATE RULES

The idea behind association rules is to examine all possible rules between items in an if–then format and select only those that are most likely to be indicators of true dependence. We use the term *antecedent* to describe the "if" part, and *consequent* to describe the "then" part. In association analysis, the antecedent and consequent are sets of items (called *item sets*) that are disjoint (do not have any items in common).

Returning to the phone faceplate purchase example, one example of a possible rule is "if red, then white," meaning that if a red faceplate is purchased, a white one is, too. Here the antecedent is *red* and the consequent is *white*. The antecedent and consequent each contain a single item in this case. Another possible rule is "if red and white, then green." Here the antecedent includes the item set {*red, white*} and the consequent is {*green*}.

The first step in affinity analysis is to generate all the rules that would be candidates for indicating associations between items. Ideally, we might want to look at all possible combinations of items in a database with p distinct items (in the phone faceplate example, $p = 6$). This means finding all combinations of single items, pairs of items, triplets of items, and so on in the transactions database. However, generating all these combinations requires a long computation time that grows exponentially in k. A practical solution is to consider only combinations that occur with higher frequency in the database. These are called *frequent item sets*.

Determining what consists of a frequent item set is related to the concept of *support*. The support of a rule is simply the number of transactions that include both the antecedent and consequent item sets. It is called a support because it measures the degree to which the data "support" the validity of the rule. The support is sometimes expressed as a percentage of the total number of records in the database. For example, the support for the item set {red,white} in the phone faceplate example is 4 ($100 \times \frac{4}{10} = 40\%$).

What constitutes a frequent item set is therefore defined as an item set that has a support that exceeds a selected minimum support, determined by the user.

The Apriori Algorithm

Several algorithms have been proposed for generating frequent item sets, but the classic algorithm is the *Apriori algorithm* of Agrawal et al. (1993). The key idea of the algorithm is to begin by generating frequent item sets with just one item (one-item sets) and to recursively generate frequent item sets with two items, then with three items, and so on until we have generated frequent item sets of all sizes.

It is easy to generate frequent one-item sets. All we need to do is to count, for each item, how many transactions in the database include the item. These transaction counts are the supports for the one-item sets. We drop one-item sets that have support below the desired minimum support to create a list of the frequent one-item sets.

To generate frequent two-item sets, we use the frequent one-item sets. The reasoning is that if a certain one-item set did not exceed the minimum support, any larger size item set that includes it will not exceed the minimum support. In general, generating k-item sets uses the frequent $(k - 1)$-item sets that were generated in the preceding step. Each step requires a single run through the database, and therefore the Apriori algorithm is very fast even for a large number of unique items in a database.

13.4   SELECTING STRONG RULES

From the abundance of rules generated, the goal is to find only the rules that indicate a strong dependence between the antecedent and consequent item sets. To measure the strength of association implied by a rule, we use the measures of *confidence* and *lift ratio*, as described below.

Support and Confidence

In addition to support, which we described earlier, there is another measure that expresses the degree of uncertainty about the if–then rule. This is known as the *confidence*[1] of the rule. This measure compares the co-occurrence of the antecedent and consequent item sets in the database to the occurrence of the antecedent item sets. Confidence is defined as the ratio of the number of transactions that include all antecedent and consequent item sets (namely, the support) to the number of transactions that include all the antecedent item sets:

$$\text{Confidence} = \frac{\text{no. transactions with both antecedent and consequent item sets}}{\text{no. transactions with antecedent item set}}.$$

For example, suppose that a supermarket database has 100,000 point-of-sale transactions. Of these transactions, 2000 include both orange juice and (over-the-counter) flu medication, and 800 of these include soup purchases. The association rule "IF orange juice and flu medication are purchased THEN soup is purchased on the same trip" has a support of 800 transactions (alternatively, $0.8\% = 800/100{,}000$) and a confidence of 40% ($= 800/2000$).

To see the relationship between support and confidence, let us think about what each is measuring (estimating). One way to think of support is that it is the (estimated) probability that a transaction selected randomly from the database will contain all items in the antecedent and the consequent:

$$P(\text{antecedent AND consequent}).$$

In comparison, the confidence is the (estimated) *conditional probability* that a transaction selected randomly will include all the items in the consequent *given* that the transaction includes all the items in the antecedent:

$$\frac{P(\text{antecedent AND consequent})}{P(\text{antecedent})} = P(\text{consequent} \mid \text{antecedent}).$$

A high value of confidence suggests a strong association rule (in which we are highly confident). However, this can be deceptive because if the antecedent

---

[1] The concept of confidence is different from and unrelated to the ideas of confidence intervals and confidence levels used in statistical inference.

and/or the consequent has a high level of support, we can have a high value for confidence even when the antecedent and consequent are independent! For example, if nearly all customers buy bananas and nearly all customers buy ice cream, the confidence level will be high regardless of whether there is an association between the items.

Lift Ratio

A better way to judge the strength of an association rule is to compare the confidence of the rule with a benchmark value, where we assume that the occurrence of the consequent item set in a transaction is independent of the occurrence of the antecedent for each rule. In other words, if the antecedent and consequent item sets are independent, what confidence values would we expect to see? Under independence, the support would be

$$P(\text{antecedent AND consequent}) = P(\text{antecedent}) \times P(\text{consequent}),$$

and the benchmark confidence would be

$$\frac{P(\text{antecedent}) \times P(\text{consequent})}{P(\text{antecedent})} = P(\text{consequent}).$$

The estimate of this benchmark from the data, called the *benchmark confidence value* for a rule, is computed by

$$\text{Benchmark confidence} = \frac{\text{no. transactions with consequent item set}}{\text{no. transactions in database}}.$$

We compare the confidence to the benchmark confidence by looking at their ratio: This is called the *lift ratio* of a rule. The lift ratio is the confidence of the rule divided by the confidence, assuming independence of consequent from antecedent:

$$\text{Lift ratio} = \frac{\text{confidence}}{\text{benchmark confidence}}.$$

A lift ratio greater than 1.0 suggests that there is some usefulness to the rule. In other words, the level of association between the antecedent and consequent item sets is higher than would be expected if they were independent. The larger the lift ratio, the greater the strength of the association.

To illustrate the computation of support, confidence, and lift ratio for the cellular phone faceplate example, we introduce a presentation of the data better suited to this purpose.

Data Format

Transaction data are usually displayed in one of two formats: a list of items purchased (each row representing a transaction), or a binary matrix in which columns are items, rows again represent transactions, and each cell has either a 1 or a 0, indicating the presence or absence of an item in the transaction. For example, Table 13.1 displays the data for the cellular faceplate purchases in item list format. We translate these into binary matrix format in Table 13.2.

Now suppose that we want association rules between items for this database that have a support count of at least 2 (equivalent to a percentage support of $2/10 = 20\%$): in other words, rules based on items that were purchased together in at least 20% of the transactions. By enumeration, we can see that only the item sets listed in Table 13.3 have a count of at least 2.

TABLE 13.2   PHONE FACEPLATE DATA IN BINARY MATRIX FORMAT

Transaction	Red	White	Blue	Orange	Green	Yellow
1	1	1	0	0	1	0
2	0	1	0	1	0	0
3	0	1	1	0	0	0
4	1	1	0	1	0	0
5	1	0	1	0	0	0
6	0	1	1	0	0	0
7	1	0	1	0	0	0
8	1	1	1	0	1	0
9	1	1	1	0	0	0
10	0	0	0	0	0	1

TABLE 13.3   ITEM SETS WITH SUPPORT COUNT OF AT LEAST TWO

Item Set	Support (Count)
{red}	6
{white}	7
{blue}	6
{orange}	2
{green}	2
{red, white}	4
{red, blue}	4
{red, green}	2
{white, blue}	4
{white, orange}	2
{white, green}	2
{red, white, blue}	2
{red, white, green}	2

The first item set {red} has a support of 6 because six of the transactions included a red faceplate. Similarly, the last item set {red, white, green} has a support of 2 because only two transactions included red, white, and green faceplates.

> In XLMiner the user can choose to input data using the *Affinity* > *Association Rules* facility in either item list format or binary matrix format.

Process of Rule Selection

The process of selecting strong rules is based on generating all association rules that meet stipulated support and confidence requirements. This is done in two stages. The first stage, described in Section 13.3, consists of finding all "frequent" item sets, those item sets that have a requisite support. In the second stage we generate, from the frequent item sets, association rules that meet a confidence requirement. The first step is aimed at removing item combinations that are rare in the database. The second stage then filters the remaining rules and selects only those with high confidence. For most association analysis data, the computational challenge is the first stage, as described in the discussion of the Apriori algorithm.

The computation of confidence in the second stage is simple. Since any subset (e.g., {red} in the phone faceplate example) must occur at least as frequently as the set it belongs to (e.g., {red, white}), each subset will also be in the list. It is then straightforward to compute the confidence as the ratio of the support for the item set to the support for each subset of the item set. We retain the corresponding association rule only if it exceeds the desired cutoff value for confidence. For example, from the item set {red, white, green} in the phone faceplate purchases, we get the following association rules:

Rule 1: {red, white} $\Rightarrow$ {green} with confidence
$$= \frac{\text{support of } \{\text{red, white, green}\}}{\text{support of } \{\text{red, white}\}} = 2/4 = 50\%.$$

Rule 2: {red, green} $\Rightarrow$ {white} with confidence
$$= \frac{\text{support of } \{\text{red, white, green}\}}{\text{support of } \{\text{red, green}\}} = 2/2 = 100\%.$$

Rule 3: {white, green} $\Rightarrow$ {red} with confidence
$$= \frac{\text{support of } \{\text{red, white, green}\}}{\text{support of } \{\text{white, green}\}} = 2/2 = 100\%.$$

Rule 4: { red} $\Rightarrow$ { white, green} with confidence
$$= \frac{\text{support of } \{\text{red, white, green}\}}{\text{support of } \{\text{red}\}} = 2/6 = 33\%.$$

Data	
Input Data	Faceplates!$B$1:$G$11
Data Format	Binary Matrix
Minimum Support	2
Minimum Confidence %	70
# Rules	6
Overall Time (secs)	2

Place the cursor on a cell in the rules table to read a rule.
Use up / down arrow keys to browse through the rules.

Rule #	Conf. %	Antecedent (a)	Consequent (c)	Support(a)	Support(c)	Support(a U c)	Lift Ratio
1	100	green=>	red, white	2	4	2	2.5
2	100	green=>	red	2	6	2	1.666667
3	100	green, white=>	red	2	6	2	1.666667
4	100	green=>	white	2	7	2	1.428571
5	100	green, red=>	white	2	7	2	1.428571
6	100	orange=>	white	2	7	2	1.428571

FIGURE 13.2   **ASSOCIATION RULES FOR PHONE FACEPLATE TRANSACTIONS: XLMINER OUTPUT**

Rule 5: $\{\text{white}\} \Rightarrow \{\text{red, green}\}$ with confidence
$$= \frac{\text{support of } \{\text{red, white, green}\}}{\text{support of } \{\text{white}\}} = 2/7 = 29\%.$$

Rule 6: $\{\text{green}\} \Rightarrow \{\text{red, white}\}$ with confidence
$$= \frac{\text{support of } \{\text{red, white, green}\}}{\text{support of } \{\text{green}\}} = 2/2 = 100\%.$$

If the desired minimum confidence is 70%, we would report only the second, third, and last rules.

We can generate association rules in XLMiner by specifying the minimum support count (2) and minimum confidence level percentage (70%). Figure 13.2 shows the output. Note that here we consider all possible item sets, not just {red, white, green} as above.

The output includes information on the support of the antecedent, the support of the consequent, and the support of the combined set [denoted by Support $(a \cup c)$]. It also gives the confidence of the rule (in percent) and the lift ratio. In addition, XLMiner has an *interpreter* that translates the rule from a certain row into English. In the snapshot shown in Figure 13.2, the first rule is highlighted (by clicking), and the corresponding English rule appears in the yellow box:

Rule 1: If item(s) green= is/are purchased, then this implies item(s) red, white is/are also purchased. This rule has confidence of 100%.

Interpreting the Results

In interpreting results, it is useful to look at the various measures. The support for the rule indicates its impact in terms of overall size: What proportion of transactions is affected? If only a small number of transactions are affected, the rule may be of little use (unless the consequent is very valuable and/or the rule is very efficient in finding it).

The lift ratio indicates how efficient the rule is in finding consequents, compared to random selection. A very efficient rule is preferred to an inefficient rule, but we must still consider support: A very efficient rule that has very low support may not be as desirable as a less efficient rule with much greater support.

The confidence tells us at what rate consequents will be found and is useful in determining the business or operational usefulness of a rule: A rule with low confidence may find consequents at too low a rate to be worth the cost of (say) promoting the consequent in all the transactions that involve the antecedent.

Statistical Significance of Rules

What about confidence in the nontechnical sense? How sure can we be that the rules we develop are meaningful? Considering the matter from a statistical perspective, we can ask: Are we finding associations that are really just chance occurrences?

Let us examine the output from an application of this algorithm to a small database of 50 transactions, where each of the 9 items is assigned randomly to each transaction. The data are shown in Table 13.4, and the association rules generated are shown in Table 13.5.

In this example, the lift ratios highlight rule 6 as most interesting, as it suggests that purchase of item 4 is almost five times as likely when items 3 and 8 are purchased than if item 4 was not associated with the item set {3,8}. Yet we know

TABLE 13.4   FIFTY TRANSACTIONS OF RANDOMLY ASSIGNED ITEMS

Transaction	Items					Transaction	Items				Transaction	Items			
1	8					18	8				35	3	4	6	8
2	3	4	8			19					36	1	4	8	
3	8					20	9				37	4	7	8	
4	3	9				21	2	5	6	8	38	8	9		
5	9					22	4	6	9		39	4	5	7	9
6	1	8				23	4	9			40	2	8	9	
7	6	9				24	8	9			41	2	5	9	
8	3	5	7	9		25	6	8			42	1	2	7	9
9	8					26	1	6	8		43	5	8		
10						27	5	8			44	1	7	8	
11	1	7	9			28	4	8	9		45	8			
12	1	4	5	8	9	29	9				46	2	7	9	
13	5	7	9			30	8				47	4	6	9	
14	6	7	8			31	1	5	8		48	9			
15	3	7	9			32	3	6	9		49	9			
16	1	4	9			33	7	9			50	6	7	8	
17	6	7	8			34	7	8	9						

TABLE 13.5   ASSOCIATION RULES OUTPUT FOR RANDOM DATA

Input Data: $A$5:$E$54
Min. Support: 2 = 4%
Min. Conf. % :   70

Rule	Confidence (%)	Anteced,, a		Conseq., c	Support (a)	Support (c)	Support $(a \cup c)$	Confidence If $P(c\|a)=$ $P(c)$ (%)	Lift Ratio (conf./ prev. col.)
1	80	2	⇒	9	5	27	4	54	1.5
2	100	5, 7	⇒	9	3	27	3	54	1.9
3	100	6, 7	⇒	8	3	29	3	58	1.7
4	100	1, 5	⇒	8	2	29	2	58	1.7
5	100	2, 7	⇒	9	2	27	2	54	1.9
6	100	3, 8	⇒	4	2	11	2	22	4.5
7	100	3, 4	⇒	8	2	29	2	58	1.7
8	100	3, 7	⇒	9	2	27	2	547	1.9
9	100	4, 5	⇒	9	2	27	2	54	1.9

there is no fundamental association underlying these data—they were generated randomly.

Two principles can guide us in assessing rules for possible spuriousness due to chance effects:

1. The more records the rule is based on, the more solid the conclusion. The key evaluative statistics are based on ratios and proportions, and we can look to statistical confidence intervals on proportions, such as political polls, for a rough preliminary idea of how variable rules might be owing to chance sampling variation. Polls based on 1500 respondents, for example, yield margins of error in the range of ±1.5%.

2. The more distinct rules we consider seriously (perhaps consolidating multiple rules that deal with the same items), the more likely it is that at least some will be based on chance sampling results. For one person to toss a coin 10 times and get 10 heads would be quite surprising. If 1000 people toss a coin 10 times apiece, it would not be nearly so surprising to have one get 10 heads. Formal adjustment of "statistical significance" when multiple comparisons are made is a complex subject in its own right and beyond the scope of this book. A reasonable approach is to consider rules from the top down in terms of business or operational applicability and not consider more than can reasonably be incorporated in a human decision-making process. This will impose a rough constraint on the dangers that arise from an automated

review of hundreds or thousands of rules in search of "something interesting."

We now consider a more realistic example, using a larger database and real transactional data.

Example 2: Rules for Similar Book Purchases

The following example (drawn from the Charles Book Club case) examines associations among transactions involving various types of books. The database includes 2000 transactions, and there are 11 different types of books. The data, in binary matrix form, are shown in Figure 13.3. For instance, the first transaction included YouthBks (youth books), DoItYBks (do-it-yourself books), and GeogBks (geography books). Figure 13.4 shows (part of) the rules generated by XLMiner's *Association Rules* on these data. We specified a minimal support of 200 transactions and a minimal confidence of 50%. This resulted in 49 rules (the first 26 rules are shown in Figure 13.4).

In reviewing these rules, we can see that the information can be compressed. First, rule 1, which appears from the confidence level to be a very promising rule, is probably meaningless. It says: "If Italian cooking books have been purchased, then cookbooks are purchased." It seems likely that Italian cooking books are simply a subset of cookbooks. Rules 2 and 7 involve the same trio of books, with different antecedents and consequents. The same is true of rules 14 and 15 and rules 9 and 10. (Pairs and groups like this are easy to track down by looking for rows that share the same support.) This does not mean that the rules are not useful. On the contrary, it can reduce the number of item sets to be considered for possible action from a business perspective.

ChildBks	YouthBks	CookBks	DoItYBks	RefBks	ArtBks	GeogBks	ItalCook	ItalAtlas	ItalArt	Florence
0	1	0	1	0	0	1	0	0	0	0
1	0	0	0	0	0	0	0	0	0	0
0	0	0	0	0	0	0	0	0	0	0
1	1	1	0	1	0	1	0	0	0	0
0	0	1	0	0	0	1	0	0	0	0
1	0	0	0	0	1	0	0	0	0	1
0	1	0	0	0	0	0	0	0	0	0
0	1	0	0	1	0	0	0	0	0	0
1	0	0	1	0	0	0	0	0	0	0
1	1	1	0	0	0	1	0	0	0	0
0	0	0	0	0	0	0	0	0	0	0

FIGURE 13.3   **SUBSET OF BOOK PURCHASE TRANSACTIONS IN BINARY MATRIX FORMAT**

Data	
Input Data	Books! $A$1:$K$2001
Data Format	Binary Matrix
Minimum Support	200
Minimum Confidence %	50
# Rules	49
Overall Time (secs)	1

Rule 1: If item(s) ItalCook= is / are purchased, then this implies item(s) CookBks is / are also purchased. This rule has confidence of 100%.

Rule #	Conf. %	Antecedent (a)	Consequent (c)	Support(a)	Support(c)	Support(a U c)	Lift Ratio
1	100	ItalCook=>	CookBks	227	862	227	2.320186
2	62.77	ArtBks, ChildBks=>	GeogBks	325	552	204	2.274247
3	54.13	CookBks, DoItYBks=>	ArtBks	375	482	203	2.246196
4	61.98	ArtBks, CookBks=>	GeogBks	334	552	207	2.245509
5	53.77	CookBks, GeogBks=>	ArtBks	385	482	207	2.230964
6	57.11	RefBks=>	ChildBks, CookBks	429	512	245	2.230842
7	52.31	ChildBks, GeogBks=>	ArtBks	390	482	204	2.170444
8	60.78	ArtBks, CookBks=>	DoItYBks	334	564	203	2.155264
9	58.4	ChildBks, CookBks=>	GeogBks	512	552	299	2.115885
10	54.17	GeogBks=>	ChildBks, CookBks	552	512	299	2.115885
11	57.87	CookBks, DoItYBks=>	GeogBks	375	552	217	2.096618
12	56.79	ChildBks, DoItYBks=>	GeogBks	368	552	209	2.057735
13	52.49	ArtBks=>	ChildBks, CookBks	482	512	253	2.050376
14	52.12	YouthBks=>	ChildBks, CookBks	495	512	258	2.035985
15	50.39	ChildBks, CookBks=>	YouthBks	512	495	258	2.035985
16	57.03	ChildBks, CookBks=>	DoItYBks	512	564	292	2.022385
17	51.77	DoItYBks=>	ChildBks, CookBks	564	512	292	2.022385
18	56.36	CookBks, GeogBks=>	DoItYBks	385	564	217	1.998711
19	52.9	ArtBks=>	GeogBks	482	552	255	1.916832
20	82.19	ArtBks, DoItYBks=>	CookBks	247	862	203	1.906873
21	53.59	ChildBks, GeogBks=>	DoItYBks	390	564	209	1.900346
22	81.89	DoItYBks, GeogBks=>	CookBks	265	862	217	1.899926
23	80.33	CookBks, RefBks=>	ChildBks	305	846	245	1.899004
24	80	ArtBks, GeogBks=>	ChildBks	255	846	204	1.891253
25	81.18	ArtBks, GeogBks=>	CookBks	255	862	207	1.883445
26	79.63	CookBks, YouthBks=>	ChildBks	324	846	258	1.882497

FIGURE 13.4   ASSOCIATION RULES FOR BOOK PURCHASE TRANSACTIONS: XLMINER OUTPUT

13.5 SUMMARY

Affinity analysis (also called market basket analysis) is a method for deducing rules on associations between purchased items from databases of transactions. The main advantage of this method is that it generates clear, simple rules of the form "IF X is purchased THEN Y is also likely to be purchased." The method is very transparent and easy to understand.

The process of creating association rules is two staged. First, a set of candidate rules based on frequent item sets is generated (the Apriori algorithm being the most popular rule-generating algorithm). Then from these candidate rules, the rules that indicate the strongest association between items are selected. We use the measures of support and confidence to evaluate the uncertainty in a rule. The user also specifies minimal support and confidence values to be used in the rule generation and selection process. A third measure, the lift ratio, compares the efficiency of the rule to detect a real association compared to a random combination.

One shortcoming of association rules is the profusion of rules that are generated. There is therefore a need for ways to reduce these to a small set of useful and strong rules. An important nonautomated method to condense the information involves examining the rules for noninformative and trivial rules as well as for rules that share the same support.

Another issue that needs to be kept in mind is that rare combinations tend to be ignored because they do not meet the minimum support requirement. For this reason it is better to have items that are approximately equally frequent in the data. This can be achieved by using higher level hierarchies as the items. An example is to use types of audio CDs rather than names of individual audio CDs in deriving association rules from a database of music store transactions.

PROBLEMS

13.1 Satellite Radio Customers. An analyst at a subscription-based satellite radio company has been given a sample of data from their customer database, with the goal of finding groups of customers that are associated with one another. The data consist of company data, together with purchased demographic data that are mapped to the company data (see Figure 13.5). The analyst decides to apply association rules to learn more about the associations between customers. Comment on this approach.

13.2 Online Statistics Courses. Consider the data in the file CourseTopics.xls, the first few rows of which are shown in Figure 13.6. These data are for purchases of online statistics courses at statistics.com. Each row represents the courses attended by a single customer.

The firm wishes to assess alternative sequencings and combinations of courses. Use association rules to analyze these data and interpret several of the resulting rules.

13.3 Cosmetics Purchases. The data shown in Figure 13.7 and the output in Figure 13.8 are from a subset of a dataset on cosmetic purchases (Cosmetics-small.xls) given in binary matrix form. The complete dataset (in the file Cosmetics.xls) contains data on the purchases of different cosmetic items at a large chain drugstore. The store wants to analyze associations

Row Id.	zipconvert _2	zipconvert _3	zipconvert _4	zipconvert _5	homeowner dummy	NUMCHLD	INCOME	gender dummy	WEALTH
17	0	1	0	0	1	1	5	1	9
25	1	0	0	0	1	1	1	0	7
29	0	0	0	1	0	2	5	1	8
38	0	0	0	1	1	1	3	0	4
40	0	1	0	0	1	1	4	0	8
53	0	1	0	0	1	1	4	1	8
58	0	0	0	1	1	1	4	1	8
61	1	0	0	0	1	1	1	0	7
71	0	0	1	0	1	1	4	0	5
87	1	0	0	0	1	1	4	1	8
100	0	0	0	1	1	1	4	1	8
104	1	0	0	0	1	1	1	1	5
121	0	0	1	0	1	1	4	1	5
142	1	0	0	0	0	1	5	0	8

FIGURE 13.5 SAMPLE OF DATA ON SATELLITE RADIO CUSTOMERS

Course Topics

Intro	Data Mining	Survey	Cat Data	Regression	Forecast	DOE	SW
1	1	0	0	0	0	0	0
0	0	1	0	0	0	0	0
0	1	0	1	1	0	0	1
1	0	0	0	0	0	0	0
1	1	0	0	0	0	0	0
0	1	0	0	0	0	0	0
1	0	0	0	0	0	0	0
0	0	0	1	0	1	1	1
1	0	0	0	0	0	0	0
0	0	0	1	0	0	0	0
1	0	0	0	0	0	0	0

FIGURE 13.6 DATA ON PURCHASES OF ONLINE STATISTICS COURSES

Trans. #	Bag	Blush	Nail Polish	Brushes	Concealer	Eyebrow Pencils	Bronzer
1	0	1	1	1	1	0	1
2	0	0	1	0	1	0	1
3	0	1	0	0	1	1	1
4	0	0	1	1	1	0	1
5	0	1	0	0	1	0	1
6	0	0	0	0	1	0	0
7	0	1	1	1	1	0	1
8	0	0	1	1	0	0	1
9	0	0	0	0	1	0	0
10	1	1	1	1	0	0	0
11	0	0	1	0	0	0	1
12	0	0	1	1	1	0	1

FIGURE 13.7   **DATA ON COSMETICS PURCHASES IN BINARY MATRIX FORM**

point-of-sale display, guidance to sales personnel in promoting cross sales, and guidance for piloting an eventual time-of-purchase electronic recommender system to boost cross sales. Consider first only the subset shown in Figure 13.7.

a. Select several values in the matrix and explain their meaning.

b. Consider the results of the association rules analysis shown in Figure 13.8 and:

  i. For the first row, explain the "Conf. %" output and how it is calculated.

  ii. For the first row, explain the "Support(a)," "Support(c)," and "Support($a \cup c$)" output and how it is calculated.

  iii. For the first row, explain the "Lift Ratio" and how it is calculated.

  iv. For the first row, explain the rule that is represented there in words.
  Now, use the complete dataset on the cosmetics purchases (in the file Cosmetics.xls).

  v. Using XLMiner, apply association rules to these data.

  vi. Interpret the first three rules in the output in words.

  vii. Reviewing the first couple of dozen rules, comment on their redundancy and how you would assess their utility.

Rule #	Conf. %	Antecedent (a)	Consequent (c)	Support(a)	Support (c)	Support (a U c)	Lift Ratio
2	60.19	Bronzer, Nail Polish=>	Brushes, Concealer	103	77	62	3.909
1	80.52	Brushes, Concealer=>	Bronzer, Nail Polish	77	103	62	3.909
4	56.36	Brushes=>	Bronzer, Concealer, Nail Polish	110	76	62	3.708
3	81.58	Bronzer, Concealer, Nail Polish	Brushes	76	110	62	3.708
6	76.36	Brushes=>	Bronzer, Nail Polish	110	103	84	3.707
5	81.55	Bronzer, Nail Polish=>	Brushes	103	110	84	3.707
8	56.88	Concealer, Nail Polish=>	Bronzer, Brushes	109	84	62	3.386
7	73.81	Bronzer, Brushes=>	Concealer, Nail Polish	84	109	62	3.386
10	70	Brushes=>	Concealer, Nail Polish	110	109	77	3.211
9	70.64	Concealer, Nail Polish=>	Brushes	109	110	77	3.211
12	50	Brushes=>	Blush, Nail Polish	110	82	55	3.049
11	67.07	Blush, Nail Polish=>	Brushes	82	110	55	3.049

FIGURE 13.8   **ASSOCIATION RULES FOR COSMETICS PURCHASES DATA**

## CHAPTER 14

Cluster Analysis

This chapter is about the popular unsupervised learning task of clustering, where the goal is to segment the data into a set of homogeneous clusters of observations for the purpose of generating insight. Clustering is used in a vast variety of business applications, from customized marketing to industry analysis. We describe two popular clustering approaches: hierarchical clustering and k-means clustering. In hierarchical clustering, observations are sequentially grouped to create clusters, based on distances between observations and distances between clusters. We describe how the algorithm works in terms of the clustering process and mention several common distance metrics used. Hierarchical clustering also produces a useful graphical display of the clustering process and results, called a dendrogram. We present dendrograms and illustrate their usefulness. k-means clustering is widely used in large dataset applications. In k-means clustering, observations are allocated to one of a prespecified set of clusters, according to their distance from each cluster. We describe the k-means clustering algorithm and its computational advantages. Finally, we present techniques that assist in generating insight from clustering results.

14.1 INTRODUCTION

Cluster analysis is used to form groups or clusters of similar records based on several measurements made on these records. The key idea is to characterize the clusters in ways that would be useful for the aims of the analysis. This idea has been

Data Mining for Business Intelligence, By Galit Shmueli, Nitin R. Patel, and Peter C. Bruce

applied in many areas, including astronomy, archaeology, medicine, chemistry, education, psychology, linguistics, and sociology. Biologists, for example, have made extensive use of classes and subclasses to organize species. A spectacular success of the clustering idea in chemistry was Mendeleev's periodic table of the elements.

One popular use of cluster analysis in marketing is for *market segmentation:* Customers are segmented based on demographic and transaction history information, and a marketing strategy is tailored for each segment. Another use is for *market structure analysis:* identifying groups of similar products according to competitive measures of similarity. In marketing and political forecasting, clustering of neighborhoods using U.S. postal zip codes has been used successfully to group neighborhoods by lifestyles. Claritas, a company that pioneered this approach, grouped neighborhoods into 40 clusters using various measures of consumer expenditure and demographics. Examining the clusters enabled Claritas to come up with evocative names, such as Bohemian Mix, Furs and Station Wagons, and Money and Brains, for the groups that captured the dominant lifestyles. Knowledge of lifestyles can be used to estimate the potential demand for products (such as sports utility vehicles) and services (such as pleasure cruises).

In finance, cluster analysis can be used for creating *balanced portfolios:* Given data on a variety of investment opportunities (e.g., stocks), one may find clusters based on financial performance variables such as return (daily, weekly, or monthly), volatility, beta, and other characteristics, such as industry and market capitalization. Selecting securities from different clusters can help create a balanced portfolio. Another application of cluster analysis in finance is for *industry analysis:* For a given industry, we are interested in finding groups of similar firms based on measures such as growth rate, profitability, market size, product range, and presence in various international markets. These groups can then be analyzed in order to understand industry structure and to determine, for instance, who is a competitor.

An interesting and unusual application of cluster analysis, described in Berry and Linoff (1997), is the design of a new set of sizes for army uniforms for women in the U.S. Army. The study came up with a new clothing size system with only 20 sizes, where different sizes fit different body types. The 20 sizes are combinations of five measurements: chest, neck, and shoulder circumference, sleeve outseam, and neck-to-buttock length [for further details, see McCullugh et al. (1998)]. This example is important because it shows how a completely new insightful view can be gained by examining clusters of records.

Cluster analysis can be applied to huge amounts of data. For instance, Internet search engines use clustering techniques to cluster queries that users submit. These can then be used for improving search algorithms. The objective of this chapter is to describe the key ideas underlying the most commonly used techniques for cluster analysis and to lay out their strengths and weaknesses.

Typically, the basic data used to form clusters are a table of measurements on several variables, where each column represents a variable and a row represents a record. Our goal is to form groups of records so that similar records are in the same group. The number of clusters may be prespecified or determined from the data.

Example: Public Utilities

Table 14.1 gives corporate data on 22 U.S. public utilities (the definition of each variable is given in the table footnote). We are interested in forming groups of similar utilities. The records to be clustered are the utilities, and the clustering will be based on the eight measurements on each utility. An example where

TABLE 14.1   DATA ON 22 PUBLIC UTILITIES[a]

Company	Fixed	RoR	Cost	Load	Demand	Sales	Nuclear	Fuel Cost
Arizona Public Service	1.06	9.2	151	54.4	1.6	9,077	0	0.628
Boston Edison Co.	0.89	10.3	202	57.9	2.2	5,088	25.3	1.555
Central Louisiana Co.	1.43	15.4	113	53	3.4	9,212	0	1.058
Commonwealth Edison Co.	1.02	11.2	168	56	0.3	6,423	34.3	0.7
Consolidated Edison Co. (NY)	1.49	8.8	192	51.2	1	3,300	15.6	2.044
Florida Power & Light Co.	1.32	13.5	111	60	−2.2	11,127	22.5	1.241
Hawaiian Electric Co.	1.22	12.2	175	67.6	2.2	7,642	0	1.652
Idaho Power Co.	1.1	9.2	245	57	3.3	13,082	0	0.309
Kentucky Utilities Co.	1.34	13	168	60.4	7.2	8,406	0	0.862
Madison Gas & Electric Co.	1.12	12.4	197	53	2.7	6,455	39.2	0.623
Nevada Power Co.	0.75	7.5	173	51.5	6.5	17,441	0	0.768
New England Electric Co.	1.13	10.9	178	62	3.7	6,154	0	1.897
Northern States Power Co.	1.15	12.7	199	53.7	6.4	7,179	50.2	0.527
Oklahoma Gas & Electric Co.	1.09	12	96	49.8	1.4	9,673	0	0.588
Pacific Gas & Electric Co.	0.96	7.6	164	62.2	−0.1	6,468	0.9	1.4
Puget Sound Power & Light Co.	1.16	9.9	252	56	9.2	15,991	0	0.62
San Diego Gas & Electric Co.	0.76	6.4	136	61.9	9	5,714	8.3	1.92
The Southern Co.	1.05	12.6	150	56.7	2.7	10,140	0	1.108
Texas Utilities Co.	1.16	11.7	104	54	−2.1	13,507	0	0.636
Wisconsin Electric Power Co.	1.2	11.8	148	59.9	3.5	7,287	41.1	0.702
United Illuminating Co.	1.04	8.6	204	61	3.5	6,650	0	2.116
Virginia Electric & Power Co.	1.07	9.3	174	54.3	5.9	10,093	26.6	1.306

[a]Fixed = fixed-charge covering ratio (income/debt); RoR = rate of return on capital; Cost = cost per kilowatt capacity in place; Load = annual load factor; Demand = peak kilowatthour demand growth from 1974 to 1975; Sales = sales (kilowatthour use per year); Nuclear = percent nuclear; Fuel Cost = total fuel costs (cents per kilowatthour).

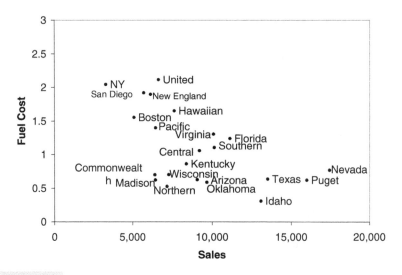

FIGURE 14.1   **SCATTERPLOT OF SALES VS. FUEL COST FOR THE 22 UTILITIES**

clustering would be useful is a study to predict the cost impact of deregulation. To do the requisite analysis, economists would need to build a detailed cost model of the various utilities. It would save a considerable amount of time and effort if we could cluster similar types of utilities and build detailed cost models for just one "typical" utility in each cluster and then scale up from these models to estimate results for all utilities.

For simplicity, let us consider only two of the measurements: Sales and Fuel Cost. Figure 14.1 shows a scatterplot of these two variables, with labels marking each company. At first glance, there appear to be two or three clusters of utilities: one with utilities that have high fuel costs, a second with utilities that have lower fuel costs and relatively low sales, and a third with utilities with low fuel costs but high sales. We can therefore think of cluster analysis as a more formal algorithm that measures the distance between records, and according to these distances (here, two-dimensional distances) forms clusters.

There are two general types of clustering algorithms for a dataset of n records:

Hierarchical Methods   Can be either agglomerative or divisive. Agglomerative methods begin with n clusters and sequentially merge similar clusters until a single cluster is left. Divisive methods work in the opposite direction, starting with one cluster that includes all observations. Hierarchical methods are especially useful when the goal is to arrange the clusters into a natural hierarchy.

Nonhierarchical Methods   For example, k-means clustering. Using a prespecified number of clusters, the method assigns cases to each cluster. These
methods are generally less computationally intensive and are therefore preferred with very large datasets.

We concentrate here on the two most popular methods: hierarchical agglomerative clustering and k-means clustering. In both cases we need to define
two types of distances: distance between two records and distance between two
clusters. In both cases there is a variety of metrics that can be used.

14.2 MEASURING DISTANCE BETWEEN TWO RECORDS

We denote by d_{ij} a *distance metric*, or *dissimilarity measure*, between records i
and j. For record i we have the vector of p measurements $(x_{i1}, x_{i2}, \ldots, x_{ip})$,
while for record j we have the vector of measurements $(x_{j1}, x_{j2}, \ldots, x_{jp})$. For
example, we can write the measurement vector for Arizona Public Service as
$[1.06, 9.2, 151, 54.4, 1.6, 9077, 0, 0.628]$.

Distances can be defined in multiple ways, but, in general, the following
properties are required:

Nonnegative $d_{ij} \geq 0$.

Self-Proximity $d_{ii} = 0$ (the distance from a record to itself is zero).

Symmetry $d_{ij} = d_{ji}$.

Triangle Inequality $d_{ij} \leq d_{ik} + d_{kj}$ (the distance between any pair cannot exceed the sum of distances between the other two pairs).

Euclidean Distance

The most popular distance measure is the *Euclidean distance*, d_{ij}, which between
two cases, i and j, is defined by

$$d_{ij} = \sqrt{(x_{i1} - x_{j1})^2 + (x_{i2} - x_{j2})^2 + \cdots + (x_{ip} - x_{jp})^2}.$$

For instance, the Euclidean distance between Arizona Public Service and Boston
Edison Co. can be computed from the raw data by

$$d_{12} = \sqrt{(1.06 - 0.89)^2 + (9.2 - 10.3)^2 + (151 - 202)^2 + \cdots + (0.628 - 1.555)^2}$$

$$= 3989.408.$$

Normalizing Numerical Measurements

The measure computed above is highly influenced by the scale of each variable,
so that variables with larger scales (e.g., Sales) have a much greater influence over

| TABLE 14.2 | ORIGINAL AND NORMALIZED MEASUREMENTS FOR SALES AND FUEL COST |

Company	Sales	Fuel Cost	NormSales	NormFuel
Arizona Public Service	9,077	0.628	0.0459	−0.8537
Boston Edison Co.	5,088	1.555	−1.0778	0.8133
Central Louisiana Co.	9,212	1.058	0.0839	−0.0804
Commonwealth Edison Co.	6,423	0.7	−0.7017	−0.7242
Consolidated Edison Co. (NY)	3,300	2.044	−1.5814	1.6926
Florida Power & Light Co.	11,127	1.241	0.6234	0.2486
Hawaiian Electric Co.	7,642	1.652	−0.3583	0.9877
Idaho Power Co.	13,082	0.309	1.1741	−1.4273
Kentucky Utilities Co.	8,406	0.862	−0.1431	−0.4329
Madison Gas & Electric Co.	6,455	0.623	−0.6927	−0.8627
Nevada Power Co.	17,441	0.768	2.4020	−0.6019
New England Electric Co.	6,154	1.897	−0.7775	1.4283
Northern States Power Co.	7,179	0.527	−0.4887	−1.0353
Oklahoma Gas & Electric Co.	9,673	0.588	0.2138	−0.9256
Pacific Gas & Electric Co.	6,468	1.4	−0.6890	0.5346
Puget Sound Power & Light Co.	15,991	0.62	1.9935	−0.8681
San Diego Gas & Electric Co.	5,714	1.92	−0.9014	1.4697
The Southern Co.	10,140	1.108	0.3453	0.0095
Texas Utilities Co.	13,507	0.636	1.2938	−0.8393
Wisconsin Electric Power Co.	7,287	0.702	−0.4583	−0.7206
United Illuminating Co.	6,650	2.116	−0.6378	1.8221
Virginia Electric & Power Co.	10,093	1.306	0.3321	0.3655
Mean	8,914.05	1.10	0.00	0.00
Standard deviation	3,549.98	0.56	1.00	1.00

the total distance. It is therefore customary to *normalize* (or *standardize*) continuous measurements before computing the Euclidean distance. This converts all measurements to the same scale. Normalizing a measurement means subtracting the average and dividing by the standard deviation (normalized values are also called z *scores*). For instance, the figure for average sales across the 22 utilities is 8914.045 and the standard deviation is 3549.984. The normalized sales for Arizona Public Service is therefore $(9077 − 8914.045)/3549.984 = 0.046$.

Returning to the simplified utilities data with only two measurements (Sales and Fuel Cost), we first normalize the measurements (see Table 14.2), and then compute the Euclidean distance between each pair. Table 14.3 gives these pairwise distanced for the first 5 utilities. A similar table can be constructed for all 22 utilities.

Other Distance Measures for Numerical Data

It is important to note that the choice of the distance measure plays a major role in cluster analysis. The main guideline is domain dependent: What exactly is being measured? How are the different measurements related? What scale should it be

TABLE 14.3   DISTANCE MATRIX BETWEEN PAIRS OF THE FIRST FIVE UTILITIES, USING EUCLIDEAN DISTANCE AND NORMALIZED MEASUREMENTS

	Arizona	Boston	Central	Commonwealth	Consolidated
Arizona	0				
Boston	2.01	0			
Central	0.77	1.47	0		
Commonwealth	0.76	1.58	1.02	0	
Consolidated	3.02	1.01	2.43	2.57	0

treated as (numerical, ordinal, or nominal)? Are there outliers? Finally, depending on the goal of the analysis, should the clusters be distinguished mostly by a small set of measurements, or should they be separated by multiple measurements that weight moderately?

Although Euclidean distance is the most widely used distance, it has three main features that need to be kept in mind. First, as mentioned above, it is highly scale dependent. Changing the units of one variable (e.g., from cents to dollars) can have a huge influence on the results. Standardizing is therefore a common solution. But unequal weighting should be considered if we want the clusters to depend more on certain measurements and less on others. The second feature of Euclidean distance is that it completely ignores the relationship between the measurements. Thus, if the measurements are in fact strongly correlated, a different distance (such as the statistical distance, described below) is likely to be a better choice. Third, Euclidean distance is sensitive to outliers. If the data are believed to contain outliers and careful removal is not a choice, the use of more robust distances (such as the Manhattan distance described below) is preferred.

Additional popular distance metrics often used (for reasons such as the ones above) are:

Correlation-Based Similarity   Sometimes it is more natural or convenient to work with a similarity measure between records rather than distance, which measures dissimilarity. A popular similarity measure is the square of the correlation coefficient, r_{ij}^2, where the correlation coefficient is defined by

$$r_{ij} \equiv \frac{\sum_{m=1}^{p} (x_{im} - \bar{x}_m)(x_{jm} - \bar{x}_m)}{\sqrt{\sum_{m=1}^{p} (x_{im} - \bar{x}_m)^2 \sum_{m=1}^{p} (x_{jm} - \bar{x}_m)^2}}.$$

Such measures can always be converted to distance measures. In the example above we could define a distance measure $d_{ij} = 1 - r_{ij}^2$.

Statistical Distance (also called *Mahalanobis distance*) This metric has an advantage over the other metrics mentioned in that it takes into account the correlation between measurements. With this metric, measurements that are highly correlated with other measurements do not contribute as much as those that are uncorrelated or mildly correlated. The statistical distance between records i and j is defined as

$$d_{i,j} = \sqrt{(\mathbf{x}_i - \mathbf{x}_j)' S^{-1} (\mathbf{x}_i - \mathbf{x}_j)},$$

where $\mathbf{x}_i$ and $\mathbf{x}_j$ are p-dimensional vectors of the measurement values for records i and j, respectively; and S is the covariance matrix for these vectors. (The notation $'$ denotes a transpose operation, which turns a column vector into a row vector.) S^{-1} is the inverse matrix of S, which is the p-dimension extension to division. For further information on statistical distance, see Chapter 12.

Manhattan Distance (*city block*) This distance looks at the absolute differences rather than squared differences, and is defined by

$$d_{ij} = \sum_{m=1}^{p} | x_{im} - x_{jm} |.$$

Maximum Coordinate Distance This distance looks only at the measurement on which records i and j deviate most. It is defined by

$$d_{ij} = \max_{m=1,2,\ldots,p} | x_{im} - x_{jm} |.$$

Distance Measures for Categorical Data

In the case of measurements with binary values, it is more intuitively appealing to use similarity measures than distance measures. Suppose that we have binary values for all the x_{ij}'s, and for records i and j we have the following 2×2 table:

		Record j		
		0	1	
Record i	0	a	b	$a + b$
	1	c	d	$c + d$
		$a + c$	$b + d$	p

where a denotes the number of predictors for which records i and j do not have that attribute, d is the number of predictors for which the two records have the attribute present, and so on. The most useful similarity measures in this situation are:

Matching Coefficient $(a + d)/p$.

Jaquard's Coefficient $d/(b + c + d)$. This coefficient ignores zero matches. This is desirable when we do not want to consider two people to be similar simply because a large number of characteristics are absent in both. For example, if Owns a Corvette is one of the variables, a matching "yes" would be evidence of similarity, but a matching "no" tells us little about whether the two people are similar.

Distance Measures for Mixed Data

When the measurements are mixed (some continuous and some binary), a similarity coefficient suggested by Gower is very useful. *Gower's similarity measure* is a weighted average of the distances computed for each variable, after scaling each variable to a [0,1] scale. It is defined as

$$s_{ij} = \frac{\sum\limits_{m=1}^{p} w_{ijm} s_{ijm}}{\sum\limits_{m=1}^{p} w_{ijm}},$$

with $w_{ijm} = 1$ subject to the following rules:

1. $w_{ijm} = 0$ when the value of the measurement is not known for one of the pair of records.

2. For nonbinary categorical measurements $s_{ijm} = 0$ unless the records are in the same category, in which case $s_{ijm} = 1$.

3. For continuous measurements, $s_{ijm} = 1 - \dfrac{|x_{im} - x_{jm}|}{\max(x_m) - \min(x_m)}$.

14.3 MEASURING DISTANCE BETWEEN TWO CLUSTERS

We define a cluster as a set of one or more records. How do we measure distance between clusters? The idea is to extend measures of *distance between records* into *distances between clusters*. Consider cluster A, which includes the m records $A_1, A_2, \ldots, A_m$ and cluster B, which includes n records $B_1, B_2, \ldots, B_n$. The most widely used measures of distance between clusters are:

Minimum Distance (Single Linkage)   The distance between the pair of records A_i and B_j that are closest:

$$\min(\text{distance}(A_i, B_j)), \quad i = 1, 2, \ldots, m; \quad j = 1, 2, \ldots, n.$$

Maximum Distance (Complete Linkage) The distance between the pair of records A_i and B_j that are farthest:

$$\max(\text{distance}(A_i, B_j)), \quad i = 1, 2, \ldots, m; \quad j = 1, 2, \ldots, n.$$

Average Distance (Average Linkage) The average distance of all possible distances between records in one cluster and records in the other cluster:

$$\text{Average}(\text{distance}(A_i, B_j)), \quad i = 1, 2, \ldots, m; \quad j = 1, 2, \ldots, n.$$

Centroid Distance The distance between the two cluster centroids. A *cluster centroid* is the vector of measurement averages across all the records in that cluster. For cluster A, this is the vector $\overline{x}_A = \left[(1/m \sum_{i=1}^{m} x_{1i}, \ldots, 1/m \sum_{i=1}^{m} x_{pi}) \right]$. The centroid distance between clusters A and B is

$$\text{distance}(\overline{x}_A, \overline{x}_B).$$

Minimum distance, maximum distance, and centroid distance are illustrated visually for two dimensions with a map of Portugal and France in Figure 14.2.

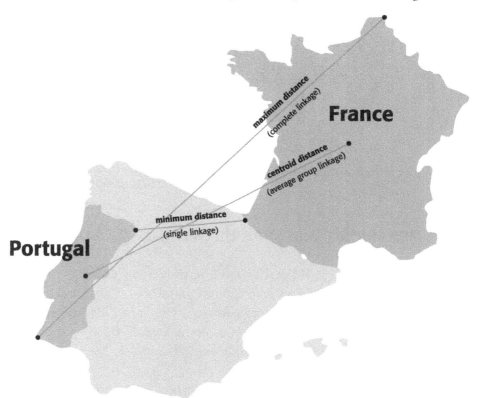

FIGURE 14.2 **TWO-DIMENSIONAL REPRESENTATION OF SEVERAL DIFFERENT DISTANCE MEASURES BETWEEN PORTUGAL AND FRANCE**

For instance, consider the first two utilities (Arizona, Boston) as cluster A, and the next three utilities (Central, Commonwealth, Consolidated) as cluster B. Using the normalized scores in Table 14.2 and the distance matrix in Table 14.3, we can compute each of the distances described above. Using Euclidean distance, we get:

- The closest pair is Arizona and Commonwealth, and therefore the minimum distance between clusters A and B is 0.76.
- The farthest pair is Arizona and Consolidated, and therefore the maximum distance between clusters A and B is 3.02.
- The average distance is $(0.77 + 0.76 + 3.02 + 1.47 + 1.58 + 1.01)/6 = 1.44$.
- The centroid of cluster A is

$$\left[\frac{0.0459 - 1.0778}{2}, \frac{-0.8537 + 0.8133}{2} \right] = [-0.516, -0.020],$$

and the centroid of cluster B is

$$\left[\frac{0.0839 - 0.7017 - 1.5814}{3}, \frac{-0.0804 - 0.7242 + 1.6926}{3} \right]$$
$$= [-0.733, 0.296].$$

The distance between the two centroids is then

$$\sqrt{(-0.516 + 0.733)^2 + (-0.020 + 0.296)^2} = 0.38.$$

In deciding among clustering methods, domain knowledge is key. If you have good reason to believe that the clusters might be chain- or sausage-like, minimum distance (single linkage) would be a good choice. This method does not require that cluster members all be close to one another, only that the new members being added be close to one of the existing members. An example of an application where this might be the case would be characteristics of crops planted in long rows, or disease outbreaks along navigable waterways that are the main areas of settlement in a region. Another example is laying and finding mines (land or marine). Single linkage is also fairly robust to small deviations in the distances. However, adding or removing data can influence it greatly.

Complete and average linkage are better choices if you know that the clusters are more likely to be spherical (e.g., customers clustered on the basis of numerous attributes). If you do not know the probable nature of the cluster these are good default choices since most clusters tend to be spherical in nature.

We now move to a more detailed description of the two major types of clustering algorithms: hierarchical (agglomerative) and nonhierarchical.

14.4 HIERARCHICAL (AGGLOMERATIVE) CLUSTERING

The idea behind hierarchical agglomerative clustering is to start with each cluster comprising exactly one record and then progressively agglomerating (combining) the two nearest clusters until there is just one cluster left at the end, which consists of all the records.

Returning to the small example of five utilities and two measures (Sales and Fuel Cost) and using the distance matrix (Table 14.3), the first step in the hierarchical clustering would join Arizona and Commonwealth, which are the closest (using normalized measurements and Euclidean distance). Next, we would recalculate a 4 × 4 distance matrix that would have the distances between these four clusters: {Arizona,Commonwealth}, {Boston}, {Central}, and {Consolidated}. At this point we use measure of distance between clusters, such as the ones described in the Section 14.3. Each of these distances (minimum, maximum, average, and centroid distance) can be implemented in the hierarchical scheme as described below.

HIERARCHICAL AGGLOMERATIVE CLUSTERING ALGORITHM

1. Start with *n* clusters (each observation = cluster).
2. The two closest observations are merged into one cluster.
3. At every step, the two clusters with the smallest distance are merged. This means that either single observations are added to existing clusters or two existing clusters are combined.

Minimum Distance (Single Linkage)

In *minimum distance clustering*, the distance between two clusters that is used is the minimum distance (the distance between the nearest pair of records in the two clusters, one record in each cluster). In our small utilities example, we would compute the distances between each of {Boston}, {Central}, and {Consolidated} with {Arizona, Commonwealth} to create the 4 × 4 distance matrix shown in Table 14.4.

TABLE 14.4   **DISTANCE MATRIX AFTER ARIZONA AND COMMONWEALTH CONSOLIDATION CLUSTER TOGETHER, USING SINGLE LINKAGE**

	Arizona–Commonwealth	Boston	Central	Consolidated
Arizona–Commonwealth	0			
Boston	min(2.01,1.58)	0		
Central	min(0.77,1.02)	1.47	0	
Consolidated	min(3.02,2.57)	1.01	2.43	0

The next step would consolidate {Central} with {Arizona, Commonwealth} because these two clusters are closest. The distance matrix will again be recomputed (this time it will be 3×3), and so on.

This method has a tendency to cluster together at an early stage records that are distant from each other because of a chain of intermediate records in the same cluster. Such clusters have elongated sausage-like shapes when visualized as objects in space.

Maximum Distance (Complete Linkage)

In *maximum distance clustering* (also called *complete linkage*) the distance between two clusters is the maximum distance (between the farthest pair of records). If we used complete linkage with the five utilities example, the recomputed distance matrix would be equivalent to Table 14.4, except that the "min" function would be replaced with a "max."

This method tends to produce clusters at the early stages with records that are within a narrow range of distances from each other. If we visualize them as objects in space, the records in such clusters would have roughly spherical shapes.

Average Distance (Average Linkage)

Average clustering is based on the average distance between clusters (between all possible pairs of records). If we used average linkage with the five utilities example, the recomputed distance matrix would be equivalent to Table 14.4, except that the "min" function would be replaced with an "average."

Note that the results of the single and complete linkage methods depend only on the order of the interrecord distances. Linear transformations of the distances (and other transformations that do not change the order) do not affect the results.

Centroid Distance (Average Group Linkage)

In clustering based on centroid distance (average group linkage clustering), clusters are represented by their mean values for each variable, a vector of means. The distance between two clusters is the distance between these two vectors. In average linkage, above, each pairwise distance is calculated, and the average of all such distances is calculated. In average group linkage or centroid distance clustering, just one distance is calculated, the distance between group means.

Ward's Method

Ward's method is also agglomerative in that it joins records and clusters together progressively to produce larger and larger clusters but operates slightly differently from the general approach described above. Ward's method considers the "loss of information" that occurs when record are clustered together. When each cluster has one record, there is no loss of information – all individual values remain available. When records are joined together and represented in clusters, information about an individual record is replaced by the information for the cluster to which it belongs. To measure loss of information, Ward's method employs a measure "error sum of squares" (ESS) that measures the difference between individual observations and a group mean.

This is easiest to see in univariate data. For example, consider the values (2, 6, 5, 6, 2, 2, 2, 2, 0, 0, 0) with a mean of 2.5. Their ESS is equal to $(2 - 2.5)^2 + (6 - 2.5)^2 + (5 - 2.5)^2 + \cdots + (0 - 2.5)^2 = 50.5$. The loss of information associated with grouping the values into a single group is therefore 50.5. Now group the observations into four groups: (0, 0, 0), (2, 2, 2, 2), (5), and (6, 6). The loss of information is the sum of the ESS's for each group, which is 0 (each observation in each group is equal to the mean for that group, so the ESS for each group is 0). Thus clustering the 10 observations into 4 clusters results in no loss of information, and this would be the first step in Ward's method. In moving to a smaller number of clusters, Ward's method would choose the configuration that results in the smallest incremental loss of information.

Ward's method tends to result in convex clusters that are of roughly equal size, which can be an important consideration in some applications (e.g., in establishing meaningful customer segments).

Dendrograms: Displaying Clustering Process and Results

A *dendrogram* is a treelike diagram that summarizes the process of clustering. At the bottom are the records. Similar records are joined by lines whose vertical length reflects the distance between the records. Figure 14.3 shows the dendrogram that results from clustering all 22 utilities using the 8 normalized measurements, Euclidean distance, and single linkage.

For any given number of clusters, we can determine the records in the clusters by sliding a horizontal line up and down until the number of vertical intersections of the horizontal line equals the number of clusters desired. For example, if we wanted to form six clusters, we would find that the clusters are:

{1, 2, 4, 10, 13, 20, 7, 12, 21, 15, 14, 19, 18, 22, 9, 3},
{8, 16} = {Idaho, Puget},
{6} = {Florida},
{17} = {San Diego},
{11} = { Nevada },
{5} = {NY}.

Note that if we wanted five clusters, they would be identical to the six, with the exception that the first two clusters would be merged into one cluster. In general, all hierarchical methods have clusters that are nested within each other

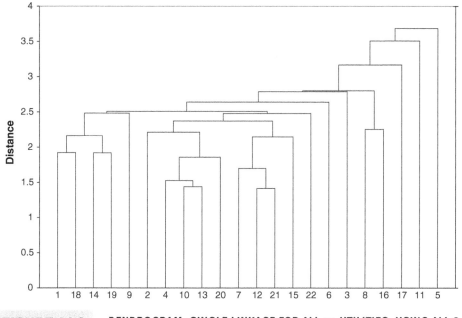

FIGURE 14.3   DENDROGRAM: SINGLE LINKAGE FOR ALL 22 UTILITIES, USING ALL 8
MEASUREMENTS

as we decrease the number of clusters. This is a valuable property for interpreting
clusters and is essential in certain applications, such as taxonomy of varieties of
living organisms.

The average linkage dendrogram is shown in Figure 14.4. If we want six
clusters using average linkage, they would be

{1, 14, 19, 18, 3, 6};{2, 4, 10, 13, 20, 22};{5};{7, 12, 9, 15, 21};{17};{8, 16, 11}.

Validating Clusters

The goal of cluster analysis is to come up with *meaningful clusters*. Since there are
many variations that can be chosen, it is important to make sure that the resulting
clusters are valid, in the sense that they really create some insight.

To see whether the cluster analysis is useful, perform the following:

1. *Cluster Interpretability*   Is the interpretation of the resulting clusters rea-
    sonable? To interpret the clusters, we explore the characteristics of each
    cluster by

    a. Obtaining summary statistics (e.g., average, min, max) from each clus-
        ter on each measurement that was used in the cluster analysis

    b. Examining the clusters for the presence of some common feature
        (variable) that was not used in the cluster analysis

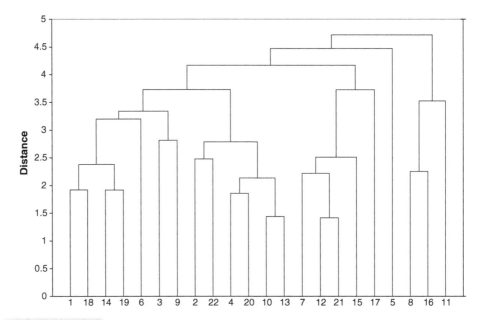

FIGURE 14.4 **DENDROGRAM: AVERAGE LINKAGE FOR ALL 22 UTILITIES, USING ALL 8 MEASUREMENTS**

    c. Cluster labeling: based on the interpretation, trying to assign a name or label to each cluster

2. *Cluster Stability* Do cluster assignments change significantly if some of the inputs were altered slightly? Another way to check stability is to partition the data and see how well clusters that are formed based on one part apply to the other part. To do this:

    a. Cluster partition A.

    b. Use the cluster centroids from A to assign each record in partition B (each record is assigned to the cluster with the closest centroid).

    c. Assess how consistent the cluster assignments are compared to the assignments based on all the data.

3. *Cluster Separation* Examine the ratio of between-cluster variation to within-cluster variation to see whether the separation is reasonable. There exist statistical tests for this task (an F ratio), but their usefulness is somewhat controversial.

    Returning to the utilities example, we notice that both methods (single and average linkage) identify {5} and {17} as singleton clusters. Also, both dendrograms imply that a reasonable number of clusters in this dataset is four. One insight that can be derived from this clustering is that clusters tend to group geographically:

A southern group: 1, 14, 19, 18, 3, 6 = Arizona, Oklahoma, Southern, Texas, Central Louisiana, Florida,

A northern group: 2,4,10,13,20 = Boston, Commonwealth, Madison, Northern States, Wisconsin,

An east/west seaboard group: 7, 12, 21, 15 = Hawaii, New England, United, Pacific.

We can further characterize each of the clusters by examining the summary statistics of their measurements.

Limitations of Hierarchical Clustering

Hierarchical clustering is very appealing in that it does not require specification of the number of clusters, and in that sense is purely data driven. The ability to represent the clustering process and results through dendrograms is also an advantage of this method, as it is easier to understand and interpret. There are, however, a few limitations to consider:

1. Hierarchical clustering requires the computation and storage of an $n \times n$ distance matrix. For very large datasets, this can be expensive and slow.

2. The hierarchical algorithm makes only one pass through the data. This means that records that are allocated incorrectly early in the process cannot be reallocated subsequently.

3. Hierarchical clustering also tends to have low stability. Reordering data or dropping a few records can lead to a very different solution.

4. With respect to the choice of distance between clusters, single and complete linkage are robust to changes in the distance metric (e.g., Euclidean, statistical distance) as long as the relative ordering is kept. Average linkage, on the other hand, is much more influenced by the choice of distance metric, and might lead to completely different clusters when the metric is changed.

5. Hierarchical clustering is sensitive to outliers.

14.5 NONHIERARCHICAL CLUSTERING: THE *k*-MEANS ALGORITHM

A nonhierarchical approach to forming good clusters is to prespecify a desired number of clusters, k, and to assign each case to one of k clusters so as to minimize a measure of dispersion within the clusters. In other words, the goal is to divide the sample into a predetermined number k of nonoverlapping clusters so that clusters are as homogeneous as possible with respect to the measurements used.

A very common measure of within-cluster dispersion is the sum of distances (or sum of squared Euclidean distances) of records from their cluster centroid. The problem can be set up as an optimization problem involving integer programming, but because solving integer programs with a large number of variables is time consuming, clusters are often computed using a fast, heuristic method that produces good (although not necessarily optimal) solutions. The k-means algorithm is one such method.

The k-means algorithm starts with an initial partition of the cases into k clusters. Subsequent steps modify the partition to reduce the sum of the distances of each record from its cluster centroid. The modification consists of allocating each record to the nearest of the k centroids of the previous partition. This leads to a new partition for which the sum of distances is smaller than before. The means of the new clusters are computed and the improvement step is repeated until the improvement is very small.

---

k-MEANS CLUSTERING ALGORITHM

1. Start with k initial clusters (user chooses k).
2. At every step, each record is reassigned to the cluster with the "closest" centroid.
3. Recompute the centroids of clusters that lost or gained a record, and repeat step 2.
4. Stop when moving any more records between clusters increases cluster dispersion.

Returning to the example with the five utilities and two measurements, let us assume that $k = 2$ and that the initial clusters are A = {Arizona, Boston} and B = {Central, Commonwealth, Consolidated}. The cluster centroids were computed in Section 14.4.

$$\bar{x}_A = [-0.516, -0.020] \text{ and } \bar{x}_B = [-0.733, 0.296].$$

The distance of each record from each of these two centroids is shown in Table 14.5. We see that Boston is closer to cluster B, and that Central and Commonwealth are each closer to cluster A. We therefore move each of these records to the other cluster and obtain

A = {Arizona, Central, Commonwealth} and B = {Consolidated, Boston}.

Recalculating the centroids gives

$$\bar{x}_A = [-0.191, -0.553] \text{ and } \bar{x}_B = [-1.33, 1.253].$$

TABLE 14.5	DISTANCE OF EACH RECORD FROM EACH CENTROID	
	Distance from Centroid A	Distance from Centroid B
Arizona	1.0052	1.3887
Boston	1.0052	0.6216
Central	0.6029	0.8995
Commonwealth	0.7281	1.0207
Consolidated	2.0172	1.6341

The distance of each record from each of the newly calculated centroids is given in Table 14.6 At this point we stop because each record is allocated to its closest cluster.

Initial Partition into *k* Clusters

The choice of the number of clusters can either be driven by external considerations (e.g., previous knowledge, practical constraints, etc.) or we can try a few different values for *k* and compare the resulting clusters. After choosing *k*, the *n* records are partitioned into these initial clusters. If there is external reasoning that suggests a certain partitioning, this information should be used. Alternatively, if there exists external information on the centroids of the *k* clusters, this can be used to allocate the records.

In many cases, there is no information to be used for the initial partition. In these cases, the algorithm can be rerun with different randomly generated starting partitions to reduce chances of the heuristic producing a poor solution. The number of clusters in the data is generally not known, so it is a good idea to run the algorithm with different values for *k* that are near the number of clusters that one expects from the data, to see how the sum of distances reduces with

TABLE 14.6	DISTANCE OF EACH RECORD FROM EACH NEWLY CALCULATED CENTROID	
	Distance from Centroid A	Distance from Centroid B
Arizona	0.3827	2.5159
Boston	1.6289	0.5067
Central	0.5463	1.9432
Commonwealth	0.5391	2.0745
Consolidated	2.6412	0.5067

(Distance from Cluster Centers are in normalized Co-ordinates)

Row Id.	Cluster id	Dist clust-1	Dist clust-2	Dist clust-3	Dist clust-4	Dist clust-5	Dist clust-6
4	1	1.3452	4.3621	3.8339	2.7494	4.2245	3.1049
10	1	1.0029	4.8342	3.5693	3.2345	4.1405	3.7869
13	1	1.5118	5.032	3.7204	4.0716	4.87	4.4756
20	1	1.3849	4.6365	3.6625	2.9824	4.4891	3.3278
22	1	1.9086	2.6776	2.7989	3.0881	3.8605	2.8665
11	2	4.7561	2.4316	3.7481	4.5463	6.613	5.0225
17	2	4.8471	2.4316	5.0856	5.3699	5.761	3.4106
8	3	3.5856	3.9739	1.6141	3.8148	5.2768	3.8129
9	3	3.2057	4.5496	2.3468	2.9569	4.5955	3.2353
16	3	4.3344	4.0545	1.6392	4.7394	5.9536	4.7098
1	4	2.8046	3.2553	2.8973	1.9144	4.2202	2.9635
3	4	3.8904	5.7882	4.174	2.2184	4.5739	4.5489
6	4	3.551	5.6974	4.6489	2.2677	4.7084	3.8032
14	4	3.5124	4.4748	4.169	1.5343	4.9322	4.3896
18	4	2.7316	3.536	2.763	1.3952	4.4402	2.5216
19	4	3.9634	4.9218	4.213	1.4988	5.2546	4.3521
5	5	4.0768	5.7051	4.9534	4.3258	0.00002	3.786
2	6	2.4426	3.6051	3.9153	3.7564	3.9435	1.9151
7	6	4.0822	4.888	3.829	3.6938	4.7044	1.9664
12	6	3.5029	3.8347	3.4951	3.3834	3.6887	0.92426
15	6	3.8236	3.448	4.1034	3.5875	4.3554	1.633
21	6	3.9527	3.4058	3.7556	4.1582	3.7299	1.2034

FIGURE 14.5 OUTPUT FOR K-MEANS CLUSTERING WITH $K = 6$ OF 22 UTILITIES (AFTER SORTING BY CLUSTER ID)

increasing values of k. Note that the clusters obtained using different values of k will not be nested (unlike those obtained by hierarchical methods).

The results of running the k-means algorithm for all 22 utilities and 8 measurements with $k = 6$ are shown in Figure 14.5. As in the results from the hierarchical clustering, we see once again that {5} is a singleton cluster and that some of the previous "geographic" clusters show up here as well.

To characterize the resulting clusters, we examine the cluster centroids [numerically in Figure 14.6 or in the line chart ("profile plot") in Figure 14.7]. We can see, for instance, that cluster 1 has the highest average Nuclear, a very high RoR, and a slow demand growth. In contrast, cluster 3 has the highest Sales, with no Nuclear, a high Demand Growth, and the highest average Cost.

We can also inspect the information on the within-cluster dispersion. From Figure 14.6 we see that cluster 2 has the highest average distance, and it includes only two records. In contrast, cluster 1, which includes five records, has the lowest within-cluster average distance. This is true for both normalized measurements (bottom left table) and original units (bottom right table). This means that cluster 1 is more homogeneous.

From the distances between clusters we can learn about the separation of the different clusters. For instance, we see that cluster 2 is very different from the other clusters except cluster 3. This might lead us to examine the possibility of

Cluster centers

Cluster	Fixed	RoR	Cost	Load_factor	Demand	Sales	Nuclear	Fuel
Cluster-1	1.112	11.480001	177.200001	55.380002	3.76	7487.399702	38.280034	0.7716
Cluster-2	0.755001	6.949994	154.500005	56.700001	7.749996	11577.49951	4.149999	1.344
Cluster-3	1.2	10.7	221.666487	57.800002	6.566652	12493.01588	-0.000008	0.597
Cluster-4	1.185	12.400001	120.833197	54.650001	0.799999	10456.00045	3.750008	0.8765
Cluster-5	1.49	8.8	192.000002	51.20002	0.999999	3300.012277	15.600001	2.044
Cluster-6	1.048	9.920001	184.600002	62.14001	2.300001	6400.400459	5.240004	1.724

Distance between cluster centers	Cluster-1	Cluster-2	Cluster-3	Cluster-4	Cluster-5	Cluster-6
Cluster-1	0	4090.309917	5005.961483	2969.338323	4187.479063	1087.549967
Cluster-2	4090.309917	0	917.995763	1122.041163	8277.584943	5177.193266
Cluster-3	5005.961483	917.995763	0	2039.52432	9193.06908	6092.733626
Cluster-4	2969.338323	1122.041163	2039.52432	0	7156.353693	4056.109579
Cluster-5	4187.479063	8277.584943	9193.06908	7156.353693	0	3100.434146
Cluster-6	1087.549967	5177.193266	6092.733626	4056.109579	3100.434146	0

Data summary

Cluster	#Obs	Average distance in cluster
Cluster-1	5	1.431
Cluster-2	2	2.432
Cluster-3	3	1.867
Cluster-4	6	1.805
Cluster-5	1	0
Cluster-6	5	1.528
Overall	22	1.64

Data summary (In Original coordinates)

Cluster	#Obs	Average distance in cluster
Cluster-1	5	1042.936117
Cluster-2	2	5863.533146
Cluster-3	3	2724.981548
Cluster-4	6	1241.097807
Cluster-5	1	0.012277017
Cluster-6	5	624.4372161
Overall	22	1622.067124

FIGURE 14.6   **CLUSTER CENTROIDS AND DISTANCES FOR *K* MEANS WITH *K* = 6**

merging the two. Cluster 5, which is a singleton cluster, appears to be very far from all the other clusters.

Finally, we can use the information on the distance between the final clusters to evaluate the cluster validity. The ratio of the sum of squared distances for a given k to the sum of squared distances to the mean of all the records ($k = 1$) is a useful measure for the usefulness of the clustering. If the ratio is near 1.0, the clustering has not been very effective, whereas if it is small, we have well-separated groups.

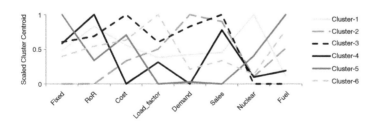

FIGURE 14.7   **VISUAL PRESENTATION OF CLUSTER CENTROIDS (SCALED TO 0–1 ON EACH MEASURE, FOR EASY COMPARISON), CREATED USING EXCEL'S LINE CHART**

PROBLEMS

14.1 **University Rankings.** The dataset on American College and University Rankings (available from www.dataminingbook.com) contains information on 1302 American colleges and universities offering an undergraduate program. For each university there are 17 measurements, including continuous measurements (such as tuition and graduation rate) and categorical measurements (such as location by state and whether it is a private or public school).

Note that many records are missing some measurements. Our first goal is to estimate these missing values from "similar" records. This will be done by clustering the complete records and then finding the closest cluster for each of the partial records. The missing values will be imputed from the information in that cluster.

a. Remove all records with missing measurements from the dataset (by creating a new worksheet).

b. For all the continuous measurements, run hierarchical clustering using complete linkage and Euclidean distance. Make sure to normalize the measurements. Examine the dendrogram: How many clusters seem reasonable for describing these data?

c. Compare the summary statistics for each cluster and describe each cluster in this context (e.g., "Universities with high tuition, low acceptance rate..."). **Hint:** To obtain cluster statistics for hierarchical clustering, use Excel's *Pivot Table* on the *Predicted Clusters* sheet.

d. Use the categorical measurements that were not used in the analysis (State and Private/Public) to characterize the different clusters. Is there any relationship between the clusters and the categorical information?

e. Can you think of other external information that explains the contents of some or all of these clusters?

f. Consider Tufts University, which is missing some information. Compute the Euclidean distance of this record from each of the clusters that you found above (using only the measurements that you have). Which cluster is it closest to? Impute the missing values for Tufts by taking the average of the cluster on those measurements.

14.2 **Pharmaceutical Industry.** An equities analyst is studying the pharmaceutical industry and would like your help in exploring and understanding the financial data collected by her firm. Her main objective is to understand the structure of the pharmaceutical industry using some basic financial measures.

Financial data gathered on 21 firms in the pharmaceutical industry are available in the file Pharmaceuticals.xls. For each firm, the following variables are recorded:

1. Market capitalization (in billions of dollars)
2. Beta
3. Price/earnings ratio
4. Return on equity
5. Return on assets
6. Asset turnover
7. Leverage
8. Estimated revenue growth

9. Net profit margin

10. Median recommendation (across major brokerages)

11. Location of firm's headquarters

12. Stock exchange on which the firm is listed

Use cluster analysis to explore and analyze the given dataset as follows:

a. Use only the quantitative variables (1–9) to cluster the 21 firms. Justify the various choices made in conducting the cluster analysis, such as weights accorded different variables, the specific clustering algorithm(s) used, the number of clusters formed, and so on.

b. Interpret the clusters with respect to the quantitative variables that were used in forming the clusters.

c. Is there a pattern in the clusters with respect to the qualitative variables (10–12) (those not used in forming the clusters)?

d. Provide an appropriate name for each cluster using any or all of the variables in the dataset.

14.3 **Customer Rating of Breakfast Cereals.** The dataset Cereals.xls includes nutritional information, store display, and consumer ratings for 77 breakfast cereals.
   Data Preprocessing. Remove all cereals with missing values.

a. Apply hierarchical clustering to the data using Euclidean distance to the standardized measurements. Compare the dendrograms from single linkage and complete linkage, and look at cluster centroids. Comment on the structure of the clusters and on their stability. **Hints:** (1) To obtain cluster centroids for hierarchical clustering, use Excel's *Pivot Table* on the *Predicted Clusters* sheet. (2) Running hierarchical clustering in XLMiner is an iterative process—run it once with a guess at the right number of clusters, then run it again after looking at the dendrogram, adjusting the number of clusters if needed.

b. Which method leads to the most insightful or meaningful clusters?

c. Choose one of the methods. How many clusters would you use? What distance is used for this cutoff? (Look at the dendrogram.)

d. The elementary public schools would like to choose a set of cereals to include in their daily cafeterias. Every day a different cereal is offered, but all cereals should support a healthy diet. For this goal you are requested to find a cluster of "healthy cereals." Should the data be standardized? If not, how should they be used in the cluster analysis?

14.4 **Marketing to Frequent Fliers.** The file EastWestAirlinesCluster.xls contains information on 4000 passengers who belong to an airline's frequent flier program. For each passenger the data include information on their mileage history and on different ways they accrued or spent miles in the last year. The goal is to try to identify clusters of passengers that have similar characteristics for the purpose of targeting different segments for different types of mileage offers.

a. Apply hierarchical clustering with Euclidean distance and Ward's method. Make sure to standardize the data first. How many clusters appear?

b. What would happen if the data were not standardized?

c. Compare the cluster centroid to characterize the different clusters and try to give each cluster a label.

   d. To check the stability of the clusters, remove a random 5% of the data (by taking a random sample of 95% of the records), and repeat the analysis. Does the same picture emerge?

   e. Use k-means clustering with the number of clusters that you found above. Does the same picture emerge?

   f. Which clusters would you target for offers, and what types of offers would you target to customers in that cluster?

## PART SIX

Forecasting Time Series

## CHAPTER 15

Handling Time Series

In this chapter we describe the context of business time series forecasting and introduce the main approaches that are detailed in the next chapters (in particular, regression-based forecasting and smoothing-based methods). Our focus is on forecasting future values of a single time series. We discuss the difference between the predictive nature of time series forecasting versus the descriptive or explanatory task of time series analysis. A general discussion of combining forecasting methods or results for added precision follows. Next, we present a time series in terms of four components (level, trend, seasonality, and noise) and present methods for visualizing the different components and for exploring time series data. We close with a discussion of data partitioning (to create training and validation sets), which is performed differently than cross-sectional data partitioning.

15.1 INTRODUCTION

Time series forecasting is performed in nearly every organization that works with quantifiable data. Retail stores use it to forecast sales. Energy companies use it to forecast reserves, production, demand, and prices. Educational institutions use it to forecast enrollment. Governments use it to forecast tax receipts and spending. International financial organizations such as the World Bank and International Monetary Fund use it to forecast inflation and economic activity. Transportation companies use time series forecasting to forecast future travel. Banks and lending institutions use it (sometimes badly!) to forecast new home purchases. And venture capital firms use it to forecast market potential and to evaluate business plans.

Previous chapters in this book deal with classifying and predicting data where time is not a factor (in the sense that it is not treated differently from other variables), and where the order of measurements in time does not matter. These are typically called cross-sectional data. In contrast, this chapter deals with a different type of data: time series.

With today's technology, many time series are recorded on very frequent time scales. Stock data are available at ticker level. Purchases at online and offline stores are recorded in real time. Although data might be available at a very frequent scale, for the purpose of forecasting it is not always preferable to use this time scale. In considering the choice of time scale, one must consider the scale of the required forecasts and the level of noise in the data. For example, if the goal is to forecast next-day sales at a grocery store, using minute-by-minute sales data are likely to be less useful for forecasting than using daily aggregates. The minute-by-minute series will contain many sources of noise (e.g., variation by peak and nonpeak shopping hours) that degrade its forecasting power, and these noise errors, when the data are aggregated to a coarser level, are likely to average out.

The focus in this part of the book is on forecasting a single time series. In some cases multiple time series are to be forecasted (e.g., the monthly sales of multiple products). Even when multiple series are being forecasted, the most popular forecasting practice is to forecast each series individually. The advantage of single-series forecasting is simplicity. The disadvantage is that it does not take into account possible relationships between series. The statistics literature contains models for multivariate time series that directly model the cross correlations between series. Such methods tend to make restrictive assumptions about the data and the cross-series structure, and they also require statistical expertise for estimation and maintenance. Econometric models often include information from one or more series as inputs into another series. However, such models are based on assumptions of causality that are based on theoretical models. An alternative approach is to capture the associations between the series of interest and external information more heuristically. An example is using the sales of lipstick to forecast some measure of the economy, based on the observation by Ronald Lauder, chairman of Estee Lauder, that lipstick sales tend to increase before tough economic times (a phenomenon called the "leading lipstick indicator").

15.2 EXPLANATORY VERSUS PREDICTIVE MODELING

As with cross-sectional data, modeling time series data is done for either explanatory or predictive purposes. In explanatory modeling, or *time series analysis*, a time series is modeled to determine its components in terms of seasonal patterns, trends, relation to external factors, and the like. These can then be used for decision making and policy formulation. In contrast, *time series forecasting* uses the information in a time series (and perhaps other information) to forecast future values of that series. The difference between the goals of time series analysis and time series forecasting leads to differences in the type of methods used and in the modeling process itself. For example, in selecting a method for explaining a time series, priority is given to methods that produce explainable results (rather than black-box methods) and to models based on causal arguments. Furthermore, explaining can be done in retrospect, while forecasting is prospective in nature. This means that explanatory models might use "future" information (e.g., averaging the values of yesterday, today, and tomorrow to obtain a smooth representation of today's value) whereas forecasting models cannot.

The focus in this chapter is on time series forecasting, where the goal is to predict future values of a time series. For information on time series analysis see Chatfield (2003).

15.3 POPULAR FORECASTING METHODS IN BUSINESS

In this part of the book we focus on two main types of forecasting methods that are popular in business applications. Both are versatile and powerful, yet relatively simple to understand and deploy. One type of forecasting tool is multiple linear regression, where the user specifies a certain model and then estimates it from the time series. The other is the more data-driven tool of smoothing, where the method learns patterns from the data. Each of the two types of tools has advantages and disadvantages, as detailed in Chapters 16 and 17. We also note that data mining methods such as neural networks and others that are intended for cross-sectional data are also sometimes used for time series forecasting, especially for incorporating external information into the forecasts.

Combining Methods

Before a discussion of specific forecasting methods in the following two chapters, it should be noted that a popular approach for improving predictive performance is to combine forecasting methods. Combining methods can be done via two-level (or multilevel) forecasters, where the first method uses the original time series to generate forecasts of future values, and the second method uses the residuals from the first model to generate forecasts of future forecast errors, thereby "correcting" the first level forecasts. We describe two-level forecasting in Section 16.4. Another combination approach is via "ensembles," where multiple methods are applied to the time series, and their resulting forecasts are averaged in some way to produce the final forecast. Combining methods can take advantage of the strengths of different forecasting methods to capture different aspects of the time series (also true in cross-sectional data). The averaging across multiple methods can lead to forecasts that are more robust and of higher precision.

15.4 TIME SERIES COMPONENTS

In both types of forecasting methods (regression models and smoothing) and in general, it is customary to dissect a time series into four components: level, trend, seasonality, and noise. The first three components are assumed to be invisible, as they characterize the underlying series, which we only observe with added noise. Level describes the average value of the series, trend is the change in the series from one period to the next, and seasonality describes a short-term cyclical behavior of the series that can be observed several times within the given series. Finally, noise is the random variation that results from measurement error or other causes not accounted for. It is always present in a time series to some degree.

In order to identify the components of a time series, the first step is to examine a time plot. In its simplest form, a time plot is a line chart of the series values over time, with temporal labels (e.g., calendar date) on the horizontal axis. To illustrate this, consider the following example.

Example: Ridership on Amtrak Trains

Amtrak, a U.S. railway company, routinely collects data on ridership. Here we focus on forecasting future ridership using the series of monthly ridership between January 1991 and March 2004. These data are publicly available at http://www.forecastingprinciples.com (click on Data, and select Series M-34 from the T-Competition Data).

A time plot for monthly Amtrak ridership series is shown in Figure 15.1. Note that the values are in thousands of riders. Looking at the time plot reveals the nature of the series components: the overall level is around 1,800,000 passengers per month. A slight U-shaped trend is discernible during this period, with pronounced annual seasonality, with peak travel during summer (July and August).

A second step in visualizing a time series is to examine it more carefully. A few tools are useful:

Zoom in: Zooming in to a shorter period within the series can reveal patterns that are hidden when viewing the entire series. This is especially important when the time series is long. Consider a series of the daily number of vehicles

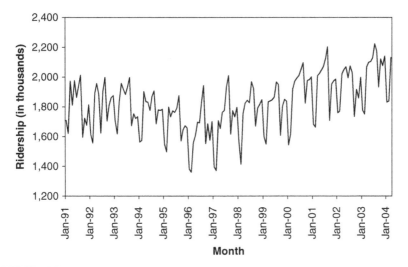

FIGURE 15.1   **MONTHLY RIDERSHIP ON AMTRAK TRAINS (IN THOUSANDS) FROM JAN-1991 TO MARCH-2004**

passing through the Baregg tunnel in Switzerland (data are available in the same location as the Amtrak Ridership data; series D028). The series from Nov. 1, 2003 to Nov. 16, 2005 is shown in the top panel of Figure 15.2. Zooming in to a 4-month period (bottom panel) reveals a strong day-of-week pattern that was not visible in the initial time plot of the complete time series.

Change Scale of Series: In order to better identify the shape of a trend, it is useful to change the scale of the series. One simple option is to change the vertical scale to a logarithmic scale (in Excel 2003 double-click on the y-axis labels and check "logarithmic scale"; in Excel 2007 select *Layout> Axes > Axes > Primary Vertical Axis* and click "logarithmic scale" in the *Format Axis* menu). If the trend on the new scale appears more linear, then the trend in the original series is closer to an exponential trend.

Add Trend Lines: Another possibility for better capturing the shape of the trend is to add a trend line (Excel 2003: *Chart> Add Trendline*; Excel 2007: *Layout> Analysis > Trendline*). By trying different trend lines one can see what type of trend (e.g., linear, exponential, cubic) best approximates the data.

Suppress Seasonality: It is often easier to see trends in the data when seasonality is suppressed. Suppressing seasonality patterns can be done by plotting the series at a cruder time scale (e.g., aggregating monthly data into years). Another popular option is to use moving-average plots. We will discuss these in Section 17.2.

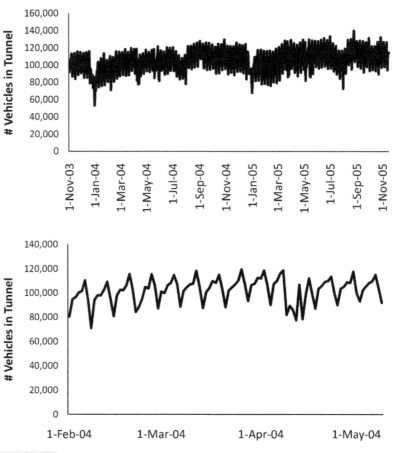

FIGURE 15.2 TIME PLOTS OF THE DAILY NUMBER OF VEHICLES PASSING THROUGH THE BAREGG TUNNEL (SWITZERLAND). THE BOTTOM PANEL ZOOMS IN TO A 4-MONTH PERIOD, REVEALING A DAY-OF-WEEK PATTERN

Continuing our example of Amtrak ridership, the plots in Figure 15.3 help make the series' components more visible.

Some forecasting methods directly model these components by making assumptions about their structure. For example, a popular assumption about trend is that it is linear or exponential over the given time period or part of it. Another common assumption is about the noise structure: Many statistical methods assume that the noise follows a normal distribution. The advantage of methods that rely on such assumptions is that when the assumptions are reasonably met, the resulting forecasts will be more robust and the models more understandable. Other forecasting methods, which are data adaptive, make fewer assumptions about the structure of these components and instead try to estimate them only from the data. Data–adaptive methods are advantageous when such assumptions are likely to be violated, or when the structure of the time series changes over

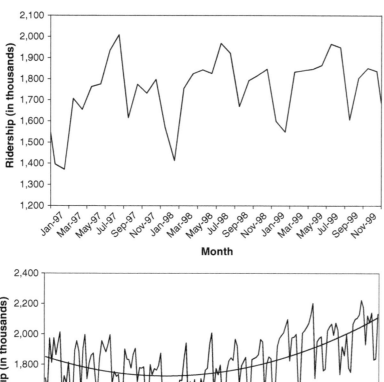

FIGURE 15.3 PLOTS THAT ENHANCE THE DIFFERENT COMPONENTS OF THE TIME SERIES. TOP: ZOOM IN TO 3 YEARS OF DATA. BOTTOM: ORIGINAL SERIES WITH OVERLAID POLYNOMIAL TREND LINE

time. Another advantage of many data-adaptive methods is their simplicity and computational efficiency.

A key criterion for choosing between model-driven and data-driven forecasting methods is the nature of the series in terms of global versus local patterns. A global pattern is one that is relatively constant throughout the series. An example is a linear trend throughout the entire series. In contrast, a local pattern is one that occurs only in a short period of the data, and then changes. An example is a trend that is approximately linear within four neighboring time points, but the trend size (slope) changes slowly over time.

Model-driven methods are generally preferable for forecasting series with global patterns, as they use all the data to estimate the global pattern. For a local pattern, a model-driven model would require specifying how and when the patterns change, which is usually impractical and often unknown. Therefore, data-driven methods are preferable for local patterns. Such methods "learn" patterns from the data, and their memory length can be set to best adapt to the rate of change in the series. Patterns that change quickly warrant a "short memory," whereas patterns that change slowly warrant a "long memory." In conclusion, the time plot should be used not only to identify the time series component but also the global/local nature of the trend and seasonality.

15.5 DATA PARTITIONING

As in the case of cross-sectional data, in order to avoid overfitting and to be able to assess the predictive performance of the model on new data, we first partition the data into a training set and a validation set (and perhaps an additional test set). However, there is one important difference between data partitioning in cross-sectional and time series data. In cross-sectional data the partitioning is usually done randomly, with a random set of observations designated as training data and the remainder as validation data. However, in time series a random partition would create two time series with "holes!" Nearly all standard forecasting methods cannot handle time series with missing values. Therefore, we partition a time series into training and validation sets differently. The series is trimmed into two periods; the earlier period is set as the training data and the later period as the validation data. Methods are then trained on the earlier periods, and their predictive performance assessed on the later period. Evaluation metrics typically use the same metrics used in cross-sectional evaluation (see Chapter 5.3) with MAPE and RMSE being the most popular metrics in practice. In evaluating and comparing forecasting methods, another important tool is visualization: Examining time plots of the actual and predicted series can shed light on performance and hint toward possible improvements.

One last important difference between cross-sectional and time series partitioning occurs when creating the actual forecasts. Before attempting to forecast future values of the series, the training and validation sets are recombined into one long series, and the chosen method/model is rerun on the complete data. This final model is then used to forecast future values. The three advantages in recombining are: (1) The validation set, which is the most recent period, usually contains the most valuable information in terms of being the closest in time to the forecasted period. (2) With more data (the complete time series, compared to only the training set), some models can be estimated more accurately. (3)

If only the training set is used to generate forecasts, then it will require forecasting farther into the future (e.g., if the validation set contains 4 time points, forecasting the next observation will require a 5-step-ahead forecast from the training set).

In XLMiner, partitioning a time series is performed within the *Time Series* menu. Figure 15.4 shows a screenshot of the time series partitioning dialog box. After the final model is chosen, the same model should be rerun on the original, unpartitioned series in order to obtain forecasts. XLMiner will only generate future forecasts if it is run on an unpartitioned series.

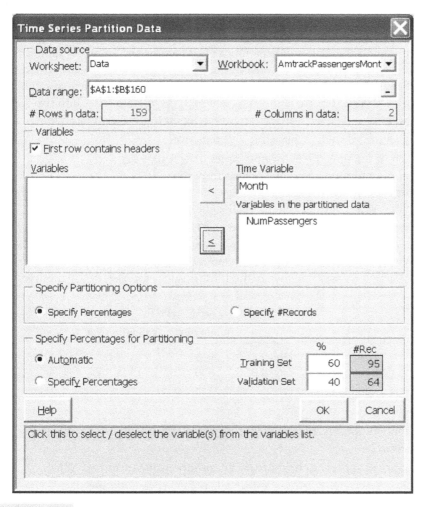

FIGURE 15.4   PARTITIONING A TIME SERIES. THE DEFAULT IN XLMINER PARTITIONS THE DATA INTO 60% TRAINING (THE EARLIEST 60% TIME POINTS) AND 40% VALIDATION (THE LATER 40% TIME POINTS)

PROBLEMS

15.1 **Impact of September 11 on Air Travel in the United States:** The Research and Innovative Technology Administration's Bureau of Transportation Statistics (BTS) conducted a study to evaluate the impact of the September 11, 2001, terrorist attack on U.S. transportation. The study report and the data can be found at http://www.bts.gov/publications/estimated_impacts_of_9_11_on_us_travel. The goal of the study was stated as follows:

The purpose of this study is to provide a greater understanding of the passenger travel behavior patterns of persons making long distance trips before and after 9/11.

The report analyzes monthly passenger movement data between January 1990 and May 2004. Data on three monthly time series are given in file Sept11Travel.xls for this period: (1) actual airline revenue passenger miles (Air), (2) rail passenger miles (Rail), and (3) vehicle miles traveled (Car).

In order to assess the impact of September 11, BTS took the following approach: Using data before September 11, it forecasted future data (under the assumption of no terrorist attack). Then, BTS compared the forecasted series with the actual data to assess the impact of the event. Our first step, therefore, is to split each of the time series into two parts: pre- and post-September 11. We now concentrate only on the earlier time series.

a. Is the goal of this study explanatory or predictive?

b. Plot each of the three preevent time series (Air, Rail, Car).
   i. What time series components appear from the plot?
   ii. What type of trend appears? Change the scale of the series, add trend lines, and suppress seasonality to better visualize the trend pattern.

15.2 **Performance on Training and Validation Data:** Two different models were fit to the same time series. The first 100 time periods were used for the training set and the last 12 periods were treated as a hold-out set. Assume that both models make sense practically and fit the data pretty well. Below are the RMSE values for each of the models:

	Training Set	Validation Set
Model A	543	20
Model B	869	14

a. Which model appears more useful for explaining the different components of this time series? Why?

b. Which model appears to be more useful for forecasting purposes? Why?

15.3 **Forecasting Department Store Sales:** The file DepartmentStoreSales.xls contains data on the quarterly sales for a department store over a 6-year period (data courtesy of Chris Albright).

a. Recreate the time plot.

b. Which of the four components (level, trend, seasonality, noise) seem to be present in this series?

FIGURE 15.5   AVERAGE ANNUAL WEEKLY HOURS SPENT BY CANADIAN
MANUFACTURING WORKERS

15.4 **Shipments of Household Appliances:** The file ApplianceShipments.xls contains the
series of quarterly shipments (in million $) of U.S. household appliances between 1985
and 1989 (data courtesy of Ken Black).

  a. Create a well-formatted time plot of the data.

  b. Which of the four components (level, trend, seasonality, noise) seem to be present in
  this series?

15.5 **Analysis of Canadian Manufacturing Workers Workhours:** The time series plot
in Figure 15.5 describes the average annual number of weekly hours spent by Canadian
manufacturing workers (data are available in CanadianWorkHours.xls—thanks to Ken
Black for these data).

  a. Reproduce the time plot.

  b. Which of the four components (level, trend, seasonality, noise) appear to be present
  in this series?

15.6 **Souvenir Sales:** The file SouvenirSales.xls contains monthly sales for a souvenir shop at a beach resort town in Queensland, Australia, between 1995 and
2001. [Source: R. J. Hyndman, Time Series Data Library, http://www.robjhyndman
.com/TSDL; accessed on December 20, 2009.]

  Back in 2001, the store wanted to use the data to forecast sales for the next 12 months
(year 2002). They hired an analyst to generate forecasts. The analyst first partitioned the
data into training and validation sets, with the validation set containing the last 12 months
of data (year 2001). She then fit a regression model to sales, using the training set.

  a. Create a well-formatted time plot of the data.

  b. Change the scale on the x axis, or on the y axis, or on both to log scale in order to
  achieve a linear relationship. Select the time plot that seems most linear.

  c. Comparing the two time plots, what can be said about the type of trend in the data?

   d. Why were the data partitioned? Partition the data into the training and validation set as explained above.

15.7   Forecasting Shampoo Sales: The file ShampooSales.xls contains data on the monthly sales of a certain shampoo over a 3-year period. (Source: R. J. Hyndman, Time Series Data Library, http://www.robjhyndman.com/TSDL; accessed on December 28, 2009.)

   a. Create a well-formatted time plot of the data.

   b. Which of the four components (level, trend, seasonality, noise) seem to be present in this series?

   c. Do you expect to see seasonality in sales of shampoo? Why?

   d. If the goal is forecasting sales in future months, which of the following steps should be taken? (choose one or more)

   - Partition the data into training and validation sets.
   - Tweak the model parameters to obtain good fit to the validation data.
   - Look at MAPE and RMSE values for the training set.
   - Look at MAPE and RMSE values for the validation set.

## CHAPTER 16

Regression-Based Forecasting

A popular forecasting tool is based on multiple linear regression models, using suitable predictors to capture trend and/or seasonality. In this chapter we show how a linear regression model can be set up to capture a time series with a trend and/or seasonality. The model, which is estimated from the training data, can then produce forecasts on future data by inserting the relevant predictor information into the estimated regression equation. We describe different types of common trends (linear, exponential, polynomial), as well as two types of seasonality (additive and multiplicative). Next, we show how a regression model can be used to quantify the correlation between neighboring values in a time series (called autocorrelation). This type of model, called an autoregressive model, is useful for improving forecast precision by making use of the information contained in the autocorrelation (beyond trend and seasonality). It is also useful for evaluating the predictability of a series (by evaluating whether the series is a "random walk"). The various steps of fitting linear regression and autoregressive models, using them to generate forecasts, and assessing their predictive accuracy, are illustrated using the Amtrak ridership series.

16.1 Model with Trend

Linear Trend

To create a linear regression model that captures a time series with a global linear trend, the output variable (Y) is set as the time series measurement or some function of it, and the predictor (X) is set as a time index. Let us consider a simple example, fitting a linear trend to the Amtrak ridership data. This type of

Data Mining for Business Intelligence, By Galit Shmueli, Nitin R. Patel, and Peter C. Bruce
Copyright © 2010 Statistics.com and Galit Shmueli

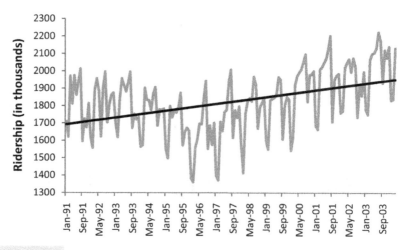

FIGURE 16.1   **LINEAR TREND FIT TO AMTRAK RIDERSHIP**

trend is shown in Figure 16.1. From the time plot it is obvious that the global trend is not linear. However, we use this example to illustrate how a linear trend is fit, and later we consider more appropriate models for this series.

To obtain a linear relationship between Ridership and Time, we set the output variable Y as the Amtrak Ridership and create a new variable that is a time index $t = 1, 2, 3, \ldots$. This time index is then used as a single predictor in the regression model:

$$Y_t = \beta_0 + \beta_1 \times t + \epsilon,$$

where Y_t is the Ridership at time point t and ϵ is the standard noise term in a linear regression. Thus, we are modeling three of the four time series components: level (β_0), trend (β_1), and noise (ε). Seasonality is not modeled. A snapshot of the two corresponding columns (Y and t) in Excel are shown in Figure 16.2.

Month	Ridership	t
Jan-91	1709	1
Feb-91	1621	2
Mar-91	1973	3
Apr-91	1812	4
May-91	1975	5
Jun-91	1862	6
Jul-91	1940	7
Aug-91	2013	8
Sep-91	1596	9
Oct-91	1725	10
Nov-91	1676	11
Dec-91	1814	12
Jan-92	1615	13
Feb-92	1557	14

FIGURE 16.2   **OUTPUT VARIABLE (MIDDLE) AND PREDICTOR VARIABLE (RIGHT) USED TO FIT A LINEAR TREND**

The Regression Model

Input variables	Coefficient	Std. Error	p-value	SS
Constant term	1713.028809	27.08552361	0	477456500
t	1.2053107	0.31751993	0.00021544	384546.3125

Training Data scoring - Summary Report

Total sum of squared errors	RMS Error	Average Error
3869551.676	162.2451256	-3.84852E-05

Validation Data scoring - Summary Report

Total sum of squared errors	RMS Error	Average Error
529326.616	210.0251207	168.8524156

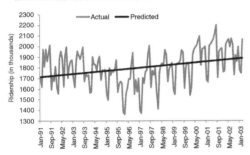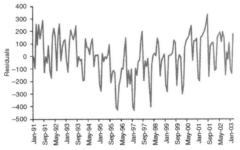

FIGURE 16.3   FITTED REGRESSION MODEL WITH LINEAR TREND. REGRESSION OUTPUT (TOP) AND TIME PLOTS OF ACTUAL AND FITTED SERIES (MIDDLE) AND RESIDUALS (BOTTOM)

After partitioning the data into training and validation sets, the next step is to fit a linear regression model to the training set, with t as the single predictor. Applying this to the Amtrak ridership data (with a validation set consisting of the last 12 months) results in the estimated model shown in Figure 16.3. The actual and fitted values and the residuals are shown in the two lower panels in time plots. Note that examining only the estimated coefficients and their statistical significance can be very misleading! In this example they would indicate that the linear fit is reasonable, although it is obvious from the time plots that the trend is not linear. The difference in the magnitude of the validation average error is also indicative of an inadequate trend shape. But an inadequate trend shape is easiest to detect by examining the series of residuals.

Exponential Trend

There are several alternative trend shapes that are useful and easy to fit via a linear regression model. Recall Excel's *Trendline* and other plots that help assess the type of trend in the data. One such shape is an exponential trend. An exponential trend implies a multiplicative increase/decrease of the series over time ($Y_t = ce^{\beta_1 t + \epsilon}$). To fit an exponential trend, simply replace the output variable Y with $\log(Y)$ and fit a linear regression ($\log(Y_t) = \beta_0 + \beta_1 t + \epsilon$). The term "log" refers to the natural logarithm (ln). In the Amtrak example, for instance, we would fit a linear regression of $\log$(Ridership) on the index variable t. Exponential trends are popular in sales data, where they reflect percentage growth.

Note: As in the general case of linear regression, when comparing the predictive accuracy of models that have a different output variable, such as a linear model trend (with Y) and an exponential model trend (with $\log(Y)$), it is essential to compare forecast or forecast errors on the same scale. An exponential trend model will produce forecasts of $\log(Y)$, and the forecast errors reported by the software will therefore be $\log(Y) - \log(\hat{Y})$. To obtain forecasts in the original units, create a new column that takes an exponent of the model forecasts. Then, use this column to create an additional column of forecast errors, by subtracting the original Y. An example is shown in Figures 16.4 and 16.5, where an exponential trend is fit to the Amtrak ridership data. Note that the performance measures for the training and validation data are not comparable to those from the linear trend model shown in Figure 16.3. Instead, we manually compute two new columns in Figure 16.5, one that gives forecasts of ridership (in thousands) and the other that gives the forecast errors in terms of ridership. To compare RMS Error or Average Error, we would now use the new forecast errors and compute their standard deviation (for RMS Error) or their average (for Average Error). These would then be comparable to the numbers in Figure 16.3.

The Regression Model

Input variables	Coefficient	Std. Error	p-value	SS
Constant term	7.44398642	0.01547452	0	8251.513672
t	0.00065125	0.00018141	0.00045169	0.11226512

Training Data scoring - Summary Report

Total sum of squared errors	RMS Error	Average Error
1.263050414	0.092694011	-5.27755E-08

Validation Data scoring - Summary Report

Total sum of squared errors	RMS Error	Average Error
0.139731707	0.107908799	0.08800547

FIGURE 16.4 OUTPUT FROM REGRESSION MODEL WITH EXPONENTIAL TREND, FIT TO TRAINING DATA

Predicted Value	Actual Value	Residual		t	Predicted Ridership	Forecast Errors
7.54037142	7.649168201	0.108796781		148	0.877392739	6.771775462
7.54102267	7.652028465	0.111005795		149	0.877430246	6.774598219
7.54167392	7.663722787	0.122048867		150	0.877467751	6.786255036
7.54232517	7.706769897	0.164444727		151	0.877505252	6.829264645
7.54297642	7.684489647	0.141513227		152	0.87754275	6.806946897
7.54362767	7.566003514	0.022375844		153	0.877580245	6.688423269
7.54427892	7.659864524	0.115585604		154	0.877617736	6.782246787
7.54493017	7.638224256	0.093294086		155	0.877655225	6.760569031
7.54558142	7.668877413	0.123295993		156	0.87769271	6.791184703
7.54623267	7.51289495	-0.03333772		157	0.877730192	6.635164759
7.54688392	7.516436567	-0.030447353		158	0.87776767	6.638668897
7.54753517	7.665024957	0.117489787		159	0.877805145	6.787219811

FIGURE 16.5 ADJUSTING FORECASTS OF log(RIDERSHIP) TO THE ORIGINAL SCALE (FIFTH COLUMN) AND COMPUTING FORECAST ERRORS IN THE ORIGINAL SCALE (RIGHT COLUMN)

Polynomial Trend

Another nonlinear trend shape that is easy to fit via linear regression is a polynomial trend, and, in particular, a quadratic relationship of the form $Y_t = \beta_0 + \beta_1 t + \beta_2 t^2 + \epsilon$. This is done by creating an additional predictor t^2 (the square of t) and fitting a multiple linear regression with the two predictors t and t^2. For the Amtrak ridership data, we have already seen a U-shaped trend in the data. We therefore fit a quadratic model, concluding from the plots of model fit and residuals (Figure 16.6) that this shape adequately captures the trend. The residuals now exhibit only seasonality.

In general, any type of trend shape can be fit as long as it has a mathematical representation. However, the underlying assumption is that this shape is applicable throughout the period of data that we have and also during the period that we

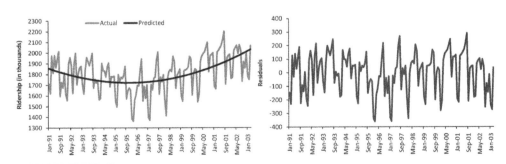

FIGURE 16.6 FITTED REGRESSION MODEL WITH QUADRATIC TREND ALONGSIDE WITH THE ACTUAL TRAINING DATA (LEFT) AND MODEL RESIDUALS (RIGHT)

are going to forecast. Do not choose an overly complex shape. Although it will fit the training data well, it will in fact be overfitting them. To avoid overfitting, always examine performance on the validation set and refrain from choosing overly complex trend patterns.

16.2   MODEL WITH SEASONALITY

A seasonal pattern in a time series means that observations that fall in some seasons have consistently higher or lower values than those that fall in other seasons. Examples are day-of-week patterns, monthly patterns, and quarterly patterns. The Amtrak ridership monthly time series, as can be seen in the time plot, exhibits strong monthly seasonality (with highest traffic during summer months).

Seasonality is modeled in a regression model by creating a new categorical variable that denotes the season for each observation. This categorical variable is then turned into dummies, which in turn are included as predictors in the regression model. To illustrate this, we created a new Month column for the Amtrak ridership data, as shown in Figure 16.7.

In order to include the season categorical variable as a predictor in a regression model for Y (e.g., Ridership), we turn it into dummies. For m seasons we create $m - 1$ dummies, which are binary variables that take on the value 1 if the record

Month	Ridership Season
Jan-91	1709 Jan
Feb-91	1621 Feb
Mar-91	1973 Mar
Apr-91	1812 Apr
May-91	1975 May
Jun-91	1862 Jun
Jul-91	1940 Jul
Aug-91	2013 Aug
Sep-91	1596 Sep
Oct-91	1725 Oct
Nov-91	1676 Nov
Dec-91	1814 Dec
Jan-92	1615 Jan
Feb-92	1557 Feb
Mar-92	1891 Mar
Apr-92	1956 Apr
May-92	1885 May

FIGURE 16.7   NEW CATEGORICAL VARIABLE (RIGHT) TO BE USED (VIA DUMMIES) AS PREDICTOR(S) IN A LINEAR REGRESSION MODEL

The Regression Model

Input variables	Coefficient	Std. Error	p-value	SS
Constant term	1855.235962	33.95079803	0	477456500
season_Aug	139.3903351	48.01367569	0.00431675	483721.3125
season_Dec	-19.82307816	48.01367569	0.68036187	33314.77734
season_Feb	-288.9631348	47.08128357	0	665331.9375
season_Jan	-251.2854462	47.08128357	0.00000034	598841.0625
season_Jul	94.34428406	48.01367569	0.05147372	187691.7656
season_Jun	-10.11090946	48.01367569	0.83352947	11869.09277
season_Mar	11.57308865	47.08128357	0.80620199	48930.94922
season_May	31.24033737	48.01367569	0.51637506	114420.9141
season_Nov	-63.96651077	48.01367569	0.18502063	3121.062012
season_Oct	-54.12883377	48.01367569	0.26158884	14579.31641
season_Sep	-193.6371613	48.01367569	0.00009163	224972.1094

Training Data scoring - Summary Report

Total sum of squared errors	RMS Error	Average Error
1867303.623	112.7064583	-5.14163E-05

Validation Data scoring - Summary Report

Total sum of squared errors	RMS Error	Average Error
841206.3548	264.765046	262.1077072

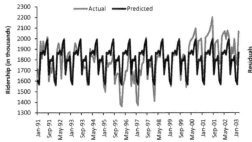
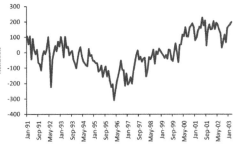

FIGURE 16.8   **FITTED REGRESSION MODEL WITH SEASONALITY. REGRESSION OUTPUT (TOP), PLOTS OF FITTED AND ACTUAL SERIES (BOTTOM LEFT), AND MODEL RESIDUALS (BOTTOM RIGHT)**

falls in that particular season, and 0 otherwise.[1] We then partition the data into training and validation sets (see Section 15.5) and fit the regression model to the training data. The top panels of Figure 16.8 show the output of a linear regression fit to Ridership (Y) on 11-month dummies (using the training data). The fitted series and the residuals from this model are shown in the lower panels. The model appears to capture the seasonality in the data. However, since we have

---

[1] We use only $m - 1$ dummies because information about the $m - 1$ seasons is sufficient. If all $m - 1$ variables are zero, then the season must be the mth season. Including the mth variable would cause redundant information and multicollinearity errors.

not included a trend component in the model (as shown in Section 16.1), the fitted values do not capture the existing trend. Therefore, the residuals, which are the difference between the actual and fitted values, clearly display the remaining U-shaped trend.

When seasonality is added as described above (create categorical seasonal variable, then create dummies from it, then regress on Y), it captures *additive seasonality*. This means that the average value of Y in a certain season is a certain amount more or less than that in another season. For example, in the Amtrak ridership, the coefficient for August (139.39) indicates that the average number of passengers in August is higher by 140,000 passengers than the average in April (the reference category). Using regression models, we can also capture *multiplicative seasonality*, where values on a certain season are on average higher or lower by a certain percentage compared to another season. To fit multiplicative seasonality, we use the same model as above, except that we use $\log(Y)$ as the output variable.

16.3 MODEL WITH TREND AND SEASONALITY

Finally, we can create models that capture both trend and seasonality by including predictors of both types. For example, from our exploration of the Amtrak Ridership data, it appears that a quadratic trend and monthly seasonality are both warranted. We therefore fit a model with 13 predictors: 11 dummies for month, and t and t^2 for trend. The output and fit from this final model are shown in Figure 16.9. This model can then be used to generate k-step-ahead forecasts (denoted by F_{t+k}) by plugging in the appropriate month and index terms.

16.4 AUTOCORRELATION AND ARIMA MODELS

When we use linear regression for time series forecasting, we are able to account for patterns such as trend and seasonality. However, ordinary regression models do not account for dependence between observations, which in cross-sectional data is assumed to be absent. Yet, in the time series context, observations in neighboring periods tend to be correlated. Such correlation, called *autocorrelation*, is informative and can help in improving forecasts. If we know that a high value tends to be followed by high values (positive autocorrelation), then we can use that to adjust forecasts. We will now discuss how to compute the autocorrelation of a series and how best to utilize the information for improving forecasting.

The Regression Model

Input variables	Coefficient	Std. Error	p-value	SS
Constant term	1932.998779	27.85863113	0	477456500
season_Aug	135.1726227	30.52143288	0.00001955	483721.3125
season_Dec	-29.65872955	30.53801155	0.33320817	33314.77734
season_Feb	-306.3078308	29.94875526	0	665331.9375
season_Jan	-267.444458	29.94642067	0	598841.0625
season_Jul	91.31225586	30.5189991	0.00330446	187691.7656
season_Jun	-12.04474545	30.51724434	0.69370645	11869.09277
season_Mar	-7.04482555	29.95207596	0.81441271	48930.94922
season_May	30.31717491	30.51618195	0.32228076	114420.9141
season_Nov	-72.26641083	30.53282547	0.01938256	3121.062012
season_Oct	-60.98049164	30.52834129	0.04781064	14579.31641
season_Sep	-199.1280975	30.52454758	0	224972.1094
t	-5.246521	0.58674908	0	398979.7188
t^2	0.0437566	0.00384071	0	725213.9375

Training Data scoring - Summary Report

Total sum of squared errors	RMS Error	Average Error
743110.0191	71.0997201	-6.05149E-05

Validation Data scoring - Summary Report

Total sum of squared errors	RMS Error	Average Error
30722.61731	50.59859789	-34.11397564

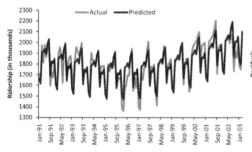
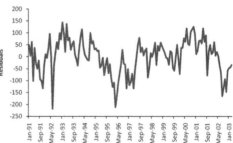

FIGURE 16.9   REGRESSION MODEL WITH MONTHLY (ADDITIVE) SEASONALITY AND QUADRATIC TREND, FIT TO AMTRAK RIDERSHIP DATA. REGRESSION OUTPUT (TOP), PLOTS OF FITTED AND ACTUAL SERIES (BOTTOM LEFT), AND MODEL RESIDUALS (BOTTOM RIGHT)

Computing Autocorrelation

Correlation between values of a time series in neighboring periods is called *autocorrelation* because it describes a relationship between the series and itself. To compute autocorrelation, we compute the correlation between the series and a lagged version of the series. A lagged series is a "copy" of the original series, which is moved forward one or more time periods. A lagged series with lag 1 is the original series moved forward one time period; a lagged series with lag 2 is the original series moved forward two time periods, and so on. Table 16.1 shows

TABLE 16.1	FIRST 24 MONTHS OF AMTRAK RIDERSHIP SERIES		
Month	**Ridership**	**Lag-1 Series**	**Lag-2 Series**
Jan-91	1709		
Feb-91	1621	1709	
Mar-91	1973	1621	1709
Apr-91	1812	1973	1621
May-91	1975	1812	1973
Jun-91	1862	1975	1812
Jul-91	1940	1862	1975
Aug-91	2013	1940	1862
Sep-91	1596	2013	1940
Oct-91	1725	1596	2013
Nov-91	1676	1725	1596
Dec-91	1814	1676	1725
Jan-92	1615	1814	1676
Feb-92	1557	1615	1814
Mar-92	1891	1557	1615
Apr-92	1956	1891	1557
May-92	1885	1956	1891
Jun-92	1623	1885	1956
Jul-92	1903	1623	1885
Aug-92	1997	1903	1623
Sep-92	1704	1997	1903
Oct-92	1810	1704	1997
Nov-92	1862	1810	1704
Dec-92	1875	1862	1810

the first 24 months of the Amtrak ridership series, the lag-1 series and the lag-2 series.

Next, to compute the lag-1 autocorrelation (which measures the linear relationship between values in consecutive time periods), we compute the correlation between the original series and the lag-1 series (e.g., via the Excel function CORREL) to be 0.08. Note that although the original series shown in Table 16.1 has 24 time periods, the lag-1 autocorrelation will only be based on 23 pairs (because the lag-1 series does not have a value for Jan-91). Similarly, the lag-2 autocorrelation (measuring the relationship between values that are two time periods apart) is the correlation between the original series and the lag-2 series (yielding −0.15).

We can use XLMiner's *ACF (autocorrelations)* utility within the *Time Series* menu, to directly compute the autocorrelations of a series at different lags. For example, the output for the 24–month ridership is shown in Figure 16.10. To display a bar chart of the autocorrelations at different lags, check the *Plot ACF* option.

A few typical autocorrelation behaviors that are useful to explore are:

Inputs

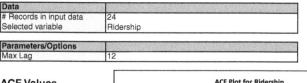

Data	
# Records in input data	24
Selected variable	Ridership

Parameters/Options	
Max Lag	12

ACF Values

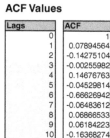

Lags	ACF
0	1
1	0.07894564
2	-0.14275104
3	-0.00255982
4	0.14676763
5	-0.04529814
6	-0.66626942
7	-0.06483612
8	0.06866533
9	0.06184223
10	-0.16368274
11	-0.05536203
12	0.32259634

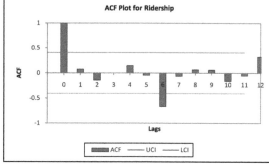

FIGURE 16.10   XLMINER OUTPUT SHOWING AUTOCORRELATION AT LAGS 1–12 FOR THE 24 MONTHS OF AMTRAK RIDERSHIP

Strong autocorrelation (positive or negative) at a lag larger than 1 typically reflects a cyclical pattern. For example, strong positive autocorrelation at lag 12 in monthly data will reflect an annual seasonality (where values during a given month each year are positively correlated).

Positive lag-1 autocorrelation (called "stickiness") describes a series where consecutive values move generally in the same direction. In the presence of a strong linear trend, we would expect to see a strong and positive lag-1 autocorrelation.

Negative lag-1 autocorrelation reflects swings in the series, where high values are immediately followed by low values and vice versa.

Examining the autocorrelation of a series can therefore help to detect seasonality patterns. In Figure 16.10, for example, we see that the strongest autocorrelation is at lag 6 and is negative. This indicates a biannual pattern in ridership, with 6-month switches from high to low ridership. A look at the time plot confirms the high-summer low-winter pattern.

In addition to looking at autocorrelations of the raw series, it is very useful to look at autocorrelations of *residual series*. For example, after fitting a regression model (or using any other forecasting method), we can examine the autocorrelation of the series of *residuals*. If we have adequately modeled the seasonal pattern, then the residual series should show no autocorrelation at the season's lag. Figure 16.11 displays the autocorrelations for the

Inputs

Data	
# Records in input data	147
Selected variable	Residual

Parameters/Options	
Max Lag	12

ACF Values

Lags	ACF
0	1
1	0.64821321
2	0.51890093
3	0.40798336
4	0.31966141
5	0.26237851
6	0.21345751
7	0.22334783
8	0.22640951
9	0.19724335
10	0.14933859
11	0.17307311
12	0.12726976

FIGURE 16.11   XLMINER OUTPUT SHOWING AUTOCORRELATION OF RESIDUAL SERIES FROM FIGURE 16.9

residuals from the regression model with seasonality and quadratic trend shown in Figure 16.9. It is clear that the 6-month (and 12-month) cyclical behavior no longer dominates the series of residuals, indicating that the regression model captured them adequately. However, we can also see a strong positive autocorrelation from lag 1 on, indicating a positive relationship between neighboring residuals. This is valuable information, which can be used to improve forecasting.

Improving Forecasts by Integrating Autocorrelation Information

In general, there are two approaches to taking advantage of autocorrelation. One is by directly building the autocorrelation into the regression model, and the other is by constructing a second-level forecasting model on the residual series.

Among regression–type models that directly account for autocorrelation are autoregressive models, or the more general class of models called ARIMA models (autoregressive integrated moving–average models). Autoregression models are similar to linear regression models, except that the predictors are the past values of the series. For example, an autoregression model of order 2 (AR(2)), can be written as

$$Y_t = \beta_0 + \beta_1 Y_{t-1} + \beta_2 Y_{t-2} + \epsilon. \tag{16.1}$$

Estimating such models is roughly equivalent to fitting a linear regression model with the series as the output, and the lagged series (at lag 1 and 2 in this example) as the predictors. However, it is better to use designated ARIMA estimation methods (e.g., those available in XLMiner's *Time Series> ARIMA* menu) over ordinary linear regression estimation, to produce more accurate results.[2] Moving from AR to ARIMA models creates a larger set of more flexible forecasting models but also requires much more statistical expertise. Even with the simpler AR models, fitting them to raw time series that contain patterns such as trends and seasonality requires the user to perform several initial data transformations and to choose the order of the model. These are not straightforward tasks. Because ARIMA modeling is less robust and requires more experience and statistical expertise than other methods, the use of such models for forecasting raw series is generally less popular in practical forecasting. We therefore direct the interested reader to classic time series textbooks [e.g., see Chapter 4 in Chatfield (2003)].

However, we do discuss one particular use of AR models that is straightforward to apply in the context of forecasting, which can provide significant improvement to short-term forecasts. This relates to the second approach for utilizing autocorrelation, which requires constructing a second-level forecasting model for the residuals, as follows:

1. Generate k-step-ahead forecast of the series (F_{t+k}), using a forecasting method.

2. Generate k-step-ahead forecast of forecast error (E_{t+k}), using an AR (or other) model.

3. Improve the initial k-step-ahead forecast of the series by adjusting it according to its forecasted error: *Improved* $F_{t+k}^* = F_{t+k} + E_{t+k}$.

In particular, we can fit low-order AR models to series of residuals (or *forecast errors*) that can then be used to forecast future forecast errors. By fitting the series of residuals, rather than the raw series, we avoid the need for initial data transformations (because the residual series is not expected to contain any trends or cyclical behavior besides autocorrelation).

To fit an AR model to the series of residuals, we first examine the autocorrelations of the residual series. We then choose the order of the AR model according to the lags in which autocorrelation appears. Often, when autocorrelation exists at lag 1 and higher, it is sufficient to fit an AR(1) model of the form

$$E_t = \beta_0 + \beta_1 E_{t-1} + \epsilon, \qquad (16.2)$$

where E_t denotes the residual (or *forecast error*) at time t. For example, although the autocorrelations in Figure 16.11 appear large from lags 1 to 10 or so, it is

---

[2] ARIMA model estimation differs from ordinary regression estimation by accounting for the dependence between observations.

likely that an AR(1) would capture all these relationships. The reason is that if neighboring values are correlated, then the relationship can propagate to values that are two periods away, then three periods away, and so forth. The result of fitting an AR(1) model to the Amtrak ridership residual series is shown in Figure 16.12. The AR(1) coefficient (0.65) is close to the lag-1 autocorrelation

Output Navigator			
Inputs	Arima Model	Residuals	ACF Plot
Elapsed Time	Forecast	Var Covar	

Inputs

Data	
# Records in input data	147
Input data	Data!$A$2:$C$148
Selected variable	Residual

Parameters/Options	
Do Not Fit Constant Term	No
AR	1
MA	0
Ordinary Differece	0
Show Forecasting Output	Yes
#Forecasts	1
Confidence Level	95
Show Residual Output	Yes

ARIMA Model

ARIMA	Coeff	StErr	p-value
Const. term	-0.00002147	4.86492062	0.99999648
AR1	0.64688748	0.06221873	0

Mean	-0.000061
-2LogL	1590.729736
S	54.047494
#Iterations	8

Lag	12	24	36	48
p-Value	0.577618	0.48778015	0.60872775	0.52462631
ChiSq	9.48063087	22.54175186	32.10258102	45.74367905
df	11	23	35	47

Forecast

Month	Forecast
Apr-03	-21.85559082

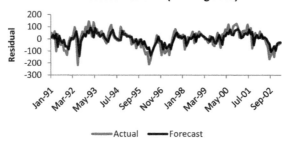

Time Plot of Actual Residual Vs Forecasted Residual (Training Data)

FIGURE 16.12   FITTING AN AR(1) MODEL TO THE RESIDUAL SERIES FROM FIGURE 16.9

that we found earlier (Figure 16.11). The forecasted residual for April 2003, given at the bottom, is computed by plugging in the most recent residual from March 2003 (equal to -33.786) into the AR(1) model: $0 + (0.647)(-33.786) = -21.866$. The negative value tells us that the regression model will produce a ridership forecast for April 2003 that is too high and that we should adjust it down by subtracting 21,866 riders. In this particular example, the regression model (with quadratic trend and seasonality) produced a forecast of 2,115,000 riders, and the improved two-stage model [regression + AR(1) correction] corrected it by reducing it to 2,093,000 riders. The actual value for April 2003 turned out to be 2,099,000 riders—much closer to the improved forecast.

Finally, from the plot of the actual versus forecasted residual series, we can see that the AR(1) model fits the residual series quite well. Note, however, that the plot is based on the training data (until March 2003). To evaluate predictive performance of the two-level model [regression + AR(1)], we would have to examine performance (e.g., via MAPE or RMSE metrics) on the validation data, in a fashion similar to the calculation that we performed for April 2003 above.

> To fit an AR model in XLMiner, use ARIMA in the *Time Series* menu. In the "Non-seasonal Parameters" set *Autoregressive (p)* to the required order, and *Moving Average (q)* to o. The *Advanced* menu will allow you to request forecasts and to display fitted values and residuals.

Finally, to examine whether we have indeed accounted for the autocorrelation in the series, and no more information remains in the series, we examine the autocorrelations of the series of residuals-of-residuals [the residuals obtained after the AR(1), which was applied to the regression residuals]. This can be seen in Figure 16.13. It is clear that no more autocorrelation remains, and that the addition of the AR(1) model has captured the autocorrelation information adequately.

We mentioned earlier that improving forecasts via an additional AR layer is useful for short-term forecasting. The reason is that an AR model of order k will usually only provide useful forecasts for the next k periods, and after that forecasts will rely on earlier forecasts rather than on actual data. For example, to forecast the residual of May 2003, when the time of prediction is March 2003, we would need the residual for April 2003. However, because that value is not available, it would be replaced by its forecast. Hence, the forecast for May 2003 would be based on the forecast for April 2003.

Evaluating Predictability

Before attempting to forecast a time series, it is important to determine whether it is predictable, in the sense that its past predicts its future. One useful way to assess

Inputs

Data	
# Records in input data	147
Selected variable	Residuals

Parameters/Options	
Max Lag	12

ACF Values

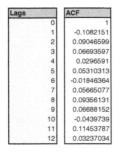

Lags	ACF
0	1
1	-0.1082151
2	0.09046599
3	0.06693597
4	0.0296591
5	0.05310313
6	-0.01846364
7	0.05665077
8	0.09356131
9	0.06688152
10	-0.0439739
11	0.11453787
12	0.03237034

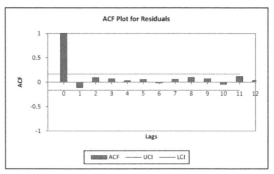

FIGURE 16.13   AUTOCORRELATIONS OF RESIDUALS-OF-RESIDUALS SERIES

predictability is to test whether the series is a random walk. A *random walk* is a series in which changes from one time period to the next are random. According to the efficient market hypothesis in economics, asset prices are random walks and therefore predicting stock prices is a game of chance.[3]

A random walk is a special case of an AR(1) model, where the slope coefficient is equal to 1:

$$Y_t = \beta_0 + Y_{t-1} + \epsilon_t. \tag{16.3}$$

We see from this equation that the difference between the values at periods $t - 1$ and t is random, hence the term random walk. Forecasts from such a model are basically equal to the most recent observed value, reflecting the lack of any other information.

To test whether a series is a random walk, we fit an AR(1) model and test the hypothesis that the slope coefficient is equal to 1 ($H_0 : \beta_1 = 1$ vs. $H_1 : \beta_1 \neq 1$). If the hypothesis is accepted (reflected by a small p-value), then the series is not a random walk, and we can attempt to predict it.

As an example, consider the AR(1) model shown in Figure 16.12. The slope coefficient (0.647) is more than 3 standard errors away from 1, indicating that this is not a random walk. In contrast, consider the AR(1) model fitted to the series of S&P500 monthly closing prices between May 1995 and August 2003 (available

---

[3] There is some controversy surrounding the efficient market hypothesis, with claims that there is slight autocorrelation in asset prices, which does make them predictable to some extent. However, transaction costs and bid–ask spreads tend to offset any prediction benefits.

Output Navigator			
Inputs	Arima Model	Residuals	ACF Plot
Elapsed Time	Forecast	Var Covar	

Inputs

Data	
# Records in input data	100
Input data	Data!$A$2:$C$101
Selected variable	Close

Parameters/Options	
Do Not Fit Constant Term	No
AR	1
MA	0
Ordinary Differece	0

ARIMA Model

ARIMA	Coeff	StErr	p-value
Const. term	15.62566853	3.68750787	0.00002261
AR1	0.98479182	0.01436355	0

FIGURE 16.14   AR(1) MODEL FITTED TO S&P500 MONTHLY CLOSING PRICES (MAY 1995–AUG 2003)

in SP500.xls, shown in Figure 16.14). Here the slope coefficient is 0.985, with a standard error of 0.015. The coefficient is sufficiently close to 1 (around one standard error away), indicating that this is a random walk. Forecasting this series using any of the methods described earlier is therefore futile.

PROBLEMS

16.1 Impact of September 11 on Air Travel in the United States: The Research and Innovative Technology Administration's Bureau of Transportation Statistics (BTS) conducted a study to evaluate the impact of the September 11, 2001, terrorist attack on U.S. transportation. The study report and the data can be found at http://www.bts.gov/publications/estimated_impacts_of_9_11_on_us_travel. The goal of the study was stated as follows:

> The purpose of this study is to provide a greater understanding of the passenger travel behavior patterns of persons making long distance trips before and after 9/11.

The report analyzes monthly passenger movement data between January 1990 and May 2004. Data on three monthly time series are given in File Sept11Travel.xls for this period: (1) actual airline revenue passenger miles (Air), (2) rail passenger miles (Rail), and (3) vehicle miles traveled (Car).

In order to assess the impact of September 11, BTS took the following approach: using data before September 11, it forecasted future data (under the assumption of no terrorist attack). Then, BTS compared the forecasted series with the actual data to assess the impact of the event. Our first step, therefore, is to split each of the time series into two parts: pre- and post-September 11. We now concentrate only on the earlier time series.

a. Plot the pre-event Air time series.

    i. Which time series components appear from the plot?

b. Figure 16.15 is a time plot of the **seasonally adjusted** pre-Sept-11 Air series. Which of the following methods would be adequate for forecasting this series?

- Linear regression model with dummies
- Linear regression model with trend
- Linear regression model with dummies and trend

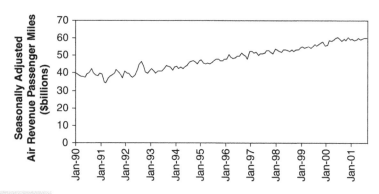

FIGURE 16.15   **SEASONALLY ADJUSTED PRE-SEPT-11 AIR SERIES**

c. Specify a linear regression model for the Air series that would produce a seasonally adjusted series similar to the one shown in (b), with multiplicative seasonality. What is the output variable? What are the predictors?

d. Run the regression model from (c). Remember to create dummy variables for the months (XLMiner will create 12 dummies; use only 11 and drop the April dummy) and to use only pre-event data.

   i. What can we learn from the statistical insignificance of the coefficients for October and September?

   ii. The actual value of Air (air revenue passenger miles) in January 1990 was 35.153577 billion. What is the residual for this month, using the regression model? Report the residual in terms of air revenue passenger miles.

e. Create an ACF (autocorrelation) plot of the regression residuals.

   i. What does the ACF plot tell us about the regression model's forecasts?

   ii. How can this information be used to improve the model?

f. Fit linear regression models to Air, Rail and Auto with additive seasonality and an appropriate trend. For Air and Rail, fit a linear trend. For Rail, use a quadratic trend. Remember to use only preevent data. Once the models are estimated, use them to forecast each of the three post-event series.

   i. For each series (Air, Rail, Auto), plot the complete pre-event and postevent actual series overlayed with the predicted series.

   ii. What can be said about the effect of the September 11 terrorist attack on the three modes of transportation? Discuss the magnitude of the effect, its time span, and any other relevant aspects.

16.2 **Analysis of Canadian Manufacturing Workers Workhours:** The time series plot in Figure 16.16 describes the average annual number of weekly hours spent by Canadian

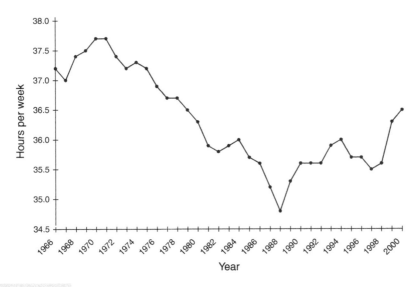

FIGURE 16.16   **AVERAGE ANNUAL WEEKLY HOURS SPENT BY CANADIAN MANUFACTURING WORKERS**

manufacturing workers (data are available in CanadianWorkHours.xls, data courtesy of Ken Black).

    a. Which one model of the following regression-based models would fit the series best?

- Linear trend model
- Linear trend model with seasonality
- Quadratic trend model
- Quadratic trend model with seasonality

    b. If we computed the autocorrelation of this series, would the lag-1 autocorrelation exhibit negative, positive, or no autocorrelation? How can you see this from the plot?

      i. Compute the autocorrelation and produce an ACF plot. Verify your answer to the previous question.

16.3 **Regression Modeling of Toys "R" Us Revenues:** Figure 16.17 is a time series plot of the quarterly revenues of Toys "R" Us between 1992 and 1995 (thanks to Chris Albright for suggesting the use of these data, which are available in ToysRUsRevenues.xls).

    a. Fit a regression model with a linear trend and seasonal dummies. Use the entire series (excluding the last two quarters) as the training set.

    b. A partial output of the regression model is shown in Figure 16.18. Use this output to answer the following questions:

      i. Mention two statistics (and their values) that measure how well this model fits the training data.

      ii. Mention two statistics (and their values) that measure the predictive accuracy of this model.

      iii. After adjusting for trend, what is the average difference between sales in Q3 and sales in Q1?

      iv. After adjusting for seasonality, which quarter (Q_1, Q_2, Q_3 or Q_4) has the highest average sales?

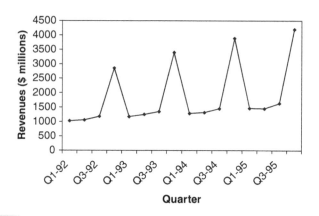

FIGURE 16.17   **QUARTERLY REVENUES OF TOYS "R" US, 1992–1995**

Regression Model

Input variables	Coefficient	Std. Error	p-value	SS
Constant term	906.749939	115.3461227	0.00002541	41669100
Trend	47.1071434	11.25662899	0.00235907	825673.5625
Quarter_2	-15.10719299	119.6596069	0.9023084	1335472
Quarter_3	89.16661835	128.6739807	0.50581801	1357627
Quarter_4	2101.726074	129.1654205	0.00000001	7514922

Residual df	9
Multiple R-squared	0.977372001
Std. Dev. estimate	168.4737854
Residual SS	255450.7656

Training Data Scoring - Summary Report

Total sum of squared errors	RMS Error	Average Error
255450.7619	135.0795432	0.000106558

Validation Data Scoring - Summary Report

Total sum of squared errors	RMS Error	Average Error
196792.9676	313.6821382	183.1429921

FIGURE 16.18   OUTPUT FOR REGRESSION MODEL FITTED TO TOYS "R" US TIME SERIES

16.4 Forecasting Wal-Mart Stock: Figures 16.19 and 16.20 show plots, summary statistics, and output from fitting an AR(1) model to the series of Wal-Mart daily closing prices between February 2001 and February 2002. (Thanks to Chris Albright for suggesting the use of these data, which are publicly available, e.g., at http://finance.yahoo.com and are in the file WalMartStock.xls.) Use all the information to answer the following questions.

a. Create a time plot of the differenced series.

b. Which of the following is/are relevant for testing whether this stock is a random walk?

- The autocorrelations of the close prices series
- The AR(1) slope coefficient
- The AR(1) constant coefficient

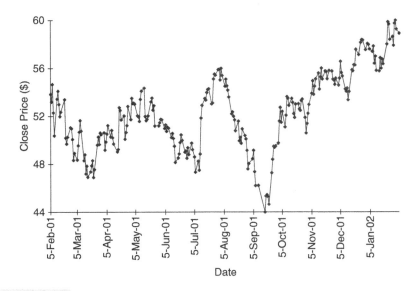

FIGURE 16.19   DAILY CLOSE PRICE OF WAL-MART STOCK, FEB 2001–2002

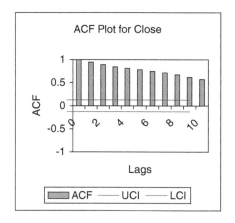

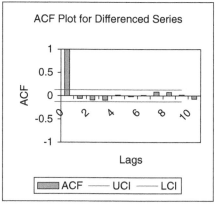

ARIMA Model for Close

ARIMA	Coeff	StErr	p-value
Const. term	2.30948	0.060417	0
AR1	0.95589	0.018672	0

ARIMA Model for Differenced Series

ARIMA	Coeff	StErr	p-value
Const. term	0.021673	0.092092	0.813947
AR1	-0.05794	0.063466	0.361313

FIGURE 16.20   **OUTPUT OF FITTING AN AR(1) MODEL TO WAL-MART STOCK SERIES**

c. Does the AR model indicate that this is a random walk? Explain how you reached your conclusion.

d. What are the implications of finding that a time series is a random walk? Choose the correct statement(s) below.

- It is impossible to obtain useful forecasts of the series.

- The series is random.

- The changes in the series from one period to the other are random.

16.5 Forecasting Department Store Sales: The time series plot shown in Figure 16.21 describes actual quarterly sales for a department store over a 6-year period. (Data are available in DepartmentStoreSales.xls, data courtesy of Chris Albright.)

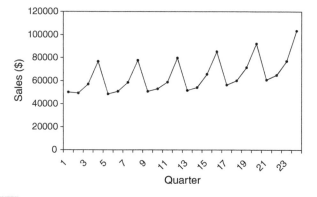

FIGURE 16.21   **DEPARTMENT STORE QUARTERLY SALES SERIES**

Input variables	Coefficient	Std. Error	p-value	SS
Constant term	10.74894524	0.01872449	0	2429.415771
Quarter	0.01108785	0.0012952	0.00000033	0.18121047
Qtr_2	0.02495589	0.02076364	0.24803306	0.11009274
Qtr_3	0.165343	0.02088447	0.00000094	0.00970232
Qtr_4	0.43374524	0.02108433	0	0.45436361

Residual df	15
Multiple R-squared	0.979125117
Std. Dev. estimate	0.03276626
Residual SS	0.01610442

FIGURE 16.22   **OUTPUT FROM REGRESSION MODEL FIT TO DEPARTMENT STORE SALES TRAINING SERIES**

a. The forecaster decided that there is an exponential trend in the series. In order to fit a regression-based model that accounts for this trend, which of the following operations must be performed?

- Take log of Quarter index
- Take log of sales
- Take an exponent of sales
- Take an exponent of Quarter index

b. Fit a regression model with an exponential trend and seasonality, using only the first 20 quarters as the training data (remember to first partition the series into training and validation series).

c. A partial output is shown in Figure 16.22. From the output, after adjusting for trend, are Q2 average sales higher, lower, or approximately equal to the average Q1 sales?

d. Use this model to forecast sales in quarters 21 and 22.

e. The plots in Figure 16.23 describe the fit (top) and forecast errors (bottom) from this regression model.

   i. Recreate these plots.

   ii. Based on these plots, what can you say about your forecasts for quarters 21 and 22? Are they likely to overforecast, underforecast, or be reasonably close to the real sales values?

f. Looking at the residual plot, which of the following statements appear true?

- Seasonality is not captured well.
- The regression model fits the data well.
- The trend in the data is not captured well by the model.

g. Which of the following solutions is adequate *and* a parsimonious solution for improving model fit?

- Fit a quadratic trend model to the residuals (with Quarter and Quarter^2.)
- Fit an AR model to the residuals.
- Fit a quadratic trend model to Sales (with Quarter and Quarter^2.)

16.6 **Souvenir Sales:** Figure 16.24 shows a time plot of monthly sales for a souvenir shop at a beach resort town in Queensland, Australia, between 1995 and 2001. [Data are available in SouvenirSales.xls, source: R. J. Hyndman, Time Series Data Library, http://www.robjhyndman.com/TSDL; accessed on December 28, 2009.] The series is presented twice, in Australian dollars and in log scale. Back in 2001, the store wanted to use the data to forecast sales for the next 12 months (year 2002). It hired an analyst to generate forecasts. The analyst first partitioned the data into training and validation sets,

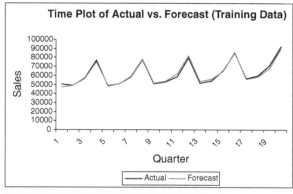

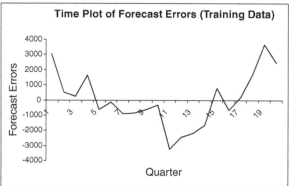

FIGURE 16.23   FIT OF REGRESSION MODEL FOR DEPARTMENT STORE SALES

with the validation set containing the last 12 months of data (year 2001). She then fit a regression model to sales, using the training set.

a. Based on the two time plots, which predictors should be included in the regression model? What is the total number of predictors in the model?

b. Run a regression model with Sales (in Australian dollars) as the output variable and with a linear trend and monthly predictors. Remember to fit only the training data. Call this model A.

    i. Examine the estimated coefficients: which month tends to have the highest average sales during the year?

    ii. The estimated trend coefficient is 245.36. What does this mean?

c. Run a regression model with log(Sales) as the output variable and with a linear trend and monthly predictors. Remember to fit only the training data. Call this model B.

    i. Fitting a model to log(Sales) with a linear trend is equivalent to fitting a model to Sales (in dollars) with what type of trend?

    ii. The estimated trend coefficient is 0.2. What does this mean?

    iii. Use this model to forecast the sales in February 2002. What is the extra step needed?

d. Compare the two regression models (A and B) in terms of forecast performance. Which model is preferable for forecasting? Mention at least two reasons based on the information in the outputs.

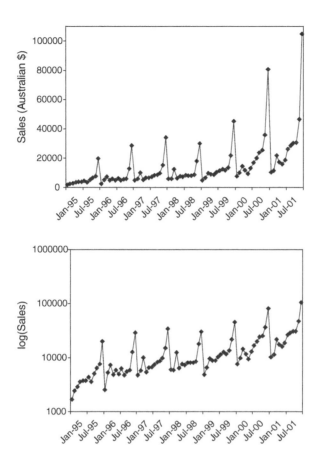

FIGURE 16.24 MONTHLY SALES AT AUSTRALIAN SOUVENIR SHOP IN DOLLARS (TOP) AND IN LOG SCALE (BOTTOM)

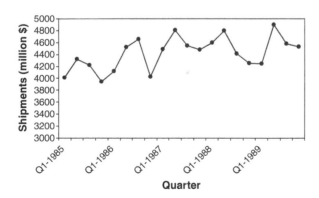

FIGURE 16.25 QUARTERLY SHIPMENTS OF U.S. HOUSEHOLD APPLIANCES OVER 5 YEARS

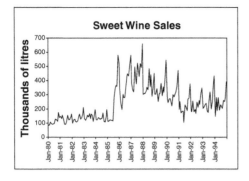

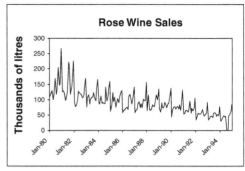

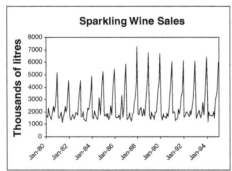

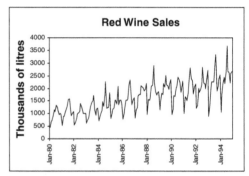

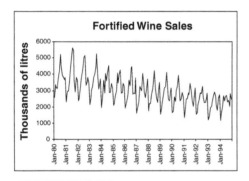

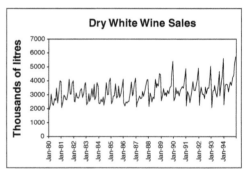

FIGURE 16.26 MONTHLY SALES OF SIX TYPES OF AUSTRALIAN WINES BETWEEN 1980 AND 1994

e. Continuing with model B [with log(Sales) as output], create an ACF plot until lag 15 for the forecast errors. Now fit an AR model with lag 2 [ARIMA(2,0,0)] to the forecast errors.

   i. Examining the ACF plot and the estimated coefficients of the AR(2) model (and their statistical significance), what can we learn about the forecasts that result from model B?

   ii. Use the autocorrelation information to compute an improved forecast for January 2002, using model B and the AR(2) model above.

f. How would you model these data differently if the goal was to understand the different components of sales in the souvenir shop between 1995 and 2001? Mention two differences.

16.7 **Shipments of Household Appliances:** The time plot in Figure 16.25 shows the series of quarterly shipments (in million dollars) of U.S. household appliances between 1985 and 1989. (Data are available in ApplianceShipments.xls; data courtesy of Ken Black.)

   a. If we compute the autocorrelation of the series, which lag (>0) is most likely to have the largest coefficient (in absolute value)? Create an ACF plot and compare with your answer.

16.8 **Forecasting Australian Wine Sales:** Figure 16.26 shows time plots of monthly sales of six types of Australian wines (red, rose, sweet white, dry white, sparkling, and fortified) for 1980–1994. [Data are available in AustralianWines.xls, source: R. J. Hyndman, Time Series Data Library, http://www.robjhyndman.com/TSDL; accessed on December 28, 2009.] The units are thousands of liters. You are hired to obtain short-term forecasts (2–3 months ahead) for each of the six series, and this task will be repeated every month.

   a. Which forecasting method would you choose if you had to choose the same method for all series except Sweet Wine? Why?

   b. Fortified wine has the largest market share of the six types of wine considered. You are asked to focus on fortified wine sales alone and produce as accurate as possible forecast for the next 2 months.

     • Start by partitioning the data using the period until December 1993 as the training set.

     • Fit a regression model to sales with a linear trend and seasonality.

       i. Comparing the "actual vs. forecast" plots for the two models, what does the similarity between the plots tell us?

       ii. Use the regression model to forecast sales in January and February 1994.

   c. Create an ACF plot for the residuals from the above model until lag 12.

       i. Examining this plot, which of the following statements are reasonable?

          • Decembers (month 12) are not captured well by the model.

          • There is a strong correlation between sales on the same calendar month.

          • The model does not capture the seasonality well.

          • We should try to fit an autoregressive model with lag 12 to the residuals.

CHAPTER 17

Smoothing Methods

In this chapter we describe popular, flexible methods for forecasting time series that rely on smoothing. Smoothing is based on averaging over multiple observations in order to reduce the noise. We start with two simple smoothers, the moving average and simple exponential smoother, which are suitable for forecasting series that contain no trend or seasonality. In both cases forecasts are averages of previous values of the series (the length of the series history that is considered and the weights that are used in the averaging differ between the methods). We also show how a moving average can be used, with a slight adaptation, for data visualization. We then proceed to describe smoothing methods that are suitable for forecasting series with a trend and/or seasonality. Smoothing methods are data driven and are able to adapt to changes in the series over time. Although highly automated, the user must specify smoothing constants, which determine how fast the method adapts to new data. We discuss the choice of such constants and their meaning. The different methods are illustrated using the Amtrak ridership series.

17.1 INTRODUCTION

A second class of methods for time series forecasting are smoothing methods. Unlike regression models, which rely on an underlying theoretical model for the components of a time series (e.g., linear model or multiplicative seasonality), smoothing methods are data driven in the sense that they estimate time series components directly from the data without a predetermined structure. Data-driven methods are especially useful in series where components change

Data Mining for Business Intelligence, By Galit Shmueli, Nitin R. Patel, and Peter C. Bruce
Copyright © 2010 Statistics.com and Galit Shmueli

over time. Here we consider a type of data-driven methods called "smoothing methods." Such methods "smooth" out the noise in a series in an attempt to uncover the patterns. Smoothing is done by averaging over multiple observations, where different smoothers differ by the number of observations averaged, how the average is computed, how many times averaging is performed, and the like. We now describe two types of smoothing methods that are popular in business applications due to their simplicity and adaptivity. These are moving average methods and exponential smoothing methods.

17.2   MOVING AVERAGE

The moving average is a simple smoother: It consists of averaging across a window of consecutive observations, thereby generating a series of averages. A moving average with window width w means averaging across each set of w consecutive values, where w is determined by the user.

In general, there are two types of moving averages: a *centered moving average* and a *trailing moving average*. Centered moving averages are powerful for visualizing trends because the averaging operation can suppress seasonality and noise, thereby making the trend more visible. In contrast, trailing moving averages are useful for forecasting. The difference between the two is where the averaging window is placed over the time series.

Centered Moving Average for Visualization

In a centered moving average, the value of the moving average at time t (MA_t) is computed by centering the window around time t and averaging across the w values within the window:

$$MA_t = \left(Y_{t-(w-1)/2} + \cdots + Y_{t-1} + Y_t + Y_{t+1} + \cdots + Y_{t+(w-1)/2} \right) / w.$$

For example, with a window of width $w = 5$, the moving average at time point $t = 3$ means averaging the values of the series at time points $1, 2, 3, 4, 5$; at time point $t = 4$ the moving average is the average of the series at time points $2, 3, 4, 5, 6$, and so on.[1] This is illustrated in the top panel of Figure 17.1.

Choosing the window width in a seasonal series is straightforward: Because the goal is to suppress seasonality for better visualizing the trend, the default choice should be the length of a seasonal cycle. Returning to the Amtrak ridership data, the annual seasonality indicates a choice of $w = 12$. Figure 17.2 shows a centered moving average line overlaid on the original series. We can see a global

---

[1] For an even window width, for example, $w = 4$, obtaining the moving average at time point $t = 3$ requires averaging across two windows: across time points 1, 2, 3, 4; across time points 2, 3, 4, 5; and finally the average of the two averages is the final moving average.

Centered window (w=5)

| $t–2$ | $t–1$ | t | $t+1$ | $t+2$ |

Trailing window (w=5)

| $t–4$ | $t–3$ | $t–2$ | $t–1$ | t |

FIGURE 17.1   SCHEMATIC OF CENTERED MOVING AVERAGE (TOP) AND TRAILING MOVING AVERAGE (BOTTOM), BOTH WITH WINDOW WIDTH $W = 5$

U-shape, but unlike the regression model that fits a strict U-shape, the moving average shows some deviation, such as the slight dip during the last year.

This plot is obtained by using Excel's *Add Trendline* (as explained earlier), and choosing the *Moving Average* option, where you can set the length of the window (see Figure 17.3).

Trailing Moving Average for Forecasting

Centered moving averages are computed by averaging across data in the past and the future of a given time point. In that sense they cannot be used for forecasting because, at the time of forecasting, the future is typically unknown. Hence, for purposes of forecasting, we use *trailing moving averages*, where the window of width w is set on the most recent available w values of the series. The k-step-ahead forecast F_{t+k} ($k = 1, 2, 3, \ldots$) is then the average of these w values (see also bottom plot in Figure 17.1):

$$F_{t+k} = (Y_t + Y_{t-1} + \cdots + Y_{t-w+1})\,/w.$$

For example, in the Amtrak ridership series, to forecast ridership in February 1992 or later months, given information until January 1992 and using a *moving average* with window width $w = 12$, we would take the average ridership during the most recent 12 months (February 1991 to January 1992).

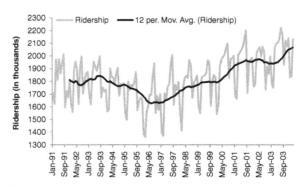

FIGURE 17.2   CENTERED MOVING AVERAGE WITH WINDOW $W = 12$, OVERLAID ON AMTRAK RIDERSHIP SERIES. THIS HELPS VISUALIZE TRENDS

FIGURE 17.3   EXCEL'S *TRENDLINE* MENU

Computing a trailing moving average can be done via XLMiner's *Moving Average* menu (within *Time Series › Smoothing*). This will yield forecasts and forecast errors for the training set and, if checked, for the validation set as well. The default window width (called *Interval* in the *Weights* box) is $w = 2$, which should be modified by the user.

Using XLMiner, we illustrate a 12-month moving average forecaster for the Amtrak ridership. We partitioned once again the Amtrak ridership series, leaving the last 12 months as the validation set. Applying a moving average forecaster with window $w = 12$, we obtained the output partially shown in Figure 17.4. Note that for the first 11 records of the training set, there is no forecast (because there are less than 12 past values to average). Also, note that the forecasts for all months in the validation set are identical (1942.73) because the method assumes that information is known only until March 2003.

In this example, it is clear that the moving average forecaster is inadequate for generating monthly forecasts because it does not capture the seasonality in

Output Navigator		
Inputs	Fitted Model	Fore cast
Elapsed Time	Error Measures (Training)	Error Measures(Validation)

Inputs

Data	
# Records in training data	147
# Records in validation data	12
Input data	Data_PartitionTS1!$B$19:$D$177
Selected variable	Ridership

Parameters/Options	
Interval	12
Season length	N.A.
Number of seasons	N.A.
Forecast	Yes
#Forecasts	12

Fitted Model

Month	Actual	Forecast	Residuals
Jan-91	1708.917	*	*
Feb-91	1620.586	*	*
Mar-91	1972.715	*	*
Apr-91	1811.665	*	*
May-91	1974.964	*	*
Jun-91	1862.356	*	*
Jul-91	1939.86	*	*
Aug-91	2013.264	*	*
Sep-91	1595.657	*	*
Oct-91	1724.924	*	*
Nov-91	1675.667	*	*
Dec-91	1813.863	*	*
Jan-92	1614.827	1809.5365	-194.7095
Feb-92	1557.088	1801.69567	-244.607667
Mar-92	1891.223	1796.40417	94.8188333
Apr-92	1955.981	1789.61317	166.367833
May-92	1884.714	1801.6395	83.0745

Time Plot of Actual vs. Forecast (Training Data)

Error Measures (Training)

MAPE	6.85518801
MAD	119.866987
MSE	21069.3208

Forecast

Month	Actual	Forecast	Error	LCI	UCI
Apr-03	2098.899	1942.73808	156.160917	1670.09809	2215.37807
May-03	2104.911	1942.73808	162.172917	1670.09809	2215.37807
Jun-03	2129.671	1942.73808	186.932917	1670.09809	2215.37807
Jul-03	2223.349	1942.73808	280.610917	1670.09809	2215.37807
Aug-03	2174.36	1942.73808	231.621917	1670.09809	2215.37807
Sep-03	1931.406	1942.73808	-11.3320833	1670.09809	2215.37807
Oct-03	2121.47	1942.73808	178.731917	1670.09809	2215.37807
Nov-03	2076.054	1942.73808	133.315917	1670.09809	2215.37807
Dec-03	2140.677	1942.73808	197.938917	1670.09809	2215.37807
Jan-04	1831.508	1942.73808	-111.230083	1670.09809	2215.37807
Feb-04	1838.006	1942.73808	-104.732083	1670.09809	2215.37807
Mar-04	2132.446	1942.73808	189.707917	1670.09809	2215.37807

Time Plot of Actual VS. Forecast (Validation Data)

Error Measures (Validation)

MAPE	7.71191897
MAD	162.040708
MSE	30531.4915

FIGURE 17.4   **PARTIAL OUTPUT FOR MOVING AVERAGE FORECASTER WITH $W = 12$ APPLIED TO AMTRAK RIDERSHIP SERIES**

the data. Hence seasons with high ridership are underforecasted, and seasons with low ridership are overforecasted. A similar issue arises when forecasting a series with a trend: The moving average "lags behind," thereby underforecasting in the presence of an increasing trend and overforecasting in the presence of a decreasing trend.

In general, the moving average can be used for forecasting *only in series that lack seasonality and trend*. Such a limitation might seem impractical. However, there are a few popular methods for removing trends (de-trending) and removing seasonality (de-seasonalizing) from a series, such as regression models. The moving average can then be used to forecast such de-trended and de-seasonalized series, and then the trend and seasonality can be added back to the forecast. For example, consider the regression model shown in Figure 16.9, which yields residuals clean of seasonality and trend (bottom plot). We can apply a moving average forecaster to that series of residuals (also called forecast errors), thereby creating a forecast for the next *forecast error*. For example, to forecast ridership in April 2003 (the first period in the validation set), assuming that we have information until March 2003, we use the regression model in Figure 16.9 to generate a forecast for April 2003 (which yields 2,115,000 riders). We then use a 12-month moving average (using the period April 2002 to March 2003) to forecast the *forecast error* for April 2003, which yields −66.33 (manually, or using XLMiner, as shown in Figure 17.5). The negative value implies that the regression model's forecast for April 2003 is too high, and therefore we should adjust it by reducing approximately 66,000 riders from the regression model's forecast of 2,115,000 riders.

Output Navigator		
Inputs	Fitted Model	Forecast
Elapsed Time	Error Measures(Training)	Error Measures(Validation)

Fitted Model

Month	Actual	Forecast	Residuals
Apr-02	19.6040031	40.2642642	-20.660261
May-02	2.95279739	36.1845692	-33.231772
Jun-02	-35.967826	31.7797777	-67.747604
Jul-02	-65.207885	18.920723	-84.128608
Aug-02	-164.72782	8.23973836	-172.96756
Sep-02	-129.88119	-12.913676	-116.96751
Oct-02	-92.549388	-17.27627	-75.273118
Nov-02	-146.91358	-26.760295	-120.15328
Dec-02	-58.825883	-43.108832	-15.717051
Jan-03	-46.758291	-48.970549	2.21225802
Feb-03	-43.925568	-54.354696	10.4291283
Mar-03	-33.785736	-63.118857	29.3331206

Error Measures (Training)

MAPE	-8.0985192
MAD	51.4815662
MSE	4098.27499

Forecast

Month	Forecast	LCI	UCI
Apr-03	-66.332197	-186.57659	53.9121966

FIGURE 17.5   APPLYING MA TO THE RESIDUALS FROM THE REGRESSION MODEL (WHICH LACK TREND AND SEASONALITY), TO FORECAST THE APRIL 2003 RESIDUAL

Choosing Window Width (*w*)

With moving average forecasting or visualization, the only choice that the user must make is the width of the window (*w*). As with other methods such as *k* nearest neighbors, the choice of the smoothing parameter is a balance between undersmoothing and oversmoothing. For visualization (using a centered window), wider windows will expose more global trends, while narrow windows will reveal local trend. Hence, examining several window widths is useful for exploring trends of differing local/global nature. For forecasting (using a trailing window), the choice should incorporate some domain knowledge in terms of relevance of past observations and how fast the series changes. Empirical predictive evaluation can also be done by experimenting with different values of *w* and comparing performance. However, care should be taken not to overfit!

17.3 SIMPLE EXPONENTIAL SMOOTHING

A popular forecasting method in business is exponential smoothing. Its popularity derives from its flexibility, ease of automation, cheap computation, and good performance. Simple exponential smoothing is similar to forecasting with a moving average, except that instead of taking a simple average over the *w* most recent values, we take a *weighted average* of all past values, such that the weights decrease exponentially into the past. The idea is to give more weight to recent information, yet not to completely ignore older information.

Like the moving average, simple exponential smoothing should only be used for forecasting *series that have no trend or seasonality*. As mentioned earlier, such series can be obtained by removing trend and/or seasonality from raw series, and then applying exponential smoothing to the series of residuals (which are assumed to contain no trend or seasonality).

The exponential smoother generates a forecast at time $t + 1$ (F_{t+1}) as follows:

$$F_{t+1} = \alpha Y_t + \alpha(1 - \alpha)Y_{t-1} + \alpha(1 - \alpha)^2 Y_{t-2} + \cdots, \tag{17.1}$$

where α is a constant between 0 and 1 called the *smoothing parameter*. The above formulation displays the exponential smoother as a weighted average of all past observations, with exponentially decaying weights.

It turns out that we can write the exponential forecaster in another way, which is very useful in practice:

$$F_{t+1} = F_t + \alpha E_t, \tag{17.2}$$

where E_t is the forecast error at time t. This formulation presents the exponential forecaster as an "active learner." It looks at the previous forecast (F_t) and how far it was from the actual value (E_t), and then corrects the next forecast based on that information. If last period the forecast was too high, the next period is adjusted down. The amount of correction depends on the value of the smoothing parameter α. The formulation in (17.2) is also advantageous in terms of data storage and computation time: It means that we need to store and use only the forecast and forecast error from the previous period, rather than the entire series. In applications where real-time forecasting is done, or many series are being forecasted in parallel and continuously, such savings are critical.

Note that forecasting further into the future yields the same as a one-step-ahead forecast. Because the series is assumed to lack trend and seasonality, forecasts into the future rely only on information until the time of prediction. Hence, the k-step-ahead forecast is equal to

$$F_{t+k} = F_{t+1}.$$

Choosing Smoothing Parameter α

The smoothing parameter α, which is set by the user, determines the rate of learning. A value close to 1 indicates fast learning (i.e., only the most recent observations have influence on forecasts), whereas a value close to 0 indicates slow learning (past observations have a large influence on forecasts). This can be seen by plugging 0 or 1 into the two equations above. Hence, the choice of α depends on the required amount of smoothing, and on how relevant the history is for generating forecasts. Default values that have been shown to work well are around 0.1–0.2. Some trial and error can also help in the choice of α: Examine the time plot of the actual and predicted series, as well as the predictive accuracy (e.g., MAPE or RMSE of the validation set). Finding the α value that optimizes predictive accuracy on the validation set can be used to determine the degree of local versus global nature of the trend (e.g., by using "optimize" in XLMiner's *Weights* box). However, beware of choosing the "best α" for forecasting, as this will most likely lead to model overfitting and low predictive accuracy on future data.

In XLMiner, forecasting using simple exponential smoothing is done via the *Exponential* menu (within *Time Series › Smoothing*). This will yield forecasts and forecast errors for both the training and validation sets. You can use the default value of $\alpha = 0.2$, set it to another value, or choose to find the optimal α in terms of minimizing RMSE of the validation data.

To illustrate forecasting with simple exponential smoothing, we return to the residuals from the regression model, which are assumed to contain no trend or seasonality. To forecast the residual on April 2003, we apply exponential smoothing to the entire period until March 2003, and use the default $\alpha = 0.2$ value.

Output Navigator		
Inputs	Fitted Model	Forecast
Elapsed Time	Error Measures(Training)	Error Measures(Validation)

Inputs

Data	
# Records in input data	159
Input data	RegResiduals!$A$2:$C$160
Selected variable	Regression Residual

Parameters/Options	
Alpha (Level)	0.2
Beta (Trend)	N.A.
Gamma (Seasonality)	N.A.
Season length	N.A.
Number of seasons	N.A.
Forecast	Yes
#Forecasts	1

Fitted Model

Month	Actual	Forecast	Residuals
Jan-91	48.5654431	*	*
Feb-91	4.21306711	48.5654431	-44.352376
Mar-91	62.1067999	39.6949679	22.4118319
Apr-91	-101.047801	44.1773343	-145.225135
May-91	36.7867358	15.1323073	21.6544285
Jun-91	-28.6941455	19.463193	-48.1573384
Jul-91	-49.8694616	9.83172528	-59.7011868
Aug-91	-15.7356564	-2.10851209	-13.6271443
Sep-91	-94.5392774	-4.83394095	-89.7053364
Oct-91	-99.0047377	-22.7750082	-76.2297294
Nov-91	-132.648186	-38.0209541	-94.627232
Dec-91	-32.8197481	-56.9464005	24.1266524
Jan-92	10.0825863	-52.12107	62.2036563
Feb-92	-4.72794809	-39.6803388	34.9523907
Mar-92	34.1216262	-32.6898606	66.8114869
Apr-92	95.7248671	-19.3275633	115.05243
May-92	-2.05675461	3.68292282	-5.73967743
Jun-92	-217.651794	2.53498733	-220.186782

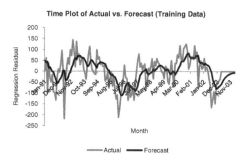

Time Plot of Actual vs. Forecast (Training Data)

Error Measures (Training)

MAPE	-3.04613294
MAD	45.0195296
MSE	3353.67331

Forecast

Month	Forecast	LCI	UCI
Apr-04	-4.00086784	-117.14875	109.147014

FIGURE 17.6 PARTIAL OUTPUT FOR SIMPLE EXPONENTIAL SMOOTHING FORECASTER WITH $\alpha = 0.2$, APPLIED TO THE SERIES OF RESIDUALS FROM THE REGRESSION MODEL (WHICH LACK TREND AND SEASONALITY). THE FORECAST FOR APRIL 2003 RESIDUAL IS AT THE BOTTOM

The output is shown in Figure 17.6. We see that the forecast for the residual is −4000 riders, implying that we should adjust the regression's forecast by reducing 4000 riders from that forecast.

Relation between Moving Average and Simple Exponential Smoothing

In both smoothing methods the user must specify a single parameter: In moving averages, the window width (w) must be set; in exponential smoothing, the smoothing parameter (α) must be set. In both cases, the parameter determines the importance of fresh information over older information. In fact, the two smoothers are approximately equal if the window width of the moving average is equal to $w = 2/\alpha - 1$.

17.4 ADVANCED EXPONENTIAL SMOOTHING

As mentioned earlier, both moving average and simple exponential smoothing should only be used for forecasting series with no trend or seasonality; series that have only a level and noise. One solution for forecasting series with trend and/or seasonality is first to remove those components (e.g., via regression models). Another solution is to use a more sophisticated version of exponential smoothing, which can capture trend and/or seasonality.

Series with a Trend

For series that contain a trend, we can use "double exponential smoothing." Unlike in regression models, the trend is not assumed to be global, but rather it can change over time. In double exponential smoothing, the local trend is estimated from the data and is updated as more data arrive. Similar to simple exponential smoothing, the level of the series is also estimated from the data, and is updated as more data arrive. The k-step-ahead forecast is given by combining the level estimate at time t (L_t) and the trend estimate at time t (T_t):

$$F_{t+k} = L_t + kT_t. \tag{17.3}$$

Note that in the presence of a trend, one- two- three-step-ahead (etc.) forecasts are no longer identical. The level and trend are updated through a pair of updating equations:

$$L_t = \alpha Y_t + (1 - \alpha)(L_{t-1} + T_{t-1}), \tag{17.4}$$

$$T_t = \beta (L_t - L_{t-1}) + (1 - \beta) T_{t-1}. \tag{17.5}$$

The first equation means that the level at time t is a weighted average of the actual value at time t and the level in the previous period, adjusted for trend (in the presence of a trend, moving from one period to the next requires factoring in the trend). The second equation means that the trend at time t is a weighted average of the trend in the previous period and the more recent information on the change in level.[2] Here there are two smoothing parameters, α and β, which determine the rate of learning. As in simple exponential smoothing, they are both constants between 0 and 1, set by the user, with higher values giving faster learning (more weight to most recent information).

---

[2] There are various ways to estimate the initial values L_1 and T_1, but the difference between these ways usually disappears after a few periods.

Series with a Trend and Seasonality

For series that contain both trend and seasonality, the "Holt–Winter's exponential smoothing" method can be used. This is a further extension of double exponential smoothing, where the k-step-ahead forecast also takes into account the season at period $t + k$. Assuming seasonality with M seasons (e.g., for weekly seasonality $M = 7$), the forecast is given by

$$F_{t+k} = (L_t + kT_t)\, S_{t+k-M}. \tag{17.6}$$

(Note that by the forecasting time t, the series must have included at least a full cycle of seasons in order to produce forecasts using this formula, i.e., $t > M$.)

Being an adaptive method, Holt–Winter's exponential smoothing allows the level, trend, and seasonality patterns to change over time. These three components are estimated and updated as more information arrives. The three updating equations are given by

$$L_t = \frac{\alpha Y_t}{S_{t-M}} + (1 - \alpha)(L_{t-1} + T_{t-1}), \tag{17.7}$$

$$T_t = \beta\,(L_t - L_{t-1}) + (1 - \beta)\,T_{t-1}, \tag{17.8}$$

$$S_t = \frac{\gamma Y_t}{L_t} + (1 - \gamma)S_{t-M}. \tag{17.9}$$

The first equation is similar to that in double exponential smoothing, except that it uses the seasonally adjusted value at time t rather than the raw value. This is done by dividing Y_t by its seasonal index, as estimated in the last cycle. The second equation is identical to double exponential smoothing. The third equation means that the seasonal index is updated by taking a weighted average of the seasonal index from the previous cycle and the current trend-adjusted value. Note that this formulation describes a multiplicative seasonal relationship, where values on different seasons differ by percentage amounts. There is also an additive seasonality version of Holt–Winter's exponential smoothing, where seasons differ by a constant amount (available also in XLMiner).

To illustrate forecasting a series with the Holt–Winter's method, consider the raw Amtrak ridership data. As we observed earlier, the data contain both a trend and monthly seasonality. Figure 17.7 shows part of XLMiner's output. We see the values of the three smoothing parameters (left at their defaults), and the chosen 12-month cycle.

Series with Seasonality (No Trend)

Finally, for series that contain seasonality but no trend, we can use a "Holt–Winter's" exponential smoothing formulation that lacks a trend term, by deleting the trend term in the forecasting equation and updating equations (in XLMiner this is called *Holt–Winter no trend*).

Output Navigator		
Inputs	Fitted Model	Forecast
Elapsed Time	Error Measures(Training)	Error Measures(Validation)

Inputs

Data	
# Records in training data	147
# Records in validation data	12
Input data	Data_PartitionTS1!$B$19:$D$177
Selected variable	Ridership

Parameters/Options	
Alpha (Level)	0.2
Beta (Trend)	0.15
Gamma (Seasonality)	0.05
Season length	12
Number of seasons	12
Forecast	Yes
#Forecasts	12

Fitted Model

Month	Actual	Forecast	Residuals
Jan-91	1708.917	1597.50951	111.407491
Feb-91	1620.586	1582.82506	37.7609411
Mar-91	1972.715	1902.76428	69.9507215
Apr-91	1811.665	1928.61318	-116.948181
May-91	1974.964	1942.29725	32.6667528
Jun-91	1862.356	1911.0349	-48.6789024
Jul-91	1939.86	2013.7081	-73.848097
Aug-91	2013.264	2047.5755	-34.3114987
Sep-91	1595.657	1699.83968	-104.182683
Oct-91	1724.924	1815.76327	-90.8392691
Nov-91	1675.667	1781.21666	-105.549661
Dec-91	1813.863	1794.56667	19.2963259
Jan-92	1614.827	1564.41763	50.4093668
Feb-92	1557.088	1526.81506	30.2729387
Mar-92	1891.223	1823.98406	67.2389446
Apr-92	1955.981	1830.72478	125.256218
May-92	1884.714	1893.90627	-9.19226749

Time Plot of Actual vs. Forecast (Training Data)

Error Measures (Training)

MAPE	2.89400095
MAD	51.3906221
MSE	4221.48184

Forecast

Month	Actual	Forecast	Error
Apr-03	2098.899	2110.41098	-11.5119757
May-03	2104.911	2157.68937	-52.7783682
Jun-03	2129.671	2123.6077	6.06329526
Jul-03	2223.349	2253.89368	-30.5446807
Aug-03	2174.36	2316.16999	-141.809986
Sep-03	1931.406	1938.84362	-7.43762261
Oct-03	2121.47	2112.97241	8.49758948
Nov-03	2076.054	2110.20872	-34.1547249
Dec-03	2140.677	2172.39818	-31.7211829
Jan-04	1831.508	1897.50816	-66.0001553
Feb-04	1838.006	1861.71438	-23.7083796
Mar-04	2132.446	2231.37186	-98.9258645

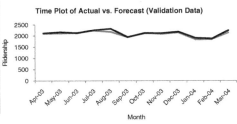

Time Plot of Actual vs. Forecast (Validation Data)

Error Measures (Validation)

MAPE	2.05679372
MAD	42.7628188
MSE	3416.88627

FIGURE 17.7 **PARTIAL OUTPUT FOR HOLT--WINTERS EXPONENTIAL SMOOTHING APPLIED TO AMTRAK RIDERSHIP SERIES**

PROBLEMS

17.1 Impact of September 11 on Air Travel in the United States: The Research and Innovative Technology Administration's Bureau of Transportation Statistics conducted a study to evaluate the impact of the September 11, 2001, terrorist attacks on U.S. transportation. The study report and the data can be found at http://www.bts.gov/publications/estimated_impacts_of_9_11_on_us_travel. The goal of the study was stated as follows:

> The purpose of this study is to provide a greater understanding of the passenger travel behavior patterns of persons making long distance trips before and after 9/11.

The report analyzes monthly passenger movement data between January 1990 and May 2004. Data on three monthly time series are given in file Sept11Travel.xls for this period: (1) actual airline revenue passenger miles (Air), (2) rail passenger miles (Rail), and (3) vehicle miles traveled (Car).

In order to assess the impact of September 11, BTS took the following approach: using data before September 11, they forecasted future data (under the assumption of no terrorist attack). Then, they compared the forecasted series with the actual data to assess the impact of the event. Our first step, therefore, is to split each of the time series into two parts: pre- and post- September 11. We now concentrate only on the earlier time series.

a. Create a time plot for the pre-event Air time series.

    i. What time series components appear from the plot?

b. Figure 17.8 is a time plot of the **seasonally adjusted** pre-September-11 Air series. Which of the following smoothing methods would be adequate for forecasting this series?

- Moving average (with what window width?)
- Simple exponential smoothing
- Double exponential smoothing
- Holt–Winter's exponential smoothing

17.2 Relation between Moving Average and Exponential Smoothing: Assume that we apply a moving average to a series, using a very short window span. If we wanted to achieve an equivalent result using simple exponential smoothing, what value should the smoothing coefficient take?

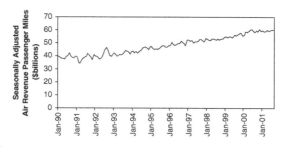

FIGURE 17.8   **SEASONALLY ADJUSTED PRE-SEPTEMBER-11 AIR SERIES**

17.3   **Forecasting with a Moving Average:** For a given time series of sales, the training set consists of 50 months. The first 5 months' data are shown below:

Month	Sales
Sept 98	27
Oct 98	31
Nov 98	58
Dec 98	63
Jan 99	59

  a. Compute the sales forecast for January 1999 based on a moving average model with span $W = 4$.

  b. Compute the forecast error for the above forecast.

17.4   **Seasonal Indexes for Monthly Series:** The seasonal indexes below were computed for a time series of monthly sales data:

Month	Jan	Feb	Mar	Apr	May	June	July	Aug	Sept	Oct	Nov	Dec
Index	0.795	0.851	0.926	1.022	1.006	1.243	1.318	1.164	0.992	0.915	0.851	0.807

  a. Are these additive or multiplicative indexes?

  b. If a moving average was applied to the raw data that were not adjusted for seasonality, would the forecasts for the month of June be on average above or below the real actual sale?

  c. The raw sales on a certain January were $139.7 million. Compute the seasonally adjusted sales for this month.

  d. Interpret the seasonal index for December.

17.5   **Optimizing Holt–Winter's Exponential Smoothing:** Figure 17.9 shows output from applying Holt–Winter's exponential smoothing to data, using "optimal" smoothing constants.

  a. In XLMiner the output smoothing constants are those that minimize (choose one of the following):

  • The MAPE of the training set

  • The MAPE of the validation set

Winters' exponential smoothing	
Smoothing constant(s)	
Level	1.000
Trend	0.000
Seasonality	0.246

FIGURE 17.9   **OPTIMIZED SMOOTHING CONSTANTS**

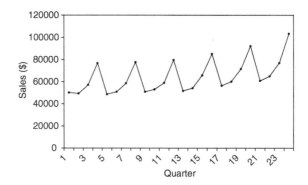

FIGURE 17.10 **DEPARTMENT STORE QUARTERLY SALES SERIES**

- The RMSE of the training set
- The RMSE of the validation set

b. The value of zero that is obtained for the trend smoothing constant means that (choose one of the following):

- There is no trend.
- The trend is estimated only from the first two points.
- The trend is updated throughout the data.
- The trend is statistically insignificant.

c. What is the danger of using the optimal smoothing constant values?

17.6 **Forecasting Department Store Sales:** The time series plot in Figure 17.10 describes actual quarterly sales for a department store over a 6-year period. (Data are available in DepartmentStoreSales.xls, data courtesy of Chris Albright.)

a. Which of the following methods would **not** be suitable for forecasting this series?

- Moving average of raw series
- Moving average of deseasonalized series
- Simple exponential smoothing of the raw series
- Holt's exponential smoothing of the raw series
- Holt–Winter's exponential smoothing of the raw series
- Regression model fit to the raw series
- Random-walk model fit to the raw series

b. The forecaster was tasked to generate forecasts for 4 quarters ahead. He therefore partitioned the data such that the last 4 quarters were designated as the validation period. The forecaster approached the forecasting task by using multiplicative Holt–Winter's exponential smoothing with smoothing parameters $\alpha = 0.2$, $\beta = 0.15$, $\gamma = 0.05$.

i. Run this method on the data.

ii. The forecasts for the validation set are given in Figure 17.11. Compute the MAPE values for the forecasts of quarters 21 and 22 for each of the two models (regression and exponential smoothing).

Quarter	Actual	Forecast	Error	LCI	UCI
21	60800	59319.9639	1480.03608	55146.12	63493.8078
22	64900	61682.6905	3217.30951	57508.8466	65856.5344
23	76997	71916.0826	5080.91737	67742.2387	76089.9266
24	103337	95159.689	8177.31103	90985.845	99333.5329

FIGURE 17.11   **FORECASTS FOR VALIDATION SERIES USING EXPONENTIAL SMOOTHING**

c. The fit and residuals from the exponential smoothing model are compared in Figure 17.12. Using all the information above, which model is more suitable for forecasting quarters 21 and 22?

17.7   Shipments of Household Appliances: The time plot in Figure 17.13 shows the series of quarterly shipments (in million dollars) of U.S. household appliances between 1985 and 1989. (Data are available in ApplianceShipments.xls, courtesy of Ken Black.)

a. Which of the following methods would be suitable for forecasting this series if applied to the raw data?

- Moving average
- Simple exponential smoothing
- Holt's exponential smoothing
- Holt–Winter's exponential smoothing

b. Apply a moving average with window span $W = 4$ to the data. Use all but that last year as the training set. Create a time plot of the moving average series.

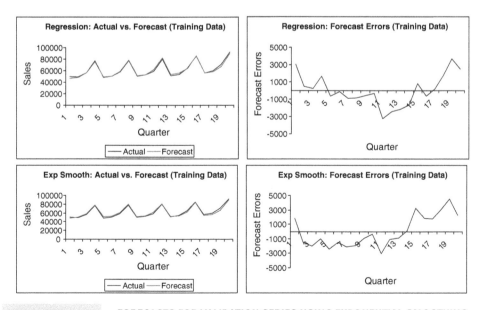

FIGURE 17.12   **FORECASTS FOR VALIDATION SERIES USING EXPONENTIAL SMOOTHING**

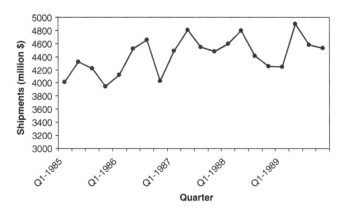

FIGURE 17.13   **QUARTERLY SHIPMENTS OF U.S. HOUSEHOLD APPLIANCES OVER 5 YEARS**

  i.   What does the MA(4) chart reveal?

 ii.   Use the MA(4) model to forecast appliance sales in Q1-1990.

iii.   Use the MA(4) model to forecast appliance sales in Q1-1991.

 iv.   Is the forecast for Q1-1990 most likely to underestimate, overestimate, or accurately estimate the actual sales on Q1-1990? Explain.

  v.   Management feels most comfortable with moving averages. The analyst therefore plans to use this method for forecasting future quarters. What else should be considered before using the MA(4) to forecast future quarterly shipments of household appliances?

c. We now focus on forecasting beyond 1989. In the following, continue to use all but the last year as the training set and the last four quarters as the validation set. First, fit a regression model to sales with a linear trend and quarterly seasonality to the training data. Next, apply Holt–Winter's exponential smoothing (with the default smoothing coefficient values) to the training data. Choose an adequate "season length."

  i.   Compute the MAPE for the validation data using the regression model.

 ii.   Compute the MAPE for the validation data using Holt–Winter's exponential smoothing.

iii.   Which model would you prefer to use for forecasting Q1-1990? Give three reasons.

 iv.   If we optimize the smoothing parameters in the Holt–Winter's method, is it likely to get values that are close to zero? Why or why not?

17.8 Forecasting Shampoo Sales: The time series plot in Figure 17.14 describes monthly sales of a certain shampoo over a 3-year period. [Data are available in ShampooSales .xls, source: R. J. Hyndman Time Series Data Library, http://www.robjhyndman .com/TSDL; accessed on December 28, 2009.]

a. Which of the following methods would be suitable for forecasting this series if applied to the raw data?

  • Moving average
  • Simple exponential smoothing

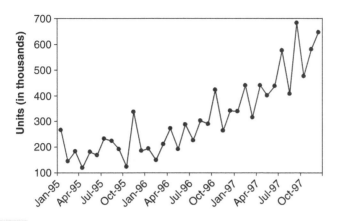

FIGURE 17.14   MONTHLY SALES OF A CERTAIN SHAMPOO

- Holt's exponential smoothing
- Holt–Winter's exponential smoothing

17.9 Forecasting Australian Wine Sales: Figure 17.15 shows time plots of monthly sales of six types of Australian wines (red, rose, sweet white, dry white, sparkling, and fortified) for 1980–1994. [Data are available in AustralianWines.xls, R. J. Hyndman, Time Series Data Library, http://www.robjhyndman.com/TSDL; accessed on December 28, 2009.] The units are thousands of liters. You are hired to obtain short-term forecasts (2–3 months ahead) for each of the six series, and this task will be repeated every month.

  a. Which forecasting method would you choose if you had to choose the same method for all series? Why?

  b. Fortified wine has the largest market share of the above six types of wine. You are asked to focus on fortified wine sales alone and produce an accurate as possible forecast for the next 2 months.

- Start by partitioning the data using the period until Dec-1993 as the training set.
- Apply Holt–Winter's exponential smoothing to sales with an appropriate season length (use the default values for the smoothing constants).

  c. Create an ACF plot for the residuals from the Holt–Winter's exponential smoothing until lag 12.

    i. Examining this plot, which of the following statements are reasonable?

- Decembers (month 12) are not captured well by the model.
- There is a strong correlation between sales on the same calendar month.
- The model does not capture the seasonality well.
- We should try to fit an autoregressive model with lag 12 to the residuals.
- We should first deseasonalize the data and then apply Holt–Winter's exponential smoothing.

    ii. How can you handle the above effect without adding another layer to your model?

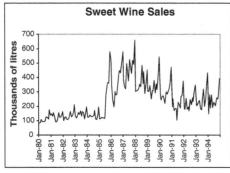

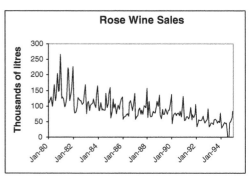

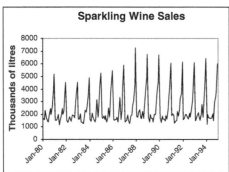

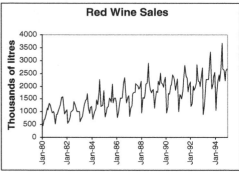

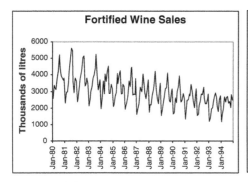

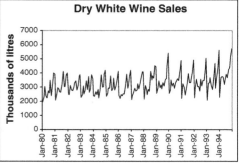

FIGURE 17.15   MONTHLY SALES OF SIX TYPES OF AUSTRALIAN WINES BETWEEN 1980 AND 1994

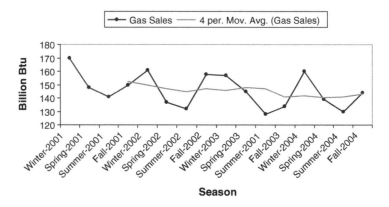

FIGURE 17.16   QUARTERLY SALES OF NATURAL GAS OVER 4 YEARS

17.10 **Natural Gas Sales:** Figure 17.16 is a time plot of quarterly natural gas sales (in billions of Btus) of a certain company, over a period of 4 years. (Data courtesy of George McCabe.) The company's analyst is asked to use a moving average model to forecast sales in Winter 2005.

    a. Reproduce the time plot with the overlaying MA(4) line (use Excel's *Add trendline*).

    b. What can we learn about the series from the MA line?

    c. Run a moving average forecaster with adequate season length. Are forecasts generated by this method expected to overforecast, underforecast, or accurately forecast actual sales? Why?

## PART SEVEN

Cases

CHAPTER 18

Cases

18.1 CHARLES BOOK CLUB

CharlesBookClub.xls is the dataset for this case study.

The Book Industry

Approximately 50,000 new titles, including new editions, are published each year in the United States, giving rise to a $25 billion industry in 2001.[1] In terms of percentage of sales, this industry may be segmented as follows:

16%	Textbooks
16%	Trade books sold in bookstores
21%	Technical, scientific, and professional books
10%	Book clubs and other mail-order books
17%	Mass-market paperbound books
20%	All other books

Book retailing in the United States in the 1970s was characterized by the growth of bookstore chains located in shopping malls. The 1980s saw increased purchases in bookstores stimulated through the widespread practice of discounting. By the 1990s, the superstore concept of book retailing gained acceptance and contributed to double-digit growth of the

---

[1] The Charles Book Club case was derived, with the assistance of Ms. Vinni Bhandari, from *The Bookbinders Club, a Case Study in Database Marketing*, prepared by Nissan Levin and Jacob Zahavi, Tel Aviv University; used with permission.

Data Mining for Business Intelligence, By Galit Shmueli, Nitin R. Patel, and Peter C. Bruce
Copyright © 2010 John Wiley & Sons Inc.

367

book industry. Conveniently situated near large shopping centers, superstores maintain large inventories of 30,000–80,000 titles and employ well-informed sales personnel. Superstores apply intense competitive pressure on book clubs and mail-order firms as well on as traditional book retailers. In response to these pressures, book clubs have sought out alternative business models that were more responsive to their customers' individual preferences.

Historically, book clubs offered their readers different types of membership programs. Two common membership programs are the continuity and negative option programs, which extended contractual relationships between the club and its members. Under a *continuity program*, a reader signs up by accepting an offer of several books for just a few dollars (plus shipping and handling) and an agreement to receive a shipment of one or two books each month thereafter at more standard pricing. The continuity program was most common in the children's book market, where parents are willing to delegate the rights to the book club to make a selection, and much of the club's prestige depends on the quality of its selections.

In a *negative option program*, readers get to select which and how many additional books they would like to receive. However, the club's selection of the month is delivered to them automatically unless they specifically mark "no" on their order form by a deadline date. Negative option programs sometimes result in customer dissatisfaction and always give rise to significant mailing and processing costs.

In an attempt to combat these trends, some book clubs have begun to offer books on a *positive option basis*, but only to specific segments of their customer base that are likely to be receptive to specific offers. Rather than expanding the volume and coverage of mailings, some book clubs are beginning to use database-marketing techniques to target customers more accurately. Information contained in their databases is used to identify who is most likely to be interested in a specific offer. This information enables clubs to design special programs carefully tailored to meet their customer segments' varying needs.

Database Marketing at Charles

The Club   The Charles Book Club (CBC) was established in December 1986 on the premise that a book club could differentiate itself through a deep understanding of its customer base and by delivering uniquely tailored offerings. CBC focused on selling specialty books by direct marketing through a variety of channels, including media advertising (TV, magazines, newspapers) and mailing. CBC is strictly a distributor and does not publish any of the books that it sells. In line with its commitment to understanding its customer base, CBC built and maintained a detailed database about its club members. Upon enrollment, readers were required to fill out an insert and mail it to CBC. Through this process, CBC created an active database of 500,000 readers; most were acquired through advertising in specialty magazines.

The Problem   CBC sent mailings to its club members each month containing the latest offerings. On the surface, CBC appeared very successful: Mailing volume was increasing, book selection was diversifying and growing, and its customer database was increasing. However, its bottom-line profits were falling. The decreasing profits led CBC to revisit its original plan of using database marketing to improve mailing yields and to stay profitable.

A Possible Solution   CBC embraced the idea of deriving intelligence from its data to allow CBC to know its customers better and enable multiple targeted campaigns where each target audience would receive appropriate mailings. CBC's management decided to focus its efforts on the most profitable customers and prospects, and to design targeted marketing strategies to best reach them. The two processes CBC had in place were:

1. Customer acquisition
   - New members would be acquired by advertising in specialty magazines, newspapers, and on TV.
   - Direct mailing and telemarketing would contact existing club members.
   - Every new book would be offered to club members before general advertising.
2. Data collection
   - All customer responses would be recorded and maintained in the database.
   - Any information not being collected that is critical would be requested from the customer.

For each new title, CBC decided to use a two-step approach:

1. Conduct a market test involving a random sample of 7000 customers from the database to enable analysis of customer responses. The analysis would create and calibrate response models for the current book offering.
2. Based on the response models, compute a score for each customer in the database. Use this score and a cutoff value to extract a target customer list for direct-mail promotion.

Targeting promotions was considered to be of prime importance. Other opportunities to create successful marketing campaigns based on customer behavior data (returns, inactivity, complaints, compliments, etc.) would be addressed by CBC at a later stage.

TABLE 18.1	LIST OF VARIABLES IN CHARLES BOOK CLUB DATASET

Variable Name	Description
Seq#	Sequence number in the partition
ID#	Identification number in the full (unpartitioned) market test dataset
Gender	0 = Male 1 = Female
M	Monetary—Total money spent on books
R	Recency—Months since last purchase
F	Frequency—Total number of purchases
FirstPurch	Months since first purchase
ChildBks	Number of purchases from the category child books
YouthBks	Number of purchases from the category youth books
CookBks	Number of purchases from the category cookbooks
DoItYBks	Number of purchases from the category do-it-yourself books
RefBks	Number of purchases from the category reference books (atlases, encyclopedias, dictionaries)
ArtBks	Number of purchases from the category art books
GeoBks	Number of purchases from the category geography books
ItalCook	Number of purchases of book title *Secrets of Italian Cooking*
ItalAtlas	Number of purchases of book title *Historical Atlas of Italy*
ItalArt	Number of purchases of book title *Italian Art*
Florence	= 1, *The Art History of Florence* was bought; = 0 if not
Related Purchase	Number of related books purchased

Art History of Florence   A new title, *The Art History of Florence*, is ready for release. CBC sent a test mailing to a random sample of 4000 customers from its customer base. The customer responses have been collated with past purchase data. The dataset has been randomly partitioned into three parts: *Training Data* (1800 customers)—initial data to be used to fit response models; *Validation Data* (1400 customers)—holdout data used to compare the performance of different response models; and *Test Data* (800 customers)—data to be used only after a final model has been selected to estimate the probable performance of the model when it is deployed. Each row (or case) in the spreadsheet (other than the header) corresponds to one market test customer. Each column is a variable, with the header row giving the name of the variable. The variable names and descriptions are given in Table 18.1.

Data Mining Techniques

Various data mining techniques can be used to mine the data collected from the market test. No one technique is universally better than another. The particular context and the particular characteristics of the data are the major factors in determining which techniques perform better in an application. For this assignment we focus on two fundamental techniques: *k*-nearest neighbors and logistic regression. We compare them with each other as well as with a standard industry practice known as *RFM (recency, frequency, monetary) segmentation*.

RFM Segmentation   The segmentation process in database marketing aims to partition customers in a list of prospects into homogeneous groups (segments) that are similar with respect to buying behavior. The homogeneity criterion we need for segmentation is the propensity to purchase the offering. But since we cannot measure this attribute, we use variables that are plausible indicators of this propensity.

In the direct marketing business the most commonly used variables are the *RFM variables*:

R *recency*, time since last purchase
F *frequency*, number of previous purchases from the company over a period
M *monetary*, amount of money spent on the company's products over a period

The assumption is that the more recent the last purchase, the more products bought from the company in the past, and the more money spent in the past buying the company's products, the more likely the customer is to purchase the product offered.

The 1800 observations in the training data and the 1400 observations in the validation data have been divided into recency, frequency, and monetary categories as follows:

Recency

0–2 months (Rcode = 1)
3–6 months (Rcode = 2)
7–12 months (Rcode = 3)
13 months and up (Rcode = 4)

Frequency

1 book (Fcode = 1)
2 books (Fcode = 2)
3 books and up (Fcode = 3)

Monetary

$0–$25 (Mcode = 1)
$26–$50 (Mcode = 2)
$51–$100 (Mcode = 3)
$101–$200 (Mcode = 4)
$201 and up (Mcode = 5)

Tables 18.2 and 18.3 display the 1800 customers in the training data cross-tabulated by these categories. Both buyers and nonbuyers are summarized. These tables are available for Excel computations in the RFM spreadsheet in the data file.

TABLE 18.2   **RFM COUNTS FOR BUYERS**

Rcode=all Sum of Florence	Mcode					
Fcode	1	2	3	4	5	Grand Total
1	2	2	10	7	17	38
2		3	5	9	17	34
3		1	1	15	62	79
Grand Total	2	6	16	31	96	151

Rcode=1 Sum of Florence	Mcode					
Fcode	1	2	3	4	5	Grand Total
1	0	0	0	2	1	3
2		1	0	0	1	2
3		1	0	0	5	6
Grand Total	0	2	0	2	7	11

Rcode=2 Sum of Florence	Mcode					
Fcode	1	2	3	4	5	Grand Total
1	1	0	1	1	5	8
2		0	3	5	5	13
3			0	4	10	14
Grand Total	1	0	4	10	20	35

Rcode=3 Sum of Florence	Mcode					
Fcode	1	2	3	4	5	Grand Total
1	1	0	1	2	5	9
2		1	1	2	4	8
3		0	0	4	31	35
Grand Total	1	1	2	8	40	52

Rcode=4 Sum of Florence	Mcode					
Fcode	1	2	3	4	5	Grand Total
1	0	2	8	2	6	18
2		1	1	2	7	11
3			1	7	16	24
Grand Total	0	3	10	11	29	53

TABLE 18.3   RFM COUNTS FOR ALL CUSTOMERS (BUYERS AND NONBUYERS)

Rcode = all
Count of Florence | Mcode | | | | |

Fcode	1	2	3	4	5	Grand Total
1	20	40	93	166	219	538
2		32	91	180	247	550
3		2	33	179	498	712
Grand Total	20	74	217	525	964	1800

Rcode = 1
Count of Florence | Mcode | | | | |

Fcode	1	2	3	4	5	Grand Total
1	2	2	6	10	15	35
2		3	4	12	16	35
3		1	2	11	45	59
Grand Total	2	6	12	33	76	129

Rcode = 2
Count of Florence | Mcode | | | | |

Fcode	1	2	3	4	5	Grand Total
1	3	5	17	28	26	79
2		2	17	30	31	80
3			3	34	66	103
Grand Total	3	7	37	92	123	262

Rcode = 3
Count of Florence | Mcode | | | | |

Fcode	1	2	3	4	5	Grand Total
1	7	15	24	51	86	183
2		12	29	55	85	181
3		1	17	53	165	236
Grand Total	7	28	70	159	336	600

Rcode = 4
Count of Florence | Mcode | | | | |

Fcode	1	2	3	4	5	Grand Total
1	8	18	46	77	92	241
2		15	41	83	115	254
3			11	81	222	314
Grand Total	8	33	98	241	429	809

Assignment

1. What is the response rate for the training data customers taken as a whole? What is the response rate for each of the $4 \times 5 \times 3 = 60$ combinations of

RFM categories? Which combinations have response rates in the training data that are above the overall response in the training data?

2. Suppose that we decide to send promotional mail only to the "above-average" RFM combinations identified in part 1. Compute the response rate in the validation data using these combinations.

3. Rework parts 1 and 2 with three segments:

> *Segment 1:* Consisting of RFM combinations that have response rates that exceed twice the overall response rate
>
> *Segment 2:* Consisting of RFM combinations that exceed the overall response rate but do not exceed twice that rate
>
> *Segment 3:* Consisting of the remaining RFM combinations

Draw the cumulative lift curve (consisting of three points for these three segments) showing the number of customers in the validation dataset on the *x* axis and cumulative number of buyers in the validation dataset on the *y* axis.

The *k*-Nearest Neighbors   The *k*-nearest neighbor technique can be used to create segments based on product proximity to similar products of the products offered as well as the propensity to purchase (as measured by the RFM variables). For *The Art History of Florence*, a possible segmentation by product proximity could be created using the following variables:

> *M:* Monetary—total money (in dollars) spent on books
>
> *R:* Recency—months since last purchase
>
> *F:* Frequency—total number of past purchases
>
> *FirstPurch:* Months since first purchase
>
> *RelatedPurch:* Total number of past purchases of related books (i.e., sum of purchases from the art and geography categories and of titles *Secrets of Italian Cooking, Historical Atlas of Italy,* and *Italian Art*).

4. Use the *k*-nearest neighbor option under the *Classify* menu choice in XLMiner to classify cases with $k = 1$, $k = 3$, and $k = 11$. Use normalized data (note the checkbox "normalize input data" in the dialog box) and all five variables.

5. Use the *k*-nearest neighbor option under the *Prediction* menu choice in XLMiner to compute a cumulative gains curve for the validation data for $k = 1$, $k = 3$, and $k = 11$. Use normalized data (note the checkbox "normalize input data" in the dialog box) and all five variables. The *k*NN prediction algorithm gives a numerical value, which is a weighted average of the values of the Florence variable for the *k*-nearest neighbors with weights that are inversely proportional to distance.

Logistic Regression   The logistic regression model offers a powerful method for modeling response because it yields well-defined purchase probabilities. (The model is especially attractive in consumer choice settings because it can be derived from the random utility theory of consumer behavior under the assumption that the error term in the customer's utility function follows a type I extreme value distribution.)

Use the training set data of 1800 observations to construct three logistic regression models with:

- The full set of 15 predictors in the dataset as independent variables and Florence as the dependent variable
- A subset that you judge to be the best
- Only the R, F, and M variables

6. Score the customers in the validation sample and arrange them in descending order of purchase probabilities.

7. Create a cumulative gains chart summarizing the results from the three logistic regression models created above, along with the expected cumulative gains for a random selection of an equal number of customers from the validation dataset.

8. If the cutoff criterion for a campaign is a 30% likelihood of a purchase, find the customers in the validation data that would be targeted and count the number of buyers in this set.

18.2 GERMAN CREDIT

GermanCredit.xls is the dataset for this case study

The German Credit dataset[2] has 30 variables and 1000 records, each record being a prior applicant for credit. Each applicant was rated as "good credit" (700 cases) or "bad credit" (300 cases).

New applicants for credit can also be evaluated on these 30 predictor variables and classified as a good or a bad credit risk based on the predictor variables. All the variables are explained in Table 18.4.

Note: The original dataset had a number of categorical variables, some of which have been transformed into a series of binary variables so that they can be handled appropriately by XLMiner. Several ordered categorical variables have been left as is, to be treated by XLMiner as numerical.

---

[2] This is available from ftp.ics.uci.edu/pub/machine-learning-databases/statlog.

| | **TABLE 18.4** | **VARIABLES FOR THE GERMAN CREDIT DATASET** | | |

Var.	Variable Name	Description	Variable Type	Code Description
1	OBS#	Observation numbers	Categorical	Sequence number in dataset
2	CHK_ACCT	Checking account status	Categorical	0 :< 0 DM
				1: 0 ⇐ ⋯ < 200 DM
				2 :⇒ 200 DM
				3: no checking account
3	DURATION	Duration of credit in months	Numerical	
4	HISTORY	Credit history	Categorical	0: no credits taken
				1: all credits at this bank paid back duly
				2: existing credits paid back duly until now
				3: delay in paying off in the past
				4: critical account
5	NEW_CAR	Purpose of credit	Binary	car (new) 0: No, 1: Yes
6	USED_CAR	Purpose of credit	Binary	car (used) 0: No, 1: Yes
7	FURNITURE	Purpose of credit	Binary	furniture/equipment 0: No, 1: Yes
8	RADIO/TV	Purpose of credit	Binary	radio/television 0: No, 1: Yes
9	EDUCATION	Purpose of credit	Binary	education 0: No, 1: Yes
10	RETRAINING	Purpose of credit	Binary	retraining 0: No, 1: Yes
11	AMOUNT	Credit amount	Numerical	
12	SAV_ACCT	Average balance in savings account	Categorical	0 :< 100 DM
				1 : 100 <= ⋯ < 500 DM
				2 : 500 <= ⋯ < 1000 DM
				3 :⇒ 1000 DM
				4 : unknown/ no savings account
13	EMPLOYMENT	Present employment since	Categorical	0 : unemployed
				1: < 1 year
				2 : 1 <= ⋯ < 4 years
				3 : 4 <= ⋯ < 7 years
				4 : >= 7 years
14	INSTALL_RATE	Installment rate as % of disposable income	Numerical	
15	MALE_DIV	Applicant is male and divorced	Binary	0: No, 1:Yes

(continued)

TABLE 18.4   (*CONTINUED*)

Var.	Variable Name	Description	Variable Type	Code Description
16	MALE_SINGLE	Applicant is male and single	Binary	0: No, 1:Yes
17	MALE_MAR_WID	Applicant is male and married or a widower	Binary	0: No, 1:Yes
18	CO-APPLICANT	Application has a coapplicant	Binary	0: No, 1:Yes
19	GUARANTOR	Applicant has a guarantor	Binary	0: No, 1:Yes
20	PRESENT_RESIDENT	Present resident since (years)	Categorical	0 :<= 1 year 1 < ··· <= 2 years 2 < ··· <= 3 years 3 :> 4 years
21	REAL_ESTATE	Applicant owns real estate	Binary	0: No, 1:Yes
22	PROP_UNKN_NONE	Applicant owns no property (or unknown)	Binary	0: No, 1:Yes
23	AGE	Age in years	Numerical	
24	OTHER_INSTALL	Applicant has other installment plan credit	Binary	0: No, 1:Yes
25	RENT	Applicant rents	Binary	0: No, 1:Yes
26	OWN_RES	Applicant owns residence	Binary	0: No, 1:Yes
27	NUM_CREDITS	Number of existing credits at this bank	Numerical	
28	JOB	Nature of job	Categorical	0 : unemployed/ unskilled— non-resident 1 : unskilled— resident 2 : skilled employee/ official 3 : management/ self-employed/ highly qualified employee/officer
29	NUM_DEPENDENTS	Number of people for whom liable to provide maintenance	Numerical	
30	TELEPHONE	Applicant has phone in his or her name	Binary	0: No, 1:Yes
31	FOREIGN	Foreign worker	Binary	0: No, 1:Yes
32	RESPONSE	Good credit rating	Binary	0: No, 1:Yes

OBS#	CHK_ACCT	DURATION	HISTORY	NEW_CAR	USED_CAR	FURNITURE	RADIO/TV	EDUCATION	RETRAINING	AMOUNT	SAV_ACCT	EMPLOYMENT	INSTALL_RATE	MALE_DIV
1	0	6	4	0	0	0	1	0	0	1169	4	4	4	0
2	1	48	2	0	0	0	1	0	0	5951	0	2	2	0
3	3	12	4	0	0	0	0	1	0	2096	0	3	2	0
4	0	42	2	0	0	1	0	0	0	7882	0	3	2	0

MALE_MAR_or_WID	CO-APPLICANT	GUARANTOR	PRESENT_RESIDENT	REAL_ESTATE	PROP_UNKN_NONE	AGE	OTHER_INSTALL	RENT	OWN_RES	NUM_CREDITS	JOB	NUM_DEPENDENTS	TELEPHONE	FOREIGN
0	0	0	4	1	0	67	0	0	1	2	2	1	1	0
0	0	0	2	1	0	22	0	0	1	1	2	1	0	0
0	0	0	3	1	0	49	0	0	1	1	1	2	0	0
0	0	1	4	0	0	45	0	0	1	1	2	2	0	0

FIGURE 18.1   DATA SAMPLE (FIRST SEVERAL ROWS)

Figure 18.1 shows the values of these variables for the first several records in the case.

The consequences of misclassification have been assessed as follows: The costs of a false positive (incorrectly saying that an applicant is a good credit risk) outweigh the benefits of a true positive (correctly saying that an applicant is a good credit risk) by a factor of 5. This is summarized in Table 18.5. The opportunity cost table was derived from the average net profit per loan as shown in Table 18.6.

Because decision makers are used to thinking of their decision in terms of net profits, we use these tables in assessing the performance of the various models.

TABLE 18.5   OPPORTUNITY COST TABLE (DEUTSCHE MARKS)

	Predicted (Decision)	
Actual	Good (Accept)	Bad (Reject)
Good	0	100
Bad	500	0

TABLE 18.6   AVERAGE NET PROFIT (DEUTSCHE MARKS)

	Predicted (Decision)	
Actual	Good (Accept)	Bad (Reject)
Good	100	0
Bad	−500	0

Assignment

1. Review the predictor variables and guess what their role in a credit decision might be. Are there any surprises in the data?

2. Divide the data into training and validation partitions, and develop classification models using the following data mining techniques in XLMiner: logistic regression, classification trees, and neural networks.

3. Choose one model from each technique and report the confusion matrix and the cost/gain matrix for the validation data. Which technique has the most net profit?

4. Let us try and improve our performance. Rather than accept XLMiner's initial classification of all applicants' credit status, use the "predicted probability of success" in logistic regression (where *success* means 1) as a basis for selecting the best credit risks first, followed by poorer risk applicants.

   a. Sort the validation on "predicted probability of success."

   b. For each case, calculate the net profit of extending credit.

   c. Add another column for cumulative net profit.

   d. How far into the validation data do you go to get maximum net profit? (Often, this is specified as a percentile or rounded to deciles.)

   e. If this logistic regression model is scored to future applicants, what "probability of success" cutoff should be used in extending credit?

18.3 TAYKO SOFTWARE CATALOGER

Tayko.xls is the dataset for this case study.

Background

Tayko is a software catalog firm that sells games and educational software.[3] It started out as a software manufacturer and later added third-party titles to its offerings. It has recently put together a revised collection of items in a new catalog, which it is preparing to roll out in a mailing.

In addition to its own software titles, Tayko's customer list is a key asset. In an attempt to expand its customer base, it has recently joined a consortium of catalog firms that specialize in computer and software products. The consortium affords members the opportunity to mail catalogs to names drawn from a pooled list of customers. Members supply their own customer lists to the pool and can "withdraw" an equivalent number of names each quarter. Members are allowed

---

[3] Resampling Stats, Inc. 2006; used with permission.

to do predictive modeling on the records in the pool so they can do a better job of selecting names from the pool.

The Mailing Experiment

Tayko has supplied its customer list of 200,000 names to the pool, which totals over 5 million names, so it is now entitled to draw 200,000 names for a mailing. Tayko would like to select the names that have the best chance of performing well, so it conducts a test—it draws 20,000 names from the pool and does a test mailing of the new catalog.

This mailing yielded 1065 purchasers, a response rate of 0.053. Average spending was $103 for each of the purchasers, or $5.46 per catalog mailed. To optimize the performance of the data mining techniques, it was decided to work with a stratified sample that contained equal numbers of purchasers and nonpurchasers. For ease of presentation, the dataset for this case includes just 1000 purchasers and 1000 nonpurchasers, an apparent response rate of 0.5. Therefore, after using the dataset to predict who will be a purchaser, we must adjust the purchase rate back down by multiplying each case's "probability of purchase" by 0.053/0.5, or 0.107.

Data

There are two response variables in this case. *Purchase* indicates whether or not a prospect responded to the test mailing and purchased something. *Spending* indicates, for those who made a purchase, how much they spent. The overall procedure in this case will be to develop two models. One will be used to classify records as Purchase or No purchase. The second will be used for those cases that are classified as *purchase* and will predict the amount they will spend.

Table 18.7 provides a description of the variables available in this case. A partition variable is used because we will be developing two different models in this case and want to preserve the same partition structure for assessing each model.

Figure 18.2 shows the first few rows of data (the top shows the sequence number plus the first 14 variables, and the bottom shows the remaining 11 variables for the same rows).

Assignment

1. Each catalog costs approximately $2 to mail (including printing, postage, and mailing costs). Estimate the gross profit that the firm could expect from the remaining 180,000 names if it selected them randomly from the pool.

TABLE 18.7 **DESCRIPTION OF VARIABLES FOR TAYKO DATASET**

Var.	Variable Name	Description	Variable Type	Code Description
1	US	Is it a U.S. address?	Binary	1: yes 0: no
2–16	Source_*	Source catalog for the record (15 possible sources)	Binary	1: yes 0: no
17	Freq.	Number of transactions in last year at source catalog	Numerical	
18	last_update_days_ago	How many days ago last update was made to customer record	Numerical	
19	1st_update_days_ago	How many days ago first update was made to customer record	Numerical	
20	RFM%	Recency–frequency–monetary percentile, as reported by source catalog (see Section 18.1)	Numerical	
21	Web_order	Customer placed at least one order via Web	Binary	1: yes 0: no
22	Gender=mal	Customer is male	Binary	1: yes 0: no
23	Address_is_res	Address is a residence	Binary	1: yes 0: no
24	Purchase	Person made purchase in test mailing	Binary	1: yes 0: no
25	Spending	Amount (dollars) spent by customer in test mailing	Numerical	
26	Partition	Variable indicating which partition the record will be assigned to	Alphabetical	t: training v: validation s: test

2. Develop a model for classification of a customer as a purchaser or non-purchaser.

   a. Partition the data into training data on the basis of the partition variable, which has 800 t's, 700 v's, and 500 s's (training data, validation data, and test data, respectively) assigned randomly to cases.

   b. Using the "best subset" option in logistic regression, implement the full logistic regression model, select the best subset of variables, and then implement a regression model with just those variables to classify the data into purchasers and nonpurchasers. (Logistic regression is used because it yields an estimated "probability of purchase," which is required later in the analysis.)

3. Develop a model for predicting spending among the purchasers.

   a. Make a copy of the data sheet (call it data2), sort by the Purchase variable, and remove the records where Purchase = 0 (the resulting spreadsheet will contain only purchasers).

sequence_number	US	source_a	source_c	source_b	source_d	source_e	source_m	source_o	source_h	source_r	source_s	source_t	source_u	source_p
1	1	0	0	1	0	0	0	0	0	0	0	0	0	0
2	1	0	0	0	0	1	0	0	0	0	0	0	0	0
3	1	0	0	0	0	0	0	0	0	0	0	1	0	0
4	1	0	1	0	0	0	0	0	0	0	0	0	0	0
5	1	0	1	0	0	0	0	0	0	0	0	0	0	0
6	1	0	0	0	0	0	0	0	0	1	0	0	0	0
7	1	0	0	0	0	0	0	0	0	0	0	0	0	0
8	1	0	0	1	0	0	0	0	0	0	0	0	0	0
9	1	1	0	0	0	0	0	0	0	0	0	0	0	0
10	1	1	0	0	0	0	0	0	0	0	0	0	0	0

source_x	source_w	Freq	last_update_days_ago	1st_update_days_ago	Web order	Gender=male	Address_is_res	Purchase	Spending	Partition
0	0	2	3662	3662	1	0	1	1	128	s
0	0	0	2900	2900	1	1	0	0	0	s
0	0	2	3883	3914	0	0	0	1	127	t
0	0	1	829	829	0	1	0	0	0	s
0	0	1	869	869	0	0	0	0	0	t
0	0	1	1995	2002	0	0	1	0	0	s
0	1	2	1498	1529	0	0	1	0	0	s
0	0	1	3397	3397	0	1	0	0	0	t
0	0	4	525	2914	1	1	0	1	489	t
0	0	1	3215	3215	0	0	0	1	174	v

FIGURE 18.2   DATA FOR FIRST 10 RECORDS

b. Partition this dataset into training and validation partitions on the basis of the partition variable.

c. Develop models for predicting spending, using:
   i. Multiple linear regression (use best subset selection)
   ii. Regression trees

d. Choose one model on the basis of its performance with the validation data.

4. Return to the original test data partition. Note that this test data partition includes both purchasers and nonpurchasers. Note also that although it contains the scoring of the chosen classification model, we have not used this partition in our analysis up to this point, so it will give an unbiased estimate of the performance of our models. It is best to make a copy of the test data portion of this sheet to work with, since we will be adding analysis to it. This copy is called *Score Analysis*.

a. Copy to this sheet the "predicted probability of success" (Success = Purchase) column from the classification of test data.

b. Score to this data sheet the prediction model chosen.

c. Arrange the following columns so that they are adjacent:
   i. Predicted probability of purchase (Success)
   ii. Actual spending (dollars)
   iii. Predicted spending (dollars)

d. Add a column for "adjusted probability of purchase" by multiplying "predicted probability of purchase" by 0.107. *This is to adjust for over-sampling the purchasers* (see above).

e. Add a column for expected spending (adjusted probability of purchase × predicted spending).

f. Sort all records on the "expected spending" column.

g. Calculate cumulative lift [= cumulative "actual spending" divided by the average spending that would result from random selection (each adjusted by 0.107)].

h. Using this cumulative lift curve, estimate the gross profit that would result from mailing to the 180,000 names on the basis of your data mining models.

Note: Although Tayko is a hypothetical company, the data in this case (modified slightly for illustrative purposes) were supplied by a real company that sells software through direct sales. The concept of a catalog consortium is based on the Abacus Catalog Alliance. Details can be found at www.doubleclick.com/us/solutions/marketers/database/catalog/.

18.4   SEGMENTING CONSUMERS OF BATH SOAP

BathSoap.xls is the dataset for this case study.

Business Situation

CRISA is an Asian market research agency that specializes in tracking consumer purchase behavior in consumer goods (both durable and nondurable).[4] In one major research project, CRISA tracks numerous consumer product categories (e.g. "detergents") and, within each category, perhaps dozens of brands. To track purchase behavior, CRISA constituted household panels in over 100 cities and towns in India, covering most of the Indian urban market. The households were carefully selected using stratified sampling to ensure a representative sample; a

---

[4] Cytel, Inc. and Resampling Stats, Inc. 2006; used with permission.

subset of 600 records is analyzed here. The strata were defined on the basis of socioeconomic status and the market (a collection of cities).

CRISA has both transaction data (each row is a transaction) and household data (each row is a household), and for the household data, maintains the following information:

- Demographics of the households (updated annually)
- Possession of durable goods (car, washing machine, etc., updated annually; an "affluence index" is computed from this information)
- Purchase data of product categories and brands (updated monthly)

CRISA has two categories of clients: (1) advertising agencies that subscribe to the database services, obtain updated data every month, and use the data to advise their clients on advertising and promotion strategies; (2) and consumer goods manufacturers, which monitor their market share using the CRISA database.

Key Problems

CRISA has traditionally segmented markets on the basis of purchaser demographics. It would now like to segment the market based on two key sets of variables more directly related to the purchase process and to brand loyalty:

1. Purchase behavior (volume, frequency, susceptibility to discounts, and brand loyalty)
2. Basis of purchase (price, selling proposition)

Doing so would allow CRISA to gain information about what demographic attributes are associated with different purchase behaviors and degrees of brand loyalty, and thus deploy promotion budgets more effectively. More effective market segmentation would enable CRISA's clients (in this case, a firm called IMRB) to design more cost-effective promotions targeted at appropriate segments. Thus, multiple promotions could be launched, each targeted at different market segments at different times of the year. This would result in a more cost-effective allocation of the promotion budget to different market segments. It would also enable IMRB to design more effective customer reward systems and thereby increase brand loyalty.

Data

The data in Table 18.8 profile each household, each row containing the data for one household.

| TABLE 18.8 | DESCRIPTION OF VARIABLES FOR EACH HOUSEHOLD |

Variable Type	Variable Name	Description
Member ID	Member id	Unique identifier for each household
Demographics	SEC	Socioeconomic class (1 = high, 5 = low)
	FEH	Eating habits (1 =vegetarian, 2 = vegetarian but eat eggs, 3 =nonvegetarian, 0 = not specified)
	MT	Native language (see table in worksheet)
	SEX	Gender of homemaker (1=male, 2=female)
	AGE	Age of homemaker
	EDU	Education of homemaker (1 = minimum, 9 = maximum)
	HS	Number of members in household
	CHILD	Presence of children in household (4 categories)
	CS	Television availability (1 = available, 2 = unavailable)
	Affluence Index	Weighted value of durables possessed
Purchase summary over the period	No. of Brands	Number of brands purchased
	Brand Runs	Number of instances of consecutive purchase of brands
	Total Volume	Sum of volume
	No. of Trans	Number of purchase transactions (multiple brands purchased in a month are counted as separate transactions
	Value	Sum of value
	Trans/ Brand Runs	Average transactions per brand run
	Vol/Trans	Average volume per transaction
	Avg. Price	Average price of purchase
Purchase within promotion	Pur Vol	Percent of volume purchased
	No Promo - %	Percent of volume purchased under no promotion
	Pur Vol Promo 6%	Percent of volume purchased under promotion code 6
	Pur Vol Other Promo %	Percent of volume purchased under other promotions
Brandwise purchase	Br. Cd. (57, 144), 55, 272, 286, 24, 481, 352, 5, and 999 (others)	Percent of volume purchased of the brand
Price categorywise purchase	Price Cat 1 to 4	Percent of volume purchased under the price category
Selling propositionwise purchase	Proposition Cat 5 to 15	Percent of volume purchased under the product proposition category

Measuring Brand Loyalty

Several variables in this case measure aspects of brand loyalty. The number of different brands purchased by the customer is one measure. However, a consumer who purchases one or two brands in quick succession, then settles on a third for a long streak, is different from a consumer who constantly switches back and forth among three brands. How often customers switch from one brand to another is another measure of loyalty. Yet a third perspective on the same issue is the proportion of purchases that go to different brands—a consumer who spends 90% of his or her purchase money on one brand is more loyal than a consumer who spends more equally among several brands.

All three of these components can be measured with the data in the purchase summary worksheet.

Assignment

1. Use k-means clustering to identify clusters of households based on:
   a. The variables that describe purchase behavior (including brand loyalty)
   b. The variables that describe the basis for purchase
   c. The variables that describe both purchase behavior and basis of purchase

   Note 1: How should k be chosen? Think about how the clusters would be used. It is likely that the marketing efforts would support two to five different promotional approaches.

   Note 2: How should the percentages of total purchases comprised by various brands be treated? Isn't a customer who buys all brand A just as loyal as a customer who buys all brand B? What will be the effect on any distance measure of using the brand share variables as is? Consider using a single derived variable.

2. Select what you think is the best segmentation and comment on the characteristics (demographic, brand loyalty, and basis for purchase) of these clusters. (This information would be used to guide the development of advertising and promotional campaigns.)

3. Develop a model that classifies the data into these segments. Since this information would most likely be used in targeting direct-mail promotions, it would be useful to select a market segment that would be defined as a *success* in the classification model.

Appendix

Although not used in the assignment, two additional datasets were used in the derivation of the summary data.

CRISA_Purchase_Data is a transaction database in which each row is a transaction. Multiple rows in this dataset corresponding to a single household were consolidated into a single household row in CRISA_Summary_Data.

The *Durables* sheet in IMRB_Summary_Data contains information used to calculate the affluence index. Each row is a household, and each column represents a durable consumer good. A 1 in the column indicates that the durable is possessed by the household; a 0 indicates that it is not possessed. This value is multiplied by the weight assigned to the durable item. For example, a 5 indicates the weighted value of possessing the durable. The sum of all the weighted values of the durables possessed equals the affluence index.

18.5 DIRECT-MAIL FUNDRAISING

Fundraising.xls and FutureFundraising.xls are the datasets used for this case study.

Background

A national veterans' organization wishes to develop a data mining model to improve the cost-effectiveness of its direct marketing campaign. The organization, with its in-house database of over 13 million donors, is one of the largest direct-mail fundraisers in the United States. According to its recent mailing records, the overall response rate is 5.1%. Out of those who responded (donated), the average donation is $13.00. Each mailing, which includes a gift of personalized address labels and assortments of cards and envelopes, costs $0.68 to produce and send. Using these facts, we take a sample of this dataset to develop a classification model that can effectively capture donors so that the expected net profit is maximized. Weighted sampling is used, underrepresenting the nonresponders so that the sample has equal numbers of donors and nondonors.

Data

The file Fundraising.xls contains 3120 data points with 50% donors (TARGET_B = 1) and 50% nondonors (TARGET_B = 0). The amount of donation (TARGET_D) is also included but is not used in this case. The descriptions for the 25 variables (including 2 target variables) are listed in Table 18.9.

Assignment

Step 1: Partitioning. Partition the dataset into 60% training and 40% validation (set the seed to 12345).

Step 2: Model Building. Follow these steps:

TABLE 18.9   DESCRIPTION OF VARIABLES FOR THE FUNDRAISING DATASET

Variable	Description
ZIP	Zip code group (Zip codes were grouped into five groups; only four are needed for analysis, since if a potential donor falls into none of the four, s/he must be in the other group. Inclusion of all five variables is redundant and will cause some methods to fail. "1" indicates that the potential donor belongs to this zip group.
	00000–19999 ⇒ 1 (omitted for reason stated above)
	20000–39999 ⇒ zipconvert_2
	40000–59999 ⇒ zipconvert_3
	60000–79999 ⇒ zipconvert_4
	80000–99999 ⇒ zipconvert_5
HOMEOWNER	1 = homeowner, 0 = not a homeowner
NUMCHLD	Number of children
INCOME	Household income
GENDER	0 = male, 1 = female
WEALTH	Wealth rating uses median family income and population statistics from each area to index relative wealth within each state. Segments are denoted 0–9 (0=lowest wealth group, 9 = highest wealth group). Each rating has a different meaning within each state.
HV	Average home value in potential donor's neighborhood in hundreds of dollars
ICmed	Median family income in potential donor's neighborhood in hundreds of dollars
ICavg	Average family income in potential donor's neighborhood in hundreds
IC15	Percent earning less than $15K in potential donor's neighborhood
NUMPROM	Lifetime number of promotions received to date
RAMNTALL	Dollar amount of lifetime gifts to date
MAXRAMNT	Dollar amount of largest gift to date
LASTGIFT	Dollar amount of most recent gift
TOTALMONTHS	Number of months from last donation to July 1998 (the last time the case was updated)
TIMELAG	Number of months between first and second gift
AVGGIFT	Average dollar amount of gifts to date
TARGET_B	Target variable: binary indicator for response (1 = donor, 0 = nondonor)
TARGET_D	Target variable: donation amount (in dollars). We will NOT use this variable for this case.

1. *Selecting classification tool and parameters.* Run the following classification tools on the data:

○ Logistic regression

○ Classification trees

○ Neural networks

Be sure to test different parameter values for each method. You may also want to run each method on a subset of the variables. Be sure NOT to include TARGET_D in your analysis.

2. *Classification under asymmetric response and cost.* What is the reasoning behind using weighted sampling to produce a training set with equal numbers of donors and nondonors? Why not use a simple random sample from the original dataset? (*Hint:* Given the actual response rate of 5.1%, how do you think the classification models will behave under simple sampling?) In this case, is classification accuracy a good performance metric for our purposes of maximizing net profit? If not, how would you determine the best model? Explain your reasoning.

3. *Calculate net profit.* For each method, calculate the lift of net profit for both the training and validation set based on the actual response rate (5.1%). Again, the expected donation, given that they are donors, is $13.00, and the total cost of each mailing is $0.68. (*Hint:* To calculate estimated net profit, we will need to undo the effects of the weighted sampling and calculate the net profit that would reflect the actual response distribution of 5.1% donors and 94.9% nondonors.)

4. *Draw lift curves.* Draw each model's net profit lift curve for the validation set onto a single graph. Do any models dominate?

5. *Best model.* From your answer in 2, what do you think is the "best" model?

Step 3: Testing. The file FutureFundraising.xls contains the attributes for future mailing candidates. Using your "best" model from step 2 (number 5), which of these candidates do you predict as donors and nondonors? List them in descending order of the probability of being a donor.

18.6 CATALOG CROSS SELLING

CatalogCrossSell.xls is the dataset for this case study.

Background

Exeter, Inc., is a catalog firm that sells products in a number of different catalogs that it owns.[5] The catalogs number in the dozens, but fall into nine basic categories:

1. Clothing
2. Housewares
3. Health
4. Automotive
5. Personal electronics

[5] Resampling Stats, Inc. 2006; used with permission.

6. Computers

7. Garden

8. Novelty gift

9. Jewelry

The costs of printing and distributing catalogs are high. By far the biggest cost of operation is the cost of promoting products to people who buy nothing. Having invested so much in the production of artwork and printing of catalogs, Exeter wants to take every opportunity to use them effectively. One such opportunity is in cross selling—once a customer has "taken the bait" and purchases one product, try to sell him or her another product while you have his or her attention.

Such cross promotion might take the form of enclosing a catalog in the shipment of a purchased product, together with a discount coupon to induce a purchase from that catalog. Or it might take the form of a similar coupon sent by e-mail, with a link to the Web version of that catalog.

But which catalog should be enclosed in the box or included as a link in the e-mail with the discount coupon? Exeter would like it to be an informed choice—a catalog that has a higher probability of inducing a purchase than simply choosing a catalog at random.

Assignment

Using the dataset CatalogCrossSell.xls, perform an association rules analysis and comment on the results. Your discussion should provide interpretations in English of the meanings of the various output statistics (lift ratio, confidence, support) and include a very rough estimate (precise calculations not necessary) of the extent to which this will help Exeter make an informed choice about which catalog to cross promote to a purchaser.

Acknowledgment The data for this case have been adapted from the data in a set of cases provided for educational purposes by the Direct Marketing Education Foundation ("DMEF Academic Data Set Two, Multi Division Catalog Company, Code: 02DMEF"); used with permission.

18.7 PREDICTING BANKRUPTCY

Bankruptcy.xls is the dataset for this case study.

DARDEN
BUSINESS PUBLISHING
UNIVERSITY *of* VIRGINIA

Predicting Corporate Bankruptcy [6]

Just as doctors check blood pressure and pulse rate as vital indicators of the health of a patient, so business analysts scour the financial statements of a corporation to monitor its financial health. Whereas blood pressure, pulse rate, and most medical vital signs, however, are measured through precisely defined procedures, financial variables are recorded under much less specific general principles of accounting. A primary issue in financial analysis, then, is how predictable is the health of a company?

One difficulty in analyzing financial report information is the lack of disclosure of actual cash receipts and disbursements. Users of financial statements have had to rely on proxies for cash flow, perhaps the simplest of which is income (INC) or earnings per share. Attempts to improve INC as a proxy for cash flow include using income plus depreciation (INCDEP), working capital from operations (WCFO), and cash flow from operations (CFFO). CFFO is obtained by adjusting income from operations for all noncash expenditures and revenues and for changes in the current asset and current liabilities accounts.

A further difficulty in interpreting historical financial disclosure information is caused whenever major changes are made in accounting standards. For example, the Financial Accounting Standards Board issued several promulgations in the middle 1970s that changed the requirements for reporting accruals pertaining to such things as equity earnings, foreign currency gain and losses, and deferred taxes. One effect of changes of this sort was that earnings figures became less reliable indicators of cash flow.

In the light of these difficulties in interpreting accounting information, just what are the important vital signs of corporate health? Is cash flow an important signal? If not, what is? If so, what is the best way to approximate cash flow? How can we predict the impending demise of a company?

To begin to answer some of these important questions, we conducted a study of the financial vital signs of bankrupt and healthy companies. We first identified 66 failed firms from a list provided by Dun and Bradstreet. These firms were in manufacturing or retailing and had financial data available on the Compustat Research tape. Bankruptcy occurred somewhere between 1970 and 1982.

For each of these 66 failed firms, we selected a healthy firm of approximately the same size (as measured by the book value of the firm's assets) from the same industry (3 digit SIC code) as a basis of comparison. This matched sample

[6] This case was prepared by Professor Mark E. Haskins and Professor Phillip E. Pfeifer. It was written as a basis for class discussion rather than to illustrate effective or ineffective handling of an administrative situation. Copyright 1988 by the University of Virginia Darden School Foundation, Charlottesville, VA. All rights reserved. To order copies, send an e-mail to sales@dardenpublishing.com. No part of this publication may be reproduced, stored in a retrieval system, used in a spreadsheet, or transmitted in any form or by any means—electronic, mechanical, photocopying, recording, or otherwise—without the permission of the Darden School Foundation.

technique was used to minimize the impact of any extraneous factors (such as industry) on the conclusions of the study.

The study was designed to see how well bankruptcy can be predicted two years in advance. A total of 24 financial ratios were computed for each of the 132 firms using data from the Compustat tapes and from Moody's Industrial Manual for the year that was two years prior to the year of bankruptcy. Table 18.10 lists the 24 ratios together with an explanation of the abbreviations used for the fundamental financial variables. All these variables are contained in a firm's annual report with the exception of CFFO. Ratios were used to facilitate comparisons across firms of various sizes.

The first four ratios using CASH in the numerator might be thought of as measures of a firm's cash reservoir with which to pay debts. The three ratios with CURASS in the numerator capture the firm's generation of current assets with which to pay debts. Two ratios, CURDEBT/DEBT and ASSETS/DEBTS, measure the firm's debt structure. Inventory and receivables turnover are measured by COGS/INV and SALES/REC, and SALES/ASSETS measures the firm's ability to generate sales. The final 12 ratios are asset flow measures.

TABLE 18.10   PREDICTING CORPORATE BANKRUPTCY: FINANCIAL VARIABLES AND RATIOS

Abbreviation	Financial Variable	Ratio	Definition
ASSETS	Total assets	R1	CASH/CURDEBT
CASH	Cash	R2	CASH/SALES
CFFO	Cash flow from operations	R3	CASH/ASSETS
COGS	Cost of goods sold	R4	CASH/DEBTS
CURASS	Current assets	R5	CFFO/SALES
CURDEBT	Current debt	R6	CFFO/ASSETS
DEBTS	Total debt	R7	CFFO/DEBTS
INC	Income	R8	COGS/INV
INCDEP	Income plus depreciation	R9	CURASS/CURDEBT
INV	Inventory	R10	CURASS/SALES
REC	Receivables	R11	CURASS/ASSETS
SALES	Sales	R12	CURDEBT/DEBTS
WCFO	Working capital from operations	R13	INC/SALES
		R14	INC/ASSETS
		R15	INC/DEBTS
		R16	INCDEP/SALES
		R17	INCDEP/ASSETS
		R18	INCDEP/DEBTS
		R19	SALES/REC
		R20	SALES/ASSETS
		R21	ASSETS/DEBTS
		R22	WCFO/SALES
		R23	WCFO/ASSETS
		R24	WCFO/DEBTS

Assignment

1. What data mining technique(s) would be appropriate in assessing whether there are groups of variables that convey the same information and how important that information is? Conduct such an analysis.

2. Comment on the distinct goals of profiling the characteristics of bankrupt firms versus simply predicting (black box style) whether a firm will go bankrupt and whether both goals, or only one, might be useful. Also comment on the classification methods that would be appropriate in each circumstance.

3. Explore the data to gain a preliminary understanding of which variables might be important in distinguishing bankrupt from nonbankrupt firms. (*Hint:* As part of this analysis, use XLMiner's boxplot option, specifying the bankrupt/not bankrupt variable as the x variable.)

4. Using your choice of classifiers, use XLMiner to produce several models to predict whether or not a firm goes bankrupt, assessing model performance on a validation partition.

5. Based on the above, comment on which variables are important in classification, and discuss their effect.

18.8   TIME SERIES CASE: FORECASTING PUBLIC TRANSPORTATION DEMAND

The dataset bicup2006.xls is used for this case study.

Background

Forecasting transportation demand is important for multiple purposes such as staffing, planning, and inventory control. The public transportation system in Santiago de Chile has gone through a major effort of reconstruction. In this context, a business intelligence competition took place in October 2006, which focused on forecasting demand for public transportation. This case is based on the competition, with some modifications.

Problem Description

A public transportation company is expecting an increase demand for its services and is planning to acquire new buses and to extend its terminals. These investments require a reliable forecast of future demand. To create such forecasts, one can use data on historic demand. The company's data warehouse has data on each 15-minute interval between 6:30 AM and 22:00, on the number of passengers arriving at the terminal. As a forecasting consultant you have been asked to create a forecasting method that can generate forecasts for the number of passengers arriving at the terminal.

Available Data

Part of the historic information is available in the file bicup2006.xls. The file contains the worksheet "Historic Information" with known demand for a 3-week period, separated into 15-minute intervals. The second worksheet ("Future") contains dates and times for a future 3-day period, for which forecasts should be generated (as part of the 2006 competition).

Assignment Goal

Your goal is to create a model/method that produces accurate forecasts. To evaluate your accuracy, partition the given historic data into two periods: a training period (the first two weeks) and a validation period (the last week). Models should be fitted only to the training data and evaluated on the validation data.

Although the competition winning criterion was the lowest Mean Absolute Error (MAE) on the future 3-day data, this is *not* the goal for this assignment. Instead, if we consider a more realistic business context, our goal is to create a model that generates reasonably good forecasts on any time/day of the week. Consider not only predictive metrics such as MAE, MAPE, and RMSE, but also look at actual and forecasted values, overlaid on a time plot.

Assignment

For your final model, present the following summary:

1. Name of the method/combination of methods.
2. A brief description of the method/combination.
3. All estimated equations associated with constructing forecasts from this method.
4. The MAPE and MAE for the training period and the validation period.
5. Forecasts for the future period (March 22–24), in 15-minute bins.
6. A single chart showing the fit of the final version of the model to the entire period (including training, validation, and future). Note that this model should be fitted using the combined training + validation data.

Tips and Suggested Steps

1. Use exploratory analysis to identify the components of this time series. Is there a trend? Is there seasonality? If so, how many "seasons" are there? Are there any other visible patterns? Are the patterns global (the same throughout the series) or local?

2. Consider the frequency of the data from a practical and technical point of view. What are some options?

3. Compare the weekdays and weekends. How do they differ? Consider how these differences can be captured by different methods.

4. Examine the series for missing values or unusual values. Think of solutions.

5. Based on the patterns that you found in the data, which models or methods should be considered?

6. Consider how to handle actual counts of zero within the computation of MAPE.

References

Agrawal, R., Imielinski, T., and Swami, A. (1993). "Mining associations between sets of items in massive databases," in *Proceedings of the 1993 ACM-SIGMOD International Conference on Management of Data* (pp. 207–216), New York: ACM Press.

Berry, M. J. A., and Linoff, G. S. (1997). *Data Mining Techniques*. New York: Wiley.

Berry, M. J. A., and Linoff, G. S. (2000). *Mastering Data Mining*. New York: Wiley.

Breiman, L., Friedman, J., Olshen, R., and Stone, C. (1984). *Classification and Regression Trees*. Boca Raton, FL: Chapman & Hall/CRC (orig. published by Wadsworth).

Chatfield, C. (2003). *The Analysis of Time Series: An Introduction*, 6th ed. Chapman & Hall/CRC.

Delmaster, R., and Hancock, M. (2001). *Data Mining Explained*. Boston: Digital Press.

Few, S. (2004). *Show Me the Numbers*. Oakland, CA, Analytics Press.

Few, S. (2009). *Now You See It*. Oakland, CA, Analytics Press.

Han, J., and Kamber, M. (2001). *Data Mining: Concepts and Techniques*. San Diego, CA: Academic.

Hand, D., Mannila, H. and Smyth, P. (2001). *Principles of Data Mining*. Cambridge, MA: MIT Press.

Hastie, T., Tibshirani, R., and Friedman, J. (2009). *The Elements of Statistical Learning*. 2nd ed. New York: Springer.

Hosmer, D. W., and Lemeshow, S. (2000). *Applied Logistic Regression*, 2nd ed. New York: Wiley-Interscience.

Jank, W., and Yahav, I. (2010). E-Loyalty Networks in Online Auctions. *Annals of Applied Statistics*, forthcoming.

Johnson, W., and Wichern, D. (2002). *Applied Multivariate Statistics*. Upper Saddle River, NJ: Prentice Hall.

Larsen, K. (2005). "Generalized Naïve Bayes Classifiers," *SIGKDD Explorations*, **7**(1), pp. 76–81.

Lyman, P., and Varian, H. R. (2003). "How much information," retrieved from http://www.sims.berkeley.edu/how-much-info-2003 on Nov. 29, 2005.

Manski, C. F. (1977). "The structure of random utility models," *Theory and Decision*, **8**, pp. 229–254.

McCullugh, C. E., Paal, B., and Ashdown, S. P. (1998). "An optimisation approach to apparel sizing," *Journal of the Operational Research Society*, **49**(5), pp. 492–499.

Pregibon, D. (1999). "2001: A statistical odyssey," invited talk at The Fifth ACM SIGKDD International Conference on Knowledge Discovery and Data Mining, ACM Press, NY, p. 4.

Trippi, R., and Turban, E. (eds.) (1996). *Neural Networks in Finance and Investing*. New York: McGraw-Hill.

Veenhoven, R., World Database of Happiness, Erasmus University Rotterdam; available at http://worlddatabaseofhappiness.eur.nl.

Index